Fundamentals of
LASER MACHINING

Fundamentals of LASER MACHINING

Mykola Stetsko
Kebra Ward

Kruger Brentt
Publishers

2025

Kruger Brentt Publishers UK. LTD.
Company Number 9728962

Regd. Office: 68 St Margarets Road, Edgware, Middlesex HA8 9UU

ISBN: 978-1-78715-420-9

For information on all our publications visit our website at http://krugerbrentt.com/

PREFACE

Laser machining has emerged as a powerful and versatile technique for precision material processing, offering unique advantages in terms of speed, accuracy, and flexibility. With the ability to deliver intense, focused beams of light, lasers can efficiently cut, drill, weld, and surface treat a wide range of materials, including metals, ceramics, composites, and semiconductors. Laser machining has revolutionized manufacturing processes across industries, enabling the fabrication of intricate components with tight tolerances and superior surface quality. Understanding the principles, techniques, and applications of laser machining is essential for engineers, researchers, and practitioners seeking to harness the full potential of this transformative technology for advanced material processing and manufacturing.

This handbook, titled "Fundamentals of Laser Machining" is a comprehensive exploration of the principles, methodologies, and applications of laser machining in the fabrication of advanced materials. Drawing upon the expertise of materials scientists, laser physicists, and manufacturing engineers, this volume offers theoretical insights, practical guidance, and case studies for understanding and utilizing laser machining techniques in various material processing applications.

From the fundamentals of laser-material interactions and beam characteristics to the latest advancements in laser system design, process optimization, and quality control, the "Fundamentals of Laser Machining" covers a wide range of topics essential for mastering the field of laser machining. Through clear explanations, experimental techniques, and industrial examples, readers will gain a deep understanding of the mechanisms and parameters that govern laser-material interactions and the factors that influence machining performance.

Moreover, this handbook explores the applications of laser machining across different materials and industries, including aerospace, automotive, electronics, medical devices, and renewable energy. By examining case studies, research findings, and technological trends, we highlight the diverse ways in which laser machining is used to fabricate complex geometries, enhance material properties, and improve manufacturing efficiency.

As we continue to push the boundaries of material science and engineering, the importance of laser machining in enabling advanced material processing has never been greater. It is our hope that the "Fundamentals of Laser Machining" handbook will serve as a valuable resource for researchers, engineers, and manufacturers in the field of advanced manufacturing. Whether you are developing laser machining processes, optimizing manufacturing workflows, or integrating laser systems into production environments, this handbook invites you to explore the capabilities and opportunities of laser machining and contribute to the advancement of material processing technologies.

Mykola Stetsko

Kebra Ward

CONTENTS

CHAPTER 1

ADVANCED PULSE LASER MACHINING TECHNOLOGY: INDUSTRIAL AND SCIENTIFIC APPLICATIONS

This book is an expeditiously evolving domain that can be customized for specific industrial and research uses. This is primarily influenced by the accessibility of high-repetition-rate laser sources and innovative beam delivery methods. Recent papers by Saraceno et al. [1], Schille and Löschner [2], and Weber and Graf [3] illustrated the evolution of ultrashortpulse laser technology during the past several decades, as depicted in the data synthesis from [2,3] in Figure 1. The average power of ultrafast lasers adheres to a variant of Moore's Law, resulting in a biannual doubling of their average power [4].Moreover, the average power achieved by lasers in research laboratories surpasses that of conventional industrial lasers by approximately a decade [3], illustrated by the horizontal distance between the two lines in Figure 1. The remarkable advancement in laser technology has resulted in a fiber-laser-based average power of 10.4 kW at a wavelength of 1.4 μm, with a pulse duration of 254 fs and a repetition rate of 80 MHz [5]. The prevailing industrial standard comprises ultrashortpulse lasers that output average powers approximately at 100 W.

For industrial use, the high output power of ultrafast lasers must be directed to the workpieces to be machined by appropriate beam guidance and deflection systems [6]. For many materials such as metals and semiconductors, moderate laser fluences (i.e., laser pulse energies) are sufficient to process the workpiece with high precision and efficiency [7,8,9]. Therefore, high laser pulse repetition rates up to the above-mentioned MHz level can be utilized to significantly reduce the processing times. This requires extremely fast beam deflection systems, possibly coupled with optics for multibeam processing. In a recent study, an area processing rate up to 3.8 m2/min was demonstrated with a single-pass raster on steel sheets with a scanning speed of the laser beam of 950 m/s [10]. The key element of the experimental setup was a polygon scanner [11] in connection with a MHz laser system.

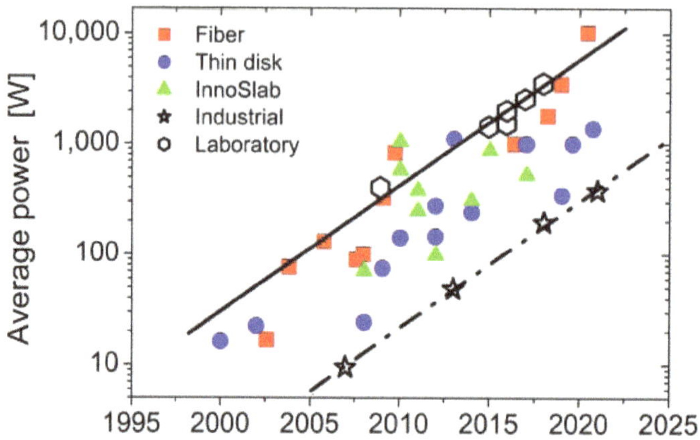

Figure 1. Progress in ultrafast laser technology featuring an exponential increase in the average output power over the past 20 years. The plot is a synthesis of data, indicating the related laser technology (fiber, thin disk, and InnoSlab) as fully colored data points [2], as well as their realization in laboratories or as commercial industrial products as open black data points [3]. The black lines guide the eye.

The Special Issue addresses not only the advantages of modern laser processing using short and ultrashort laser pulses, but also limitations caused by unwanted secondary hazards such as X-ray emissions. This phenomenon is not completely new and was described for laser machining applications using ultrashort laser pulses with repetition rates of the order of 1 kHz two decades ago [**12,13,14**]. However, the use of high pulse repetition rates in the multi-100 kHz range and burst pulses has recently exacerbated the problem [**15,16,17,18,19,20,21**]. Therefore, it is a pleasant fact that five publications in this Special Issue discuss this problem area in detail and, thus, make an important contribution to the field of combined laser and radiation protection for improving work safety aspects [**22,23,24,25,26**].

The application of laser pulse bursts facilitates novel processing regimes for metals, enhancing structuring rates and surface quality of machined samples.Förster et al. [27] evaluate the outcomes of both experimental and numerical investigations in this area. Machining solutions employing pulse bursts with intraburst repetition frequencies in the MHz to GHz range must account for an elevated risk of secondary Xray emissions associated with this operational mode [21,25,26].This Special Issue comprises one review paper [27], 1 perspective article [28], and fourteen original research articles [22,23,24,25,26,29,30,31,32,33,34,35,36,37], all concentrating on recent advancements in surface and volume laser material processing. Topics include laser-induced forward transfer and laser printing technologies [28], spatial and temporal beam shaping [29,30,31,32], Bessel-beam structuring of high-aspect-ratio void channels within glass [34], direct laser interference patterning [35], pulse burst machining [27], waveguide writing [36], femtosecond pulse laser-induced amorphization and recrystallization of single-crystalline silicon [37], and a comparative analysis of various beam shuttering technologies [33], a frequently overlooked yet practically significant aspect.

REFERENCES

1. Saraceno, J.C.; Sutter, D.; Metzger, T.; Ahmed, M.A. The amazing progress of high-power ultrafast thin-disk lasers. *J. Eur. Opt. Soc. Rapid publ.* **2019**, *15*, 15. **[Google Scholar] [CrossRef] [Green Version]**
2. Schille, J.; Löschner, U. Ultrashort pulse lasers in high-rate laser micro processing—Quo vadis? *Adv. Opt. Technol.* **2021**, *10*, 233–237. **[Google Scholar] [CrossRef]**
3. Weber, R.; Graf, T. The challenges of productive materials processing with ultrafast lasers. *Adv. Opt. Technol.* **2021**, *10*, 239–245. **[Google Scholar] [CrossRef]**
4. Han, M.; Smith, D.; Ng, S.H.; Anand, V.; Katkus, T.; Juodkazis, S. Ultra-short-pulse lasers—Materials—Applications. *Eng. Proc.* **2021**, *11*, 44. **[Google Scholar]**
5. Müller, M.; Aleshire, C.; Klenke, A.; Haddad, E.; Légaré, F.; Tünnermann, A.; Limpert, J. 10.4 kW coherently combined ultrafast fiber laser. *Opt. Lett.* **2020**, *45*, 3083–3086. **[Google Scholar] [CrossRef] [PubMed]**
6. Račiukaitis, G. Ultra-short pulse lasers for microfabrication: A review. *IEEE J. Sel. Top. Quantum Electron.* **2021**, *27*, 1100112. **[Google Scholar] [CrossRef]**
7. Kautek, W.; Krüger, J. Femtosecond pulse laser ablation of metallic, semiconducting, ceramic, and biological materials. *Proc. SPIE* **1994**, *2207*, 600–611. **[Google Scholar]**
8. Račiukaitis, G.; Brikas, M.; Gecys, P.; Voisiat, B.; Gedvilas, M. Use of high repetition rate and high power lasers in microfabrication: How to keep the efficiency high? *J. Laser Micro Nanoeng.* **2009**, *4*, 186–191. **[Google Scholar] [CrossRef]**
9. Neuenschwander, B.; Jäggi, B.; Schmid, M.; Hennig, G. Surface structuring with ultra-short laser pulses: Basics, limitations and needs for high throughput. *Phys. Procedia* **2014**, *56*, 1047–1058. **[Google Scholar] [CrossRef] [Green Version]**
10. Schille, J.; Schneider, L.; Mauersberger, S.; Szokup, S.; Höhn, S.; Pötschke, J.; Reiß, F.; Leidich, E.; Löschner, U. High-rate laser surface texturing for advanced tribological functionality. *Lubricants* **2020**, *8*, 33. **[Google Scholar] [CrossRef] [Green Version]**
11. De Loor, R. Polygon scanner system for ultra short pulsed laser micro-machining applications. *Phys. Procedia* **2013**, *41*, 544–551. **[Google Scholar] [CrossRef] [Green Version]**
12. Thogersen, J.; Borowiec, A.; Haugen, H.K.; McNeill, F.E.; Stronach, I.M. X-ray emission from femtosecond laser micromachining. *Appl. Phys. A* **2001**, *73*, 361–363. **[Google Scholar] [CrossRef]**
13. Bunte, J.; Barcikowski, S.; Püster, T.; Burmester, T.; Brose, M.; Ludwig, T. Secondary hazards: Particle and X-ray emission. *Topics Appl. Phys.* **2004**, *96*, 309–321. **[Google Scholar]**
14. Prieto-Pena, J.; Gómez, F.; González-Castaño, D.M.; Flores-Arias, M.T.; Arines, J.; Bao-Varela, C.; Cambronero-López, F.; Martínez Maqueira, A. X-ray emission from femtosecond laser micromachining. *J. Radiol. Prot.* **2018**, *38*, 716–730. **[Google Scholar] [CrossRef]**
15. Legall, H.; Schwanke, C.; Pentzien, S.; Dittmar, G.; Bonse, J.; Krüger, J. X-ray emission as a potential hazard during ultrashort pulse laser material processing. *Appl. Phys. A* **2018**, *124*, 407. **[Google Scholar] [CrossRef] [Green Version]**
16. Behrens, R.; Pullner, B.; Reginatto, M. X-ray emission from materials processing lasers. *Radiat. Prot. Dosim.* **2019**, *183*, 361–374. **[Google Scholar] [CrossRef]**
17. Weber, R.; Giedl-Wagner, R.; Förster, D.J.; Pauli, A.; Graf, T.; Balmer, J.E. Expected X-ray dose rates resulting from industrial ultrafast laser applications. *Appl. Phys. A* **2019**, *125*, 635. **[Google Scholar] [CrossRef] [Green Version]**
18. Legall, H.; Schwanke, C.; Bonse, J.; Krüger, J. X-ray radiation protection aspects during ultrashort laser processing. *J. Laser Appl.* **2020**, *32*, 022004. **[Google Scholar] [CrossRef] [Green Version]**
19. Freitag, C.; Giedl-Wagner, R. X-ray protection in an industrial production environment. *PhotonicsViews* **2020**, *17*, 37–41. **[Google Scholar] [CrossRef]**

20. Legall, H.; Bonse, J.; Krüger, J. Review of X-ray exposure and safety issues arising from ultra-short pulse laser material pro-cessing. *J. Radiol. Prot.* **2021**, *41*, R28–R42. **[Google Scholar]** **[CrossRef]**
21. Metzner, D.; Olbrich, M.; Lickschat, P.; Horn, A.; Weißmantel, S. X-ray generation by laser ablation using MHz to GHz pulse bursts. *J. Laser Appl.* **2021**, *33*, 032014. **[Google Scholar]** **[CrossRef]**
22. Mosel, P.; Sankar, P.; Düsing, J.F.; Dittmar, G.; Püster, T.; Jäschke, P.; Vahlbruch, J.-W.; Morgner, U.; Kovacev, M. X-ray dose rate and spectral measurements during ultrafast laser machining using a calibrated (high-sensitivity) novel X-ray detector. *Materials* **2021**, *14*, 4397. **[Google Scholar]** **[CrossRef]**
23. Schille, J.; Kraft, S.; Pflug, T.; Scholz, C.; Clair, M.; Horn, A.; Löschner, U. Study on X-ray emission using ultrashort pulsed lasers in materials processing. *Materials* **2021**, *14*, 4537. **[Google Scholar]** **[CrossRef]**
24. Stolzenberg, U.; Schmitt Rahner, M.; Pullner, B.; Legall, H.; Bonse, J.; Kluge, M.; Ortner, A.; Hoppe, B.; Krüger, J. X-ray emission hazards from ultrashort pulsed laser material processing in an industrial setting. *Materials* **2021**, *14*, 7163. **[Google Scholar]** **[CrossRef]**
25. Schille, J.; Kraft, S.; Kattan, D.; Löschner, U. Enhanced X-ray emissions arising from high pulse repetition frequency ultrashort pulse laser materials processing. *Materials* **2022**, *15*, 2748. **[Google Scholar]** **[CrossRef]** **[PubMed]**
26. Böttcher, K.; Schmitt Rahner, M.; Stolzenberg, U.; Kraft, S.; Bonse, J.; Feist, C.; Albrecht, D.; Pullner, B.; Krüger, J. Worst-case X-ray photon energies in ultrashort pulse laser processing. *Materials* **2022**, *15*, 8996. **[Google Scholar]** **[CrossRef]**
27. Förster, D.J.; Jäggi, B.; Michalowski, A.; Neuenschwander, B. Review on experimental and theoretical investigations of ultra-short pulsed laser ablation of metals with burst pulses. *Materials* **2021**, *14*, 3331. **[Google Scholar]** **[CrossRef]**
28. Florian, C.; Serra, P. Printing via laser-induced forward transfer and the future of digital manufacturing. *Materials* **2023**, *16*, 698. **[Google Scholar]** **[CrossRef]**
29. Hauschwitz, P.; Stoklasa, B.; Kuchařík, J.; Turčičová, H.; Písařík, M.; Brajer, J.; Rostohar, D.; Mocek, T.; Duda, M.; Lucianetti, A. Micromachining of invar with 784 Beams using 1.3 ps laser source at 515 nm. *Materials* **2020**, *13*, 2962. **[Google Scholar]** **[CrossRef]**
30. Wenisch, C.; Engel, S.; Gräf, S.; Müller, F.A. Dual laser beam processing of semiconducting thin films by excited state absorption. *Materials* **2021**, *14*, 1256. **[Google Scholar]** **[CrossRef]** **[PubMed]**
31. Zhou, T.; Kraft, S.; Perrie, W.; Schille, J.; Löschner, U.; Edwardson, S.; Dearden, G. Backward flux re-deposition patterns during multi-spot laser ablation of stainless steel with picosecond and femtosecond pulses in air. *Materials* **2021**, *14*, 2243. **[Google Scholar]** **[CrossRef]** **[PubMed]**
32. Schlutow, H.; Fuchs, U.; Müller, F.A.; Gräf, S. Squared focal intensity distributions for applications in laser material processing. *Materials* **2021**, *14*, 4981. **[Google Scholar]** **[CrossRef]** **[PubMed]**
33. Aboud, D.G.K.; Wood, M.J.; Zeppetelli, G.; Joy, N.; Kietzig, A.-M. A practical comparison of beam shuttering technologies for pulsed laser micromachining applications. *Materials* **2022**, *15*, 897. **[Google Scholar]** **[CrossRef]** **[PubMed]**
34. Vetter, C.; Giust, R.; Furfaro, L.; Billet, C.; Froehly, L.; Courvoisier, F. High aspect ratio structuring of glass with ultrafast Bessel beams. *Materials* **2021**, *14*, 6749. **[Google Scholar]** **[CrossRef]**
35. El-Khoury, M.; Voisiat, B.; Kunze, T.; Lasagni, A.F. Prediction of optimum process parameters fabricated by direct laser interference patterning based on central composite design. *Materials* **2020**, *13*, 4101. **[Google Scholar]** **[CrossRef]**
36. Macias-Montero, M.; Moreno-Zárate, P.; Muñoz, F.; Sotillo, B.; Garcia-Pardo, M.; Serna, R.; Fernandez, P.; Solis, J. Competition effects during femtosecond laser induced element redistribution in Ba- and La-migration based laser written waveguides. *Materials* **2021**, *14*, 3185. **[Google Scholar]** **[CrossRef]**
37. Florian, C.; Fischer, D.; Freiberg, K.; Duwe, M.; Sahre, M.; Schneider, S.; Hertwig, A.; Krüger, J.; Rettenmayr, M.; Beck, U.; et al. Single femtosecond laser-pulse-induced superficial amorphization and re-crystallization of silicon. *Materials* **2021**, *14*, 1651. **[Google Scholar]** **[CrossRef]**

CHAPTER 2

INVESTIGATIONS OF ULTRASHORT PULSED LASER ABLATION OF METALS USING BURST PULSES: EXPERIMENTAL AND THEORETICAL APPROACHES

1. INTRODUCTION: AN OVERVIEW

The advancement of ultra-short pulsed lasers for materials processing has consistently focused on achieving greater laser powers to enhance throughput. In recent years, ultrafast lasers with average output powers above 1 kW [1,2,3,4,5] and even reaching 10 kW [6] have been exhibited. Lasers accessible in industry currently offer pulse energies of up to approximately 100 μJ, whereas academic laser systems give several mJ. Consequently, the energy provided per pulse frequently exceeds the energy required for individual pulse ablation. To achieve the maximum ablated volume for a specified energy input, the energy may be allocated across multiple pulses within a pulse package, referred to as a pulse burst, which comprises a minimum of two pulses and can extend to several hundred or thousand pulses. [7,8,9,10,11,12,13,14,15,16,17,18,19,20,21,22,23,24].Each burst package has a substantial amount of energy suitable for laser machining, while the cumulative energy of all pulses approaches the laser's maximum attainable pulse energy. In recent years, ultra-short pulsed laser ablation utilizing burst pulses for micromachining has been extensively studied through both experimental and numerical methods. This article aims to deliver a thorough analysis of the physics underlying ultra-short pulsed laser ablation of metals using burst pulses, along with the applications derived from the associated physical processes. The subsequent subsection presents the processing strategies and terminologies employed in this article. The second section examines the physics of ablation utilizing ultra-short burst pulses on metals, initially outlining the principles of single pulse laser ablation, and subsequently addressing the interactions between two consecutive ultra-short laser pulses. Subsequently, particular emphasis is placed on machining utilizing triple pulses, followed by two segments focused on the physics associated with processing using multiple pulses at intra-burst repetition rates of MHz and GHz. Section 3 summarizes and reviews several MHz and GHz burst processing applications, including punching, drilling, line scribing, milling, surface structuring, and several specialized applications. The paper concludes with findings and a perspective.

Processing Strategies and Definitions

In material processing, different strategies can be applied depending on the application. Using a stationary beam, the so-called punching process is realized. In this process, a certain number of consecutive laser pulses N is focused on the same spot and each pulse ablates a thin layer of material. Using a Gaussian beam with a $1/e21/2$-radius of w, this results in a parabolic dimple, as shown in Figure 1a. If the applied number of pulses and the fluence are high enough, the material can be drilled through. This drilling strategy is called percussion drilling and represents only one of the possible drilling strategies used in ultra-short pulse laser machining [25].

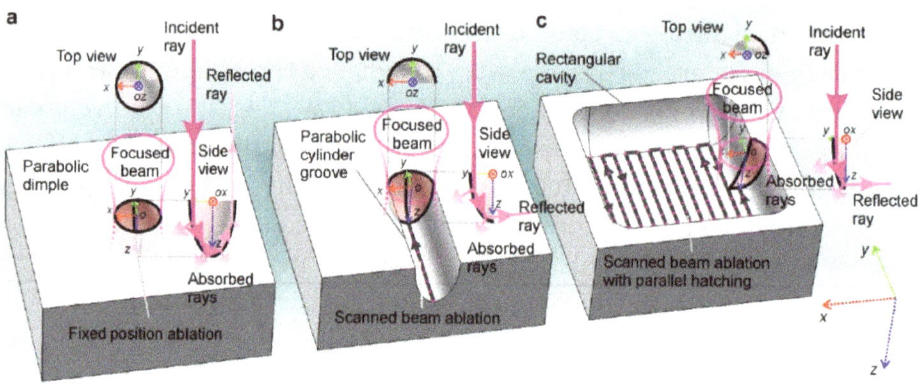

Figure 1. Schematic illustration of the processes and machining results: (a) dimple with the shape of a paraboloid ablated by multiple laser pulses with a stationary beam, resulting out of the punching process; (b) groove with the shape of a parabolic cylinder ablated by a scanned laser beam (line scan), representing the scribing process; (c) rectangular-shaped cavity ablated by a scanned laser beam and parallel line hatching (bidirectional meandering scan) resulting out of laser milling. The 3D Cartesian coordinate system x (red arrow), y (green arrow), and z (blue arrow) holds for all sketches (a–c). The center positions of the focused Gaussian beams are indicated by o. The multiple reflections of incident rays from the interior walls are given as side views (yz planes). Incident and multiple reflected rays are also given as side views (yz planes). The interaction area (cf. xy and yz planes) of laser and material decreases from (a–c). Reproduced under the terms of a Creative Commons Attribution 4.0 International License, (https://creativecommons.org/licenses/by/4.0/ (accessed on 31 May 2021)) [26].

In contrast to using a stationary beam, scribing can be realized if the laser beam is moved along one lateral direction. This results on the workpiece in a parabolic-shaped groove if a Gaussian beam is used (cf. Figure 1b). The linear movement of the beam can be realized using a linear stage or a galvanometric scanner for example. In the scribing process, consecutive pulses overlap each other. The overlap *oline* can be calculated using the moving speed of the axis *vscan*, the radius *w*, and the used pulse repetition rate of the laser *frep* by

$$o_{line} = 1 - \frac{v_{scan}}{f_{rep} \cdot 2 \cdot w} \quad (1)$$

If a deeper groove is demanded, the laser beam is moved several times along the same line path. These repetitions are the number of machined layers or overscans *NL*. If the ratio between material thickness, laser parameters, and repetitions is chosen correctly, the material can even be cut through using this line scribing technique. The third ablation process, called milling (cf. Figure 1c), is realized by machining several parallel lines with a defined line spacing. This line spacing is called the hatch distance *dh*. In the milling process, in addition to the overlap of two consecutive pulses along the scan direction, there is also an overlapping in the perpendicular direction to the machined lines *ocross*:

$$o_{cross} = 1 - \frac{d_h}{2 \cdot w} \quad (2)$$

In milling applications, the laser spot must be moved in two lateral directions. The straight-forward approach is to use a galvanometric scanner to perform this movement [**26**]. Additionally, solutions using a rotational axis in combination with different technologies such as linear stages [**27**], galvanometric scanners [**28**], or acousto-optic deflectors (AODs) [**29**] are used to allow such 2D movement. In the z-direction, the material is removed layer by layer, resulting in a so-called 2.5D process [**27**]. Using this strategy, 3D structures can be engraved by changing the processing region from layer to layer, as shown in **Figure 2**a. In addition, the direction of the hatch pattern (direction of the parallel lines) can be rotated from layer to layer by the hatch angle to avoid scanning the same lines multiple times, which could lead to line irregularities. If a central unprocessed square in the middle of a process region is enlarged linearly layer by layer (making the scanned area per layer smaller), the result is a square pyramid as a subtractively generated geometry (cf. **Figure 2**b).

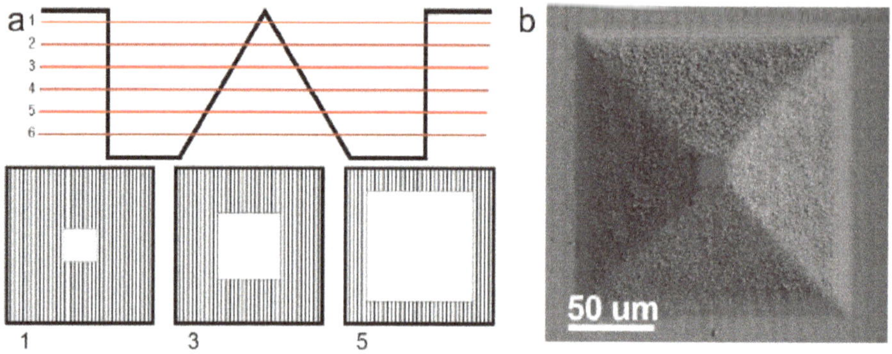

Figure 2. 2.5D processing; (a) slicing of the 3D structure and three corresponding 2D slices with the hatch pattern (no rotation of the hatch pattern from layer to layer); (b) SEM image of a 3D structure machined in copper. Reprinted from [27].

The described processing strategies have been used principally with single pulses, which are emitted at a given repetition rate from the used laser system. In burst processing, the single pulse is replaced by a pulse package including several pulses with a temporal distance much shorter than the pulse repetition rate (intra-burst delay t_B cf. **Figure 3**). This package of pulses is called a burst. Depending on the intra-burst delay time, it can be distinguished between MHz bursts (with an intra-burst delay of tens of ns) and GHz bursts (with an intra-burst delay of hundreds of ps). The interval between burst packages is referred to as the inter-burst delay, which aligns with the pulse repetition rate in single pulse operation. In the processes of scribing and milling, it may be inferred that for such time delays, the burst package is applied inside the confines of a single location, functioning akin to a stationary beam.

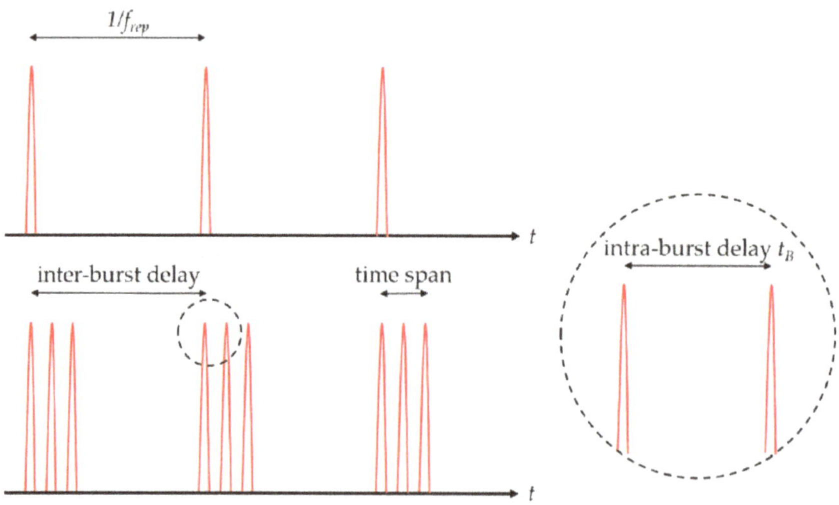

Figure 3. Upper row: single pulse operation with a defined repetition rate *frep*; lower row: burst operation with display of the two delay times inter-burst delay and intra-burst delay t_B as well as the time span.

2. PHYSICS OF ULTRA-SHORT BURST PULSE ABLATION OF METALS

This article concentrates on metal processing. For laser processing of various materials, including semiconductors [30,31,32], dielectrics [30,31,33,34,35], and ceramics [31,36], the reader is directed to the relevant articles. This discussion focuses on the physical mechanics of laser-material interactions pertinent to metal machining. Section 2.1 discusses and summarizes the mechanisms involved during and after the processing of metals using single ultra-short laser pulses, specifically for repetitive processing characterized by inter-pulse delays of tens of microseconds or repetition rates below 100 kHz. Despite the existence of numerous works addressing this topic, particularly concerning the micromachining of various metals in this regime [30,31,37,38,39,40], an overview of the fundamental effects related to single pulse operation is provided for completeness and as a foundation for the subsequent sections. Following the introduction and summary of the relevant physical mechanisms for single pulse laser ablation of metals, Section 2.2 addresses the interactions between two consecutive pulses. This section concentrates on the interactions between two pulses with inter-pulse delays ranging from several hundred femtoseconds to several tens of nanoseconds, as the experimental and theoretical results significantly diverge from those observed in single pulse metal processing. Section 2.3 examines the physical mechanics and contemporary research related to triple pulse processing with intra-burst delays of several tens of nanoseconds. Sections 2.4 and 2.5 discuss machining utilizing bursts of numerous pulses with intra-burst repetition rates in the MHz and GHz ranges.

2.1. Single Pulse Laser Ablation

When light strikes a metal surface, photons engage with free electrons, transferring energy to the electronic system through inverse Bremsstrahlung. Consequently, the electronic system transmits energy to the lattice (ions) and stimulates vibrations within the ionic system, causing the metal surface to heat and resulting in a temperature increase. For laser pulses with durations significantly surpassing the electron–phonon coupling period, the excitation of the electronic and ionic systems can be regarded as occurring in thermal equilibrium. Consequently, traditional heat conduction and physical processes such as melting and evaporation occur. In ultra-short pulsed processing, brief heating durations enable the transition of physical states across the phase diagram (e.g., the solid–liquid boundary) within picoseconds, generating tremendous pressures (tens of GPa and beyond) in thin surface layers within tens of picoseconds. For a comprehensive understanding, the specifics of the relevant physical mechanisms are available in various sources, including theoretical and numerical modeling [37,41,42,43,44,45,46]. The primary criterion for differentiating between the ultra-short pulse domain and classical thermal heating in equilibrium is the electron–phonon coupling time $\tau el\text{-}phon$, which serves as a characteristic duration for heating the ionic lattice. Initially, this value is considered constant; nevertheless, it fundamentally relies on the deposited energy, and the relaxation time may be multiples of that characteristic time [23]. The estimation can be expressed as $\tau el\text{-}phon=CP/\gamma$, where CP denotes the heat capacity of the ionic system and γ represents the electron-phonon coupling constant [37]. Both values are contingent upon the substance. In metals and their alloys, the electron–phonon coupling time ranges from several hundred femtoseconds to over 100 picoseconds; for instance, pure gold has a coupling period of approximately 115 picoseconds. Table 1 presents exemplary values. In alloys, the electron–phonon coupling times may vary from those of the constituent elements; for instance, AISI 304 stainless steel exhibits an electron–phonon coupling time of 0.5 ps, while pure iron has a coupling time of 1.3 ps.

Table 1. Electron–phonon coupling times for various metals at room temperature.

Material	Ag	Al	Au	Cu	Fe	Mb	Ni	Pt	Ti	W
$\tau_{el\text{-}phon}$ in ps	84.3 [1]	4.5 [1]	115.5 [1]	57.5 [2]	1.3 [3]	2.2 [4]	1.1 [4]	2.2 [5]	1.9 [5]	12.1 [5]

[1] calc. from C_l [**48**], g [**49**]; [2] [**50**]; [3] calc. from C_l [**48**], γ [**51**]; [4] calc. from C_l [**48**], γ [**52**]; [5] calc. from C_l [**48**], γ [**53**].

Another crucial quantity for the description of different regimes of laser processing with ultra-short pulses is the laser fluence ϕ. It is defined as the energy per irradiated area, and for a Gaussian beam the peak fluence ϕ_0 equals

$$\phi_0 = 2 \cdot \phi_{av} = 2 \cdot \frac{E_P}{\pi \cdot w^2} \quad (3)$$

E_P corresponds to the pulse energy of the incoming laser pulse, w to the laser spot radius (1/e^2), and ϕ_{av} is the average fluence. The fluence can be increased by increasing the laser pulse energy or decreasing the irradiated area (spot size). If the fluence is increased above the ablation threshold ϕ_{th}, material is removed from the surface (ablation). With increasing fluence, the ultra-short pulsed laser interaction with metals can be characterized by the following physical processes [**41,42,43,44,45,46,54**]:

heating of the metal surface;

- melting of the metal surface;
- spallation of surface layers (rupture of material when the fluence is above the ablation threshold);
- phase explosion (explosive boiling of the surface with the generation of vapor and plasma).

A microscopic view of the occurring processes for ablation of aluminum is given in **Figure 4** [46]. When a laser pulse hits the target with a fluence below the ablation threshold, the material is heated and melting of the surface may also take place.

Furthermore, two sorts of pressure waves propagate into the material from the surface. A shockwave (SW) leading to a compression of material (higher density, darker orange color) is followed by a rarefaction wave (RW), leading to a dilution of material (lower density, lighter orange color). When the fluence is increased above the ablation threshold, the intensity of both pressure waves is increased and, as a result of the lower density of the rarefaction wave, a fragmentation of surface layers takes place (spallation regime). The travel velocity of such dense layers evolving from the surface for metals lies in the range of several 100 m/s [**45**,**46**,**54**,**55**,**56**,**57**]. If the fluence is increased above values of approximately 10 times the ablation threshold, not only spallation takes place but the upper surface layers (in this representation on the left) disintegrate and show a transition into the gaseous phase. Particles and clusters in the gaseous phase then travel principally with the speed of sound (of the initial bulk material), which lies in the range of several km/s [**45**,**46**,**54**,**55**,**56**,**57**].

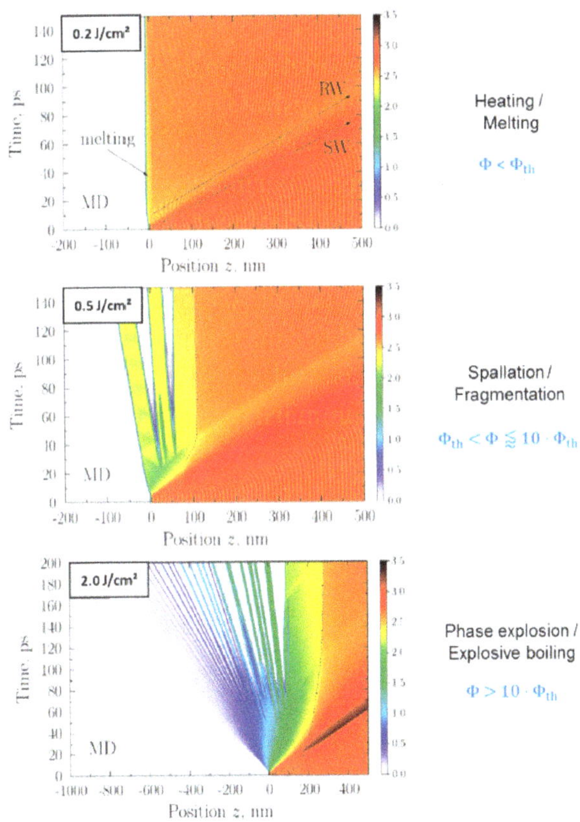

Figure 4. Results from molecular dynamics (MD) simulations for single pulse irradiation of aluminum. The *x*-axis corresponds to spatial coordinates with z = 0 nm being the initial surface of the material. The *y*-axis corresponds to the simulation time with t = 5 ps being the maximum of the laser pulse. The color of the heat maps corresponds to the density of aluminum. With increasing fluence (from top to bottom), the material passes different regimes, ranging from melting of the surface (no ablation) to phase explosion

with an excessive development of gaseous and ionized matter. In the spallation regime (center), ablation is characterized by a fragmentation of the material, leading to high surface qualities and well-defined ablation depths during laser processing. Adapted from [46].

Although not displayed here, atoms in the gaseous phase also become ionized for higher fluences, and some regions reach temperatures of several 1000 K or more [45,46,54,55,56,57]. If successive pulses have inter-pulse delays of several tens of microseconds or more (corresponding to repetition rates of $frep$= 100 kHz or less), a subsequent pulse does not significantly interact with the ablation plume of its preceding pulse [58]. For laser material processing, the different ablation regimes result in differently efficient micromachining processes. One quantity for the description of an efficient ablation process is the energy specific volume $\Delta V/\Delta E\Delta$. For Gaussian laser beams it holds [14]:

$$\frac{\Delta V}{\Delta E} = \frac{1}{2}\cdot\frac{\delta}{\phi_0}\cdot\ln^2\left(\frac{\phi_0}{\phi_{th}}\right) \quad (4)$$

where δ corresponds to the effective penetration depth of energy, $\phi 0$ is the incident peak fluence, and ϕth is the ablation threshold. Experimental data and, for the sake of clarity, only one fit function (brass, 400 fs, cyan curve) is given in Figure 5. The values were obtained in laser milling experiments by ablating cavities (cf. Figure 1c), measuring the ablated volume, and relating it to the incident total laser energy. The values follow the trend of Formula (4), having a maximum that can be extracted from the data. The corresponding values of δ, ϕth, and the fluence $\phi 0,opt$ where the maximum value was achieved and the maximum energy specific volume $\Delta V/\Delta E$ max (maximum values in Figure 5) as well as the experimental conditions such as wavelength and pulse duration are summarized in Table 2.

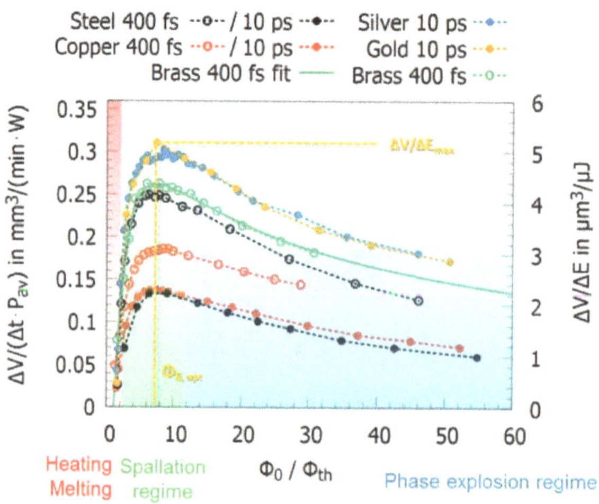

Figure 5. Energy specific volume for single pulse laser ablation of different metals processed in the near infrared. The fluence along the *x*-axis is normalized to the ablation threshold since different metals have different ablation thresholds. The corresponding values are given in Table 2. Original data taken from [59,60].

Table 2. Material parameters corresponding to the plots in Figure 5. Some of the fit parameters were calculated from original data.

Quantity		Brass [1]	Copper [1]	Copper [2]	Gold [2]	Silver [2]	Steel [1]	Steel [2]
λ	nm	1030	1030	1064	1064	1064	1030	1064
τ	ps	0.4	0.4	10	10	10	0.4	10
δ (fit)	nm	62.3	55.7	29.1	64.8	67.2	13.9	4.5
ϕ_{th}(fit)	J/cm^2	0.39	0.49	0.35	0.35	0.36	0.09	0.06
$\Delta V/\Delta E_{max}$	mm^3/W/min	0.26	0.19	0.14	0.30	0.30	0.25	0.13
$\Delta V/\Delta E_{max}$	µm^3/µJ	4.3	3.1	2.3	5.0	5.0	4.2	2.2
$\phi_{0,opt}$	J/cm^2	2.7	4.5	2.2	2.4	2.7	0.5	0.5

[1] [**60**], [2] [**59**].

For single pulse operation, this formula resembles well the experimental findings. The different metals show different maximum energy specific volumes and ablation thresholds, but the general trend is the same. Principally, such curves are found for metals when processed with pulses with a pulse duration between a few 100 fs and several ns [**60,61**].

Independent of the metals investigated, the ablation process when using ultra-short pulses is most efficient in the spallation regime, since the greatest amount of deposited energy is used for ablation. For lower fluences, only heating of the material takes place, whereas for much higher fluences the amount of energy being used for heating and ionization increases [**62**].

Beside the ablated volume per energy, the ablated volume per time (a measure for the productivity) is a very important parameter for micromachining processes such as milling or drilling. It increases for higher fluences, while the ablation process itself becomes less energetically efficient ($\Delta V/\Delta E$ decreases). At the same time, low fluences close to the ablation threshold (or several times the ablation threshold) correlate with low surface roughness and less spatter occurrence [**14,63,64**]. For industrial micromachining processes, it is often important to maintain a smooth surface while keeping the ablated volume per time high. Since the available laser pulse energy and average power of industrial laser systems have been significantly increasing within the last decade, the available energy per pulse is often much higher than the energy needed for ablation. Among other strategies such as defocusing of the processing beam to lower the fluence while keeping the productivity at a high level [**18,65**], the concept of laser burst pulses was introduced, allowing the principle use of the totally available laser energy by splitting it into several pulses with lower energy each [**66,67**]. The resulting limitations of this concept will be further discussed in the following sections.

Beside the ablated volume per energy and time, the residual energy is another important quantity influencing micromachining processes, mainly in terms of surface quality, e.g., surface roughness. If a sufficiently high amount of energy is coupled into the machined sample as heat, the surface structure and physical mechanism of material removal may change. For example, a temporary molten surface can lead to higher roughness and spatter production, decreasing the surface quality. During ablation with an ultra-short pulse, a certain amount of energy remains as heat in the bulk material. Depending on the irradiated metal and the used fluence, the amount of residual energy per pulse lies between 10% and 50% [**16,17,68,69,70,71,72,73,74,75,76**]. If high repetition rates are used, the heat accumulates and can affect the surface quality, the drilling time, and the amount of occurring melt [**14,73,77,78,79,80,81,82,83,84**]. Additionally, the amount of residual energy can increase when the geometry is changed, e.g., in hole drilling [**68,70,72,74**], or when surface structures like laser-induced periodic surface structures (LIPSS) or bumps evolve [**69,84**].

2.2. Double Pulses: Pulse-to-Pulse Interaction

The initial studies on laser treatment of metals using ultra-short double pulses date back to the early 21st century, focusing on the amplification of spectroscopic signals, now commonly referred to as laser-induced breakdown spectroscopy (LIBS), utilizing ultra-short double pulses [7,85,86,87]. Although signal enhancement was demonstrated, the depths of the ablation craters exhibited an anomalous behavior; specifically, the crater depths were limited to those produced by a single pulse or were even shallower, which is counterintuitive given that double the energy was employed to generate these craters. The subsequent summary will elucidate the interdependence between effect signal increase from the irradiated plume and the formation of shallower craters. Many experimental [**7,8,14,16,19,57,58,59,66,88,89,90,91,92,93,94,95,96,97,98,99,100,101,102,103,104,105,106,107,108,109,110,111,112,113,114**] and numerical [**9,10,12,54,93,115,116,117,118,119**] studies have been performed since the first observations about metal ablation with ultra-short double pulses were published. The following paragraphs summarize the findings to provide an overview of the effects involved in ultra-short double pulse laser ablation.

When a double pulse with short inter-pulse delay times in the range of the electron-phonon coupling time is used, the ablated volume of such a double pulse equals the volume ablated by a single pulse with doubled pulse energy. With increasing inter-pulse delay, several physical effects can occur. For delays in the range of 10 ps to a few 100 ps, rarefaction wave interference takes place. As shown in Figure 4 (center), the ablation process of single ultra-short pulses is characterized by the occurrence of a shock wave and a following rarefaction wave. The intensity of the rarefaction wave is a critical parameter of a successful ablation event. If a rarefaction wave is too weak, no ablation occurs (cf. Figure 4 (top)). For double pulse ablation, weakening of such rarefaction waves is an important process. Exemplarily, calculated density and pressure distributions for different regimes of double pulse laser ablation are given in Figure 6 [10]. For the pressure distributions given in Figure 6, red regions denote regions with compressive pressures, whereas blue regions denote regions of tension. Figure 6a shows the result of double pulse irradiation with an inter-pulse delay of Δt=Δ= 0 ps (corresponding to a single pulse of double pulse energy). An intense shockwave is followed by an intense rarefaction wave. From the regions of highest tension (blue), ablation is initiated. If the inter-pulse delay is increased to around Δt=Δ= 20 ps, the first shockwave is weaker, since this wave is only initiated by the first pulse. The following rarefaction wave is less pronounced because the second pulse increases the pressure and, hence, lowers tension. As a result, less material is removed and the ablation depth is decreased (cf. Figure 6b). This effect of "rarefaction interference" is the reason for a lowering of the ablation depth for metals if the pulses are separated by several tens of picoseconds (blue region in Figure 7) [10,120]. If the inter-pulse delay is increased to Δt=Δ= 50 ps, the second pulse interacts with material that has already left the initial metal surface and no physical connection for the exchange of pressure remains. Instead, the second pulse is shielded by the ablation cloud induced by the first pulse [9,10,54,57,118,119,120]. The ablation cloud evolving from the surface after the first pulse is heated and partially disintegrates, leading to an extreme increase in temperature and plasma formation.

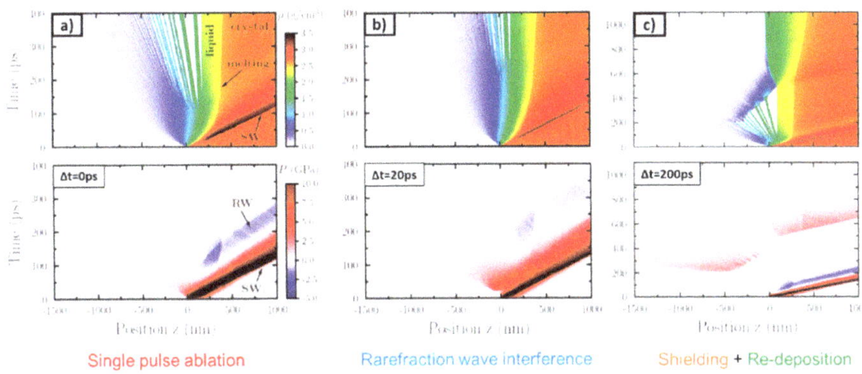

Figure 6. (a–c) Results from MD simulations showing the density (top) and pressure (bottom) distributions in space and time for different inter-pulse delays of double pulses. Material: aluminum, pulse duration 100 fs, wavelength 800 nm. Adapted with permission from Ref. [10].

The latter is an important effect beneficial for LIBS measurements and allows a better characterization of material due to an increased plasma temperature and, hence, a more pronounced spectrum (cf. Section 3.5.2). Depending on the fluence, not only the second pulse is shielded (and is not contributing to further material removal) but a re-deposition of the material ablated by the first pulse can take place (cf. Figure 6c). As a result, the ablated volume of a double pulse can be lower than the volume ablated only by the first pulse (valley of death) [57,75,95,103,109,111,121,122,

123]. This effect is material- and fluence-dependent. For example, the valley of death for stainless steel ranges between 100 and 200 ps at a fluence of 0.5 J/cm^2 [75] and shows a minimum at around 50 ps for 5 J/cm^2 [121]. Copper shows a valley of death for inter-pulse delays between 100 ps and 2 ns at fluences of 0.69 J/cm^2 [57] and 5 J/cm2 [121]. Aluminum shows a valley of death between 100 and 300 ps for an incident fluence of 5 J/cm^2 [121].

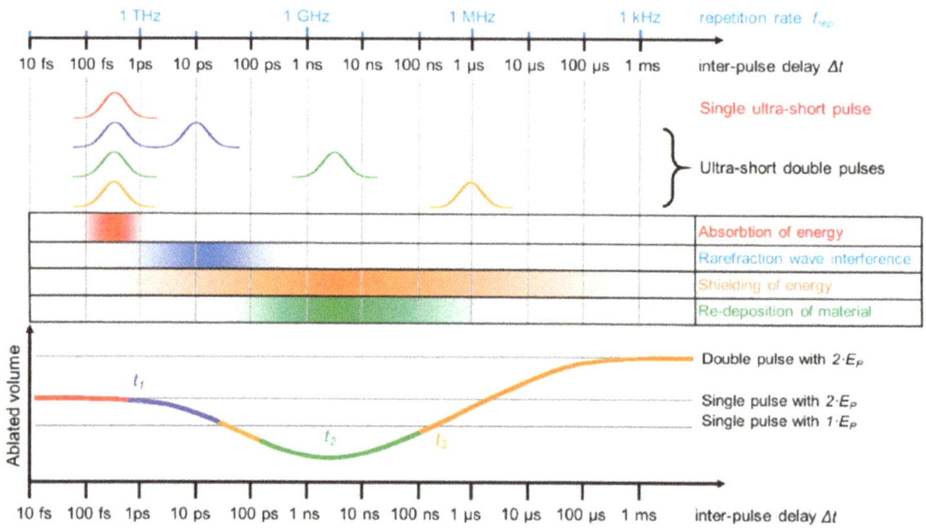

Figure 7. Interaction mechanisms of metals and ultra-short single and double pulses for different inter-pulse delays.

Titanium and platinum show valleys of death between 100 and 500 ps for incident fluences close to the ablation threshold [123]. For steel, a broader study with different time delays and fluences was also performed, and it was reported that the highest suppression of ablation takes place for steel at around a 1 ns time delay. The highest suppression ratio was close to 10, so only 10% of the volume ablated by a single pulse was finally removed using a double pulse (parameters: wavelength 1028 nm, fluence approx. 2.3 J/cm^2) [103]. If the inter-pulse delay is increased to several tens of nanoseconds, the shielding and re-deposition becomes less pronounced, since the ablation cloud formed after the first pulse dilutes with increasing time delay. For inter-pulse delays of $\Delta t > \Delta$ 10 µs, no significant interaction of the second pulse and the plume induced by the first pulse occurs anymore [58]. The discussed principal physical effects are summarized in Figure 7. Of course, the exact time delays, fluences, and pulse durations when these effects occur are material-dependent. The most significant differences between the materials are the time delay t_1, when the decrease of volume starts (blue region in Figure 7), the time delay t_2, when the minimum of ablated volume is reached, and the time delay t_3, when the re-deposition is not taking place anymore and the volume ablated by a single pulse is reached again (transition from the green to the yellow region in Figure 7).

For short delay times in the range of the electron–phonon coupling time (red color), the volume ablated by a double pulse equals the volume ablated by a single pulse with doubled pulse energy and the ablation process the same as for a single pulse, cf. **Figure 4** and **Figure 6**a. For delays in the range of 10 ps to a few 100 ps, rarefaction wave interference takes place, cf. **Figure 6**b. For inter-pulse delays between a few 100 ps and several 100 ns, a shielding of the second pulse by the ablated material of the first pulse takes place. For inter-pulse delays of a few nanoseconds, re-deposition of material ablated by the first pulse is another effect that can occur, cf. **Figure 6**c. The shielding effect diminishes for longer inter-pulse delays of several tens of nanoseconds due to the dilution of the plume initially induced by the first pulse. For delays of several tens of microseconds and longer, the second pulse does not interact with the plume, and the volume ablated by a double pulse is twice the volume ablated by a single pulse.

Although it is still the subject of current research, it seems that theses characteristic delay times are dependent on the electron–phonon coupling time. If the second pulse is reaching the target before the transfer of energy between the electronic and ionic system is finished, the material mainly reacts similar to an irradiation with a single pulse of double energy. For example, steel has a much shorter electron–phonon coupling time than copper (approx. 50 times lower), and the decrease in ablated volume starts from shorter time delays (t_1 ≈ ≈ 1 ps for steel, $_{t1}$ ≈ 10 ps for copper) [**75,103**]. For all metals investigated in the literature and as discussed before, t_2 lies between time delays of several 100 ps and several nanoseconds. For steel and copper, the difference is significant, e.g., for the fluences: 0.5 J/cm^2 for steel and 0.69 J/cm^2 for copper in the infrared, and $t2$ for steel amounts to ~200 ps, whereas for copper $t2$ amounts to ~2 ns [**75**]. Additionally, the delay time $t3$, when the volume ablated by the double pulse again is the same as for a single pulse, differs for both metals in this fluence regime. For steel, $t3$ amounts to approximately 1–3 ns [**75,103,122**], whereas for copper $t3$> 10 ns [**75,122**]. For aluminum, it was also reported that $t3$> 10 ns [**122**].

2.3. Triple Pulses with Intra-Burst Delays in the Range of Tens of Nanoseconds

As described in the previous section, the characteristic delay times $t1$, $t2$,, and $t3$ strongly depend on the material, especially on the electron–phonon coupling time, which also has a dominating influence onto the removal rates when a third pulse is added to the double pulse. If $t1$ and $t2$ are short compared to the intra-burst delay and $t3$ shows a value in the range of this delay, the energy specific volume obtained for a triple pulse only slightly differs from the one of single pulses, e.g., steel 1.4301 (AISI 304) shows such short delay times and was intensively investigated. Different reported maximum energy specific volumes for single, double, and triple pulses are summarized in **Figure 8**a [14,19,23,24,102,107]. For pulses with a pulse duration of 10 ps, an intra-burst delay of 12 ns, and wavelengths of 1064 nm and 532 nm, the maximum energy specific volume slightly decreased for a triple pulse compared to a single pulse [14,102,107]. An almost identical behavior was observed for a 2 ps pulse duration but with higher ablated volumes due to the usage of shorter pulses [19]. Similar results with even higher energy specific volumes have been reported for 300 fs and 350 fs pulse duration in the near infrared (NIR) [23,60]. In contrast to the rest of the literature, in [24], a significant drop by a factor of two of the maximum energy specific volume for steel (from about 0.4 mm³/min/W for single pulses down to less than 0.2 mm³/min/W, **Figure 8**a, blue empty squares) for a triple pulse burst was reported for a 210 fs pulse duration for a wavelength of 1030 nm. Here, in contrast to most other reported experiments by different authors, the fluence was varied by defocusing. This leads to different spot sizes at the fluences where the maximum energy specific volume was detected for the three situations. It was found for single pulse operation that there is a dependence of the threshold fluence, the energy penetration depth, and the maximum energy specific volume for steel 1.4301 (AISI 304) and copper when using different spot sizes [124]. This could serve as a possible explanation for the atypical behavior reported in [24] in contrast to all other studies in this field. The influence of the spot size on laser milling in general, but especially for bursts, still needs to be clarified by further studies.

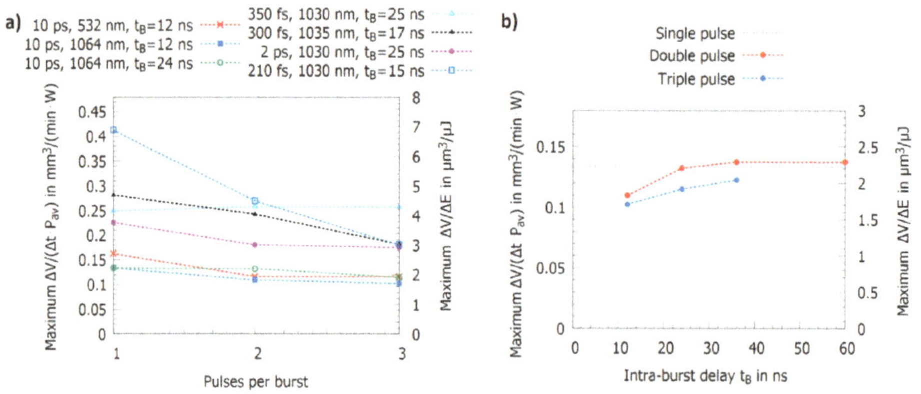

Figure 8. Maximum measured energy specific volumes for single, double, and triple pulses for steel 1.4301 (AISI 304) for (a) different pulse durations, wavelengths, and intra-burst delays and (b) as a function of the intra-burst delay for double and triple burst pulses with a 10 ps pulse duration and a 1064 nm wavelength. Data taken from [14,19,23,24,102,107].

Figure 8b shows the maximum energy specific volume for steel 1.4301 (AISI 304) using pulses with a pulse duration of 10 ps at a wavelength of 1064 nm and an intra-burst delay increasing from 12 ns to 36 ns for triple pulses and 60 ns for double pulses [14]. For the shortest intra-burst delay of 12 ns, the maximum volumes are smaller than the value for single pulse operation. For double pulses with an intra-burst delay of 24 ns and more, the same maximum value as for single pulse operation is reached (deviations are in the range of the measurement error). For triple pulses, the maximum energy specific volume also increases with increasing intra-burst delay but does not yet reach the value of single pulse operation with a 36 ns pulse delay.

The situation completely changes for copper (having longer delay times $t1$, $t2$,, and $t3$) as shown in **Figure 9**a,b. For double pulses, the aforementioned valley of death with a reduction in the maximum energy specific volume to less than 50% of the value for single pulses due to shielding effects is observed [**14,24,59,60,102**]. On the other hand, for triple pulse processing, the value strongly increases again and may even exceed the value for single pulses [**14,24,59,60,102**]. As can be seen in **Figure 9**a, the maximum energy specific volumes are functions of the pulse duration (an increase for shorter pulse durations is observed) and wavelength (for copper, green radiation is beneficial and results in higher maximum energy specific volumes, which is mainly due to higher absorptance [**16**]). For triple pulses, in the first place, it seems that after the described re-deposition effect (cf. **Section 2.2.**) induced by the second pulse, the third pulse should simply contribute again to the ablated volume. However, assuming only a complete shielding and no re-deposition of the second pulse (no contribution to the ablated volume) and a full contribution of the third pulse to the ablated volume, the maximum energy specific volume should not exceed two-thirds of the value of single pulses. This is not the case. Indeed, for triple pulse processing, the maximum energy specific volume is the same or even higher than for single pulse processing.

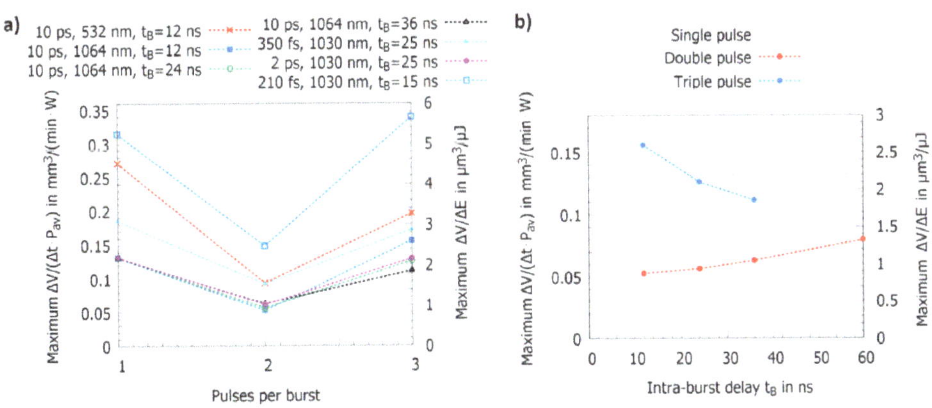

Figure 9. Maximum measured energy specific volumes for single, double, and triple pulses for (a) different pulse durations, wavelengths, and intra-burst delays for copper and (b) as a function of the intra-burst delay for double and triple burst pulses on copper with 10 ps pulse duration and 1064 nm wavelength. Data taken from [14,19,23,24,102,107].

The contribution of the second pulse was further investigated in [102,125], where the energy of the second pulse was varied compared to the first and third pulse, i.e., the relative energies were chosen to be (1,η,1). The pulses had a pulse duration of 10 ps, a wavelength of 1064 nm, and an intra-burst delay of 12 ns. For the situation (1, 0.25, 1)(1, 0.25, 1), the energy specific volume does not differ from the one of a double pulse with an intra-burst delay of 24 ns. When the energy of the second pulse is then raised to (1, 0.5, 1)(1, 0.5, 1), the maximum energy specific volume is almost doubled and even exceeds that for single pulses for the (1, 0.75, 1)(1, 0.75, 1) and (1, 1, 1)(1, 1, 1) situations. This shows that a minimum energy for the second pulse is demanded to start its clearing effect and that it saturates at a maximum energy. However, the high maximum energy specific volume for a triple pulse cannot be explained only by this effect.

Therefore, it was stated that additional effects influence the ablation process in case of triple burst pulses on copper to allow an enhanced ablation. In [**16**,**17**], results from calorimetric measurements with 10 ps pulse duration were obtained from two different calorimeter set-ups. For copper, it is shown that the residual energy coefficient, resembling the part of the applied laser energy remaining and heating up in the material only slightly varies between single, double, and triple pulses processing. For both measurement setups, the residual energy coefficient was measured to be slightly higher (an increase in the range of 2%) for the double pulse, but the difference was in the order of the measurement error. However, the absorptance of the machined surface changes significantly when using lasers in the NIR. The corresponding absorptance increased from 1.5% of the initial surface (high reflectivity) to 15.7%, 17.9%, and 26.8% for single, double, and triple pulses, respectively. This increase was obtained from the first machined layer and rests almost unchanged when the number of layers was increased to 48. For a triple pulse, the first pulse in the burst sequence could be absorbed much more strongly compared to single pulses, since 26.8% of incident energy was absorbed instead of 15.7% [**16**]. This could explain the further increase in the maximum energy specific volume for triple pulses on copper. This assumption is further supported by the results obtained for a wavelength of 532 nm, where the absorptance of the machined surface again increases from 35.7% for single pulses to 49.5% and 58.5% for double and triple pulses, respectively. This increase was lower than for 1064 nm, which is in accordance with the smaller increase in the energy specific volume for 532 nm and the triple pulse burst shown in **Figure 9**a. Hence, the absorptance seems to play a crucial role, and, due to significantly increased absorptance after the second pulse, the energy specific volume of copper might be increased to or even above the value of the single pulse operation. It should be kept in mind that these absorptances were measured on a cold surface ca. 10 mm out of the focal plane and therefore with a spot diameter between 400 µm and 700 µm, whereas, during machining, the surface is hot, and the spot diameter is in the range of 25 µm to 40 µm. Furthermore, the absorptance of a surface can also vary during irradiation with an ultra-short pulse, as shown in [**126**]. How this behavior affects burst pulse processing is unknown today and still a subject to further investigation.

In **Figure 9**b, the maximum energy specific volumes for double and triple pulses for pulses with a pulse duration of 10 ps at a wavelength of 1064 nm are shown as a function of the intra-burst delay tB [**17**]. Due to the longer delay times $t1$, $t2$,, and $t3$ of copper, the values seem to increase for intra-burst delays exceeding 60 ns in contrast to steel, where the single pulse value is already obtained for an intra-burst delay of 24 ns, as illustrated in **Figure 8**b. However, in contrast to steel, the maximum energy specific volume decreases for triple pulses when the intra-burst delay is increasing (cf. **Figure 9**b). This is a surprising behavior as the shielding effect of the second pulse is definitively reduced. A possible explanation is that since less material is redeposited, more material remains above the workpiece in a less dense ablation cloud, leading then to a shielding of the third pulse. However, this is only a hypothesis and further investigations are needed to clarify the physical processes involved during triple pulse ablation.

Figure 10a shows the maximum energy specific volumes for brass [**60**], aluminum [**23,107**], silver [**59**], gold [**59**], and molybdenum [**23**]. Except for molybdenum, a strong shielding effect for the double pulse burst followed by a significant increase in the maximum energy specific volume for the triple burst pulse is observed. Following Formula (4), the energy specific volume for a Gaussian beam shows a maximum value at the optimum peak fluence $\phi 0,pt=e2\cdot\phi th$ [**61**]. Introducing this optimum peak fluence into the expression for the energy specific volume (Formula (4)) leads to the ablated volume per pulse at the optimum fluence, which does not depend on the threshold fluence but directly scales with the energy penetration depth δ [**65,127**], i.e., the higher the energy penetration depth, the higher the volume ablated per pulse. The value of δ can be obtained by a least square fit of the formula to the data for single pulses reported in [**14,23,59,60,107,128**] following Formula (4). A larger ablated volume should lead to a stronger shielding effect for double pulses and therefore the ratio should drop for higher energy penetration depths. This is confirmed in **Figure 10**b where this ratio is plotted as a function of the energy penetration depth δ. The circles denote situations where a clear shielding for the double burst pulse followed by a significant increase of the energy specific volume for the triple burst pulse is observed, whereas for the triangles, the maximum energy specific volumes for the three situations only slightly differ. The figure indicates a trend of ratios near 1 for short energy penetration depths in the range of 10 nm towards values below 0.5 for energy penetration depths of 40 nm and higher. Hence, the removed volume per single pulse could also be a factor describing the ablation behavior for double and triple pulses. Again, further investigations are needed to gain a clearer picture and to be able to distinguish between the influence of parameters such as the spot size, the intra-burst delay, and the energy penetration depth.

$$\left(\frac{\Delta V}{\Delta E}\right)_{max,double\ pulse} / \left(\frac{\Delta V}{\Delta E}\right)_{max,single\ pulse} \tag{5}$$

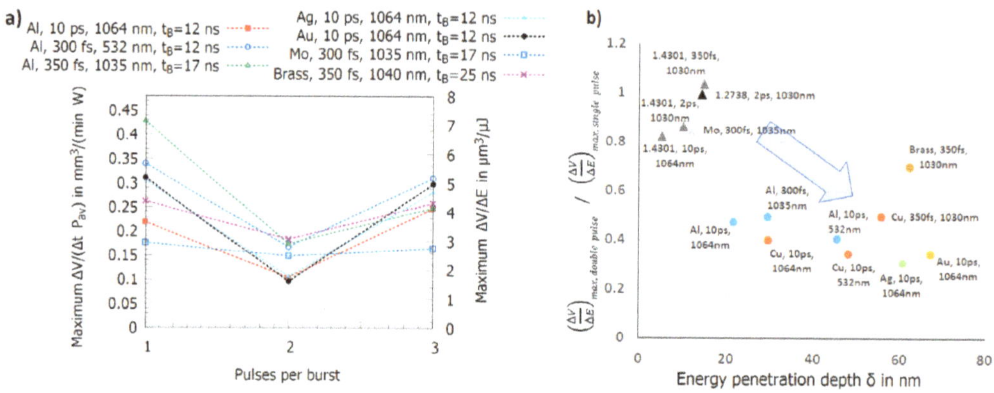

Figure 10. Maximum energy specific volumes for single, double, and triple pulses for (a) brass, aluminum, silver, gold, and molybdenum and (b) the ratio of the maximum energy specific volume of a double burst pulse and single pulses as a function of the corresponding energy penetration depth for different metals. The triangles denote metals where the energy specific volume of a double and triple pulse burst does not show a shielding effect for the double pulse burst followed by an increase in the energy specific volume for the third pulse within a triple pulse. Data taken from [19,23,24,60,128,129].

2.4. Multi-Pulse Bursts with Intra-Burst Repetition Rates in the MHz Range

The results for multi-pulse bursts for copper, brass, and aluminum are summarized in **Figure 11**a [**19,23,24,60**]. Copper, brass, and aluminum show an alternating behavior of the maximum energy specific volumes with high values for odd numbers of pulses in the burst and low values for even numbers of pulses. This can be explained as follows: the second pulse is fully or partially shielded but its energy clears the ablation cloud such that the third pulse can be absorbed by the surface. The third pulse again generates an ablation cloud, and the fourth pulse is therefore shielded again. The fourth pulse again clears the plume of the third pulse so that the fifth pulse can again contribute to the ablation process etc. For copper and a 210 fs pulse duration [**24**], aluminum and a 300 fs pulse duration [**23**], and brass and a 350 fs pulse duration [**60**], this oscillation seems to be damped.

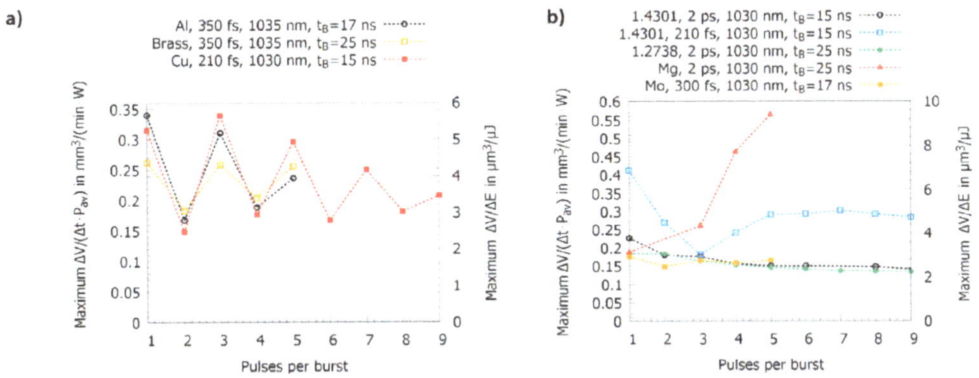

Figure 11. Maximum energy specific volumes as a function of the number of pulses per burst for (a) copper, brass, and aluminum as well as for (b) stainless steel, molybdenum, and magnesium alloy. Data extracted from [19,23,24,60,128,129].

In contrast to this oscillating behavior, this alternating behavior is not observed for steel 1.4301 (AISI 304) [**19,24**], steel 1.2738 [**130**], molybdenum [**23**], and magnesium (cf. **Figure 11**b). For steel and molybdenum, the maximum energy specific volume shows a tendency to slightly lower values when the number of pulses per burst is increased. Steel 1.4301 (AISI 304) machined with 210 fs pulses [**24**] shows an atypical behavior up to 4 pulses per burst, but these results were obtained by defocusing to vary the peak fluence. Hence, different effects cannot be distinguished, as has been explained before. For magnesium alloy [**129**], the threshold fluence and the energy penetration depth amounts to 0.06 J/cm² and 6.5 nm, respectively. Therefore, a similar behavior as for steel 1.4301 (AISI 304) could be expected, but, as can be seen, the maximum energy specific volume increases by about a factor of three when the number of pulses per burst is raised from one to five. This is in contrast to the other metals and is similar to the behavior of silicon [**17**], where the maximum energy specific volume also increases with the number of pulses per burst. For silicon, this increase of energy specific volume is mainly due to changes in absorptance due to multi-photon absorption and stable liquid surface layers (having a higher absorptance) during processing [**17**].

Recently, the behavior of stainless steel was investigated for pulses at a wavelength of 1030 nm and with pulse durations of 270 fs, 1 ps, and 10 ps for single pulses for up to 9 pulses per burst with an intra-burst delay of 15.4 ns. The results show that the energy specific volumes for peak fluences of 0.5 J/cm², 1.5 J/cm², and 2.5 J/cm² mainly depend on the pulse duration, but follow identical trends when the number of pulses per burst is increased. Depending on the pulse duration and the peak fluence, regimes where a melt film is formed leading to smooth surfaces can be identified [**131**]. Similar smoothing effects were already reported for steel 1.4301 (AISI 304) in [**14,60**] and for cobalt and titanium alloys in [**78,132**]. Neither for copper nor for brass could a smoothing effect be observed, and the surfaces only start to oxidize when they are machined with pulse bursts [**14,60**].

In contrast to double pulses, the theory behind triple pulse ablation still shows an ambiguous picture. To allow for a better understanding of the physical mechanisms taking place during processing with triple pulses, further investigations with varying intra-burst delays need to be performed.

2.5. Multi-Pulse Bursts with Intra-Burst Repetition Rates in the GHz Range

In 2016, a new type of laser emitting ultra-short burst pulses with intra-burst repetition rates in the GHz range, and, hence, with intra-burst delays of several nanoseconds and down to a few 100 ps, gained attraction for material processing due to a report showing that the efficiency of the ablation process was significantly improved for a variety of materials [**15**]. However, the reported physical mechanism influencing the ablation process, the "ablation cooling effect," appears highly questionable and has been partially disproven for metal processing. Since it can be mainly attributed to heat accumulation [**133,134,135**], the publication was the starting point for materials processing with a completely new class of laser systems emitting pulse bursts with intra-burst delays of several 100 ps or at intra-burst repetition rates of several GHz, which have since then been built and studied by various research groups. In general, it can be concluded that drilling processes are always more efficient than milling processes in this processing regime [**135,136**], which was explained by different melt flows in those two regimes [**133**]. Furthermore, the increase in energy specific volume is highly dependent on the time span of the used burst train and increases with longer time spans [**22,136,137,138,139**]. Moreover, the used inter-burst delay plays a crucial role. For short time delays (corresponding to higher repetition rates) between the burst trains, heat accumulation is also more pronounced and contributes to an increase in energy specific volume [**112**].

It has been reported by several groups that both the ablated volume per time $\Delta V/\Delta t$ and the energy specific volume $\Delta V/\Delta E\Delta$ can be increased by several factors when processing material with ultra-short burst pulses with intra-burst repetition rates in the GHz regime compared to single pulse operation. For the milling of copper, it was reported that the maximum energy specific volume $\Delta V/\Delta Emax\Delta$ can increase by a factor of three [**135**], whereas for steel it can increase by a factor of two when using GHz burst pulses instead of single pulses [**112,135,137**]. For the drilling of copper, it was reported that the maximum energy specific volume $\Delta V/\Delta Emax$ can increase by a factor of 3.5; for steel, it can increase by a factor of 4.6; and, for aluminum, it can increase by a factor of 5.8 when using GHz burst pulses instead of single pulses [**136**]. All these effects diminish and are less pronounced when during experimental design the spot diameters on the samples investigated are also changed for optimization [**24**].

A thorough overview of data from many authors about laser milling using GHz laser sources is shown in Figure 12 (top) to accommodate diverse study designs. In the application of double pulses, the previously mentioned shielding effect prevails, resulting in a fivefold reduction in the highest attainable energy specific volume for intra-burst delays spanning several hundred picoseconds. Moreover, when the quantity of burst pulses escalates, the reduction in maximum attainable energy specific volume diminishes further, being inhibited by about 93% for steel when employing 5 to 25 pulses in a pulse burst [23,24]. After around 30 pulses per burst, the energy-specific volume increases once more, and for several hundred or thousands of pulses per burst, it surpasses the value observed for single pulse ablation in both copper and steel. A limited number of pulses inside a burst creates a region of increased density in the interaction zone between the laser and the target, enhancing the shielding effect significantly [134]. In the case of a moderate quantity of pulses (ranging from several tens to hundreds), the ablation process is primarily governed by heat accumulation, facilitating the removal of material in a gaseous state, even for pulses with a fluence significantly below the ablation threshold. This phenomenon occurs because the total available laser pulse energy is predominantly distributed among multiple pulses within the burst. In instances involving a substantial quantity of pulses (ranging from several hundred to several thousand) within a burst, the ablation process resembles that of nanosecond-duration pulses, leading to elevated energy-specific volumes and a deterioration of surface quality due to melt expulsion, as observed in GHz laser ablation [64,135,136]. Figure 12 (bottom) presents a straightforward yet equitable comparison of the maximum attainable energy specific volumes relative to the duration of burst packages, juxtaposed with the energy specific volumes of nanosecond and microsecond laser processes. The maximum energy specific volumes of GHz laser processes align completely with those of ns and μs lasers within the same temporal domain. This applies not only to steel but also to copper (not illustrated here). Consequently, it may be inferred that GHz processing is analogous to ns laser processing for pulses of equivalent duration in metals. Undoubtedly, further comprehensive investigations are required to obtain a complete understanding of this subject. The analyzed research indicate that surface roughness reaches a minimum with a limited number of burst pulses (associated with brief burst durations), which can be considerably lower than that of single pulses [21,23,24,64,124,135,141,142,143,144]. Over extended durations and increased pulse counts, heat accumulation predominantly influences laser processing, resulting in elevated energy-specific volumes and much greater roughness. This phenomenon is also observed with longer pulses in the nanosecond and microsecond regimes due to the onset of melting.

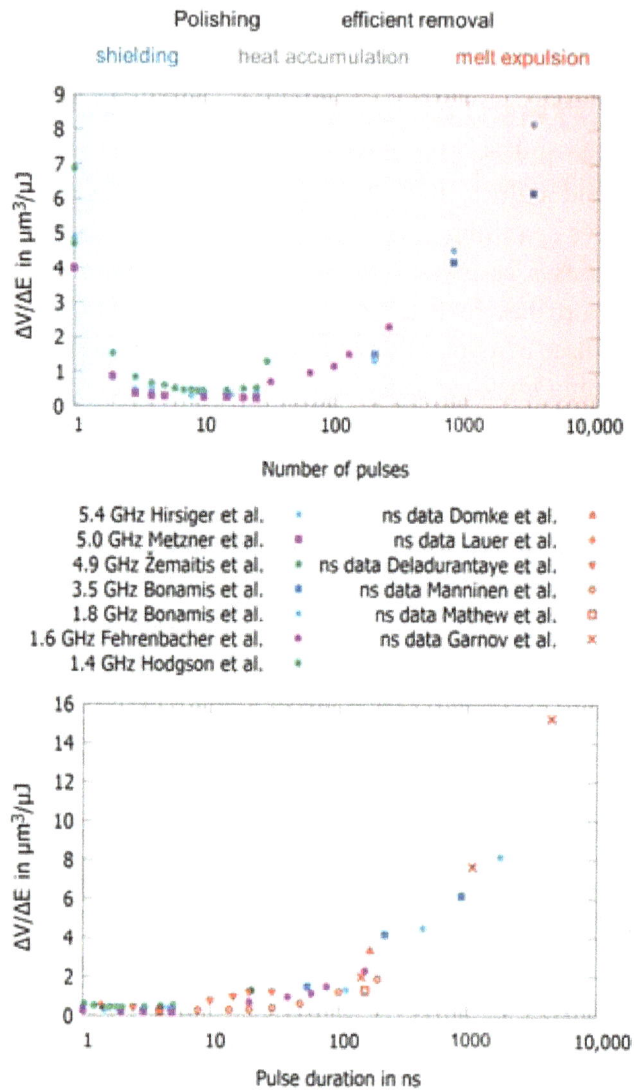

Figure 12. Maximum achievable energy specific volumes as a function of applied laser pulses within GHz bursts (top) and maximum energy specific volumes as a function of the duration of GHz bursts and durations of ns and µs pulses (bottom). Data from laser milling of stainless steel, data extracted from [21,23,24,64,98,124,135,141,142,143,144,145,146,147].

In addition to materials processing utilizing ultra-short burst pulses with intra-burst repetition rates in the MHz and GHz ranges, research is also focused on processing with THz intra-burst repetition rates, which correspond to delay durations between successive pulses of several picoseconds. In this regime, matter responds to stimulation with successive pulses in a manner analogous to its reaction to prolonged pulses within the same temporal domain. In addition to the dependence of the incubation effect (a material's degradation) on burst spacing [149,153], the material response is analogous when employing multiple consecutive ultra-short pulses over a specified duration compared to a single pulse with an equivalent duration. For instance, a burst of 10 pulses administered over a 10 ps interval exhibits effects comparable to a single pulse lasting 10 ps. Consequently, distinguishing physical impacts across these time spans is difficult. From the perspective of material processing, the application of burst pulses with THz intra-burst repetition rates is of lesser significance, however it has been utilized in the creation of particle beams and X-rays (see Section 3.5.3).

3. APPLICATIONS USING BURST PULSES

The subsequent subsections summarize and explore applications related to pulse burst processing of metals. The primary emphasis is on micromachining applications, with a summary of applications from other domains utilizing pulse bursts included in the final subsection.

3.1. Punching and Drilling

Ultrashort pulsed laser light has long been a prevalent approach for creating precise holes in diverse materials through drilling. A substantial quantity of laser pulses and adequately elevated pulse energy are generally necessary to create blind and through holes with high aspect ratios of 1:10 or above. The explanation resides in the characteristics of the drilling process, wherein the irradiated region often expands with increased drilling depth and pulse count, resulting in a gradual decline of fluence within the boreholes over time [74,154,155]. This need fundamentally opposes the division of pulse energy during burst processing, which is intended to enhance laser drilling efficiency and reduce drilling time. An increase in the pulse repetition rate also results in more significant shielding effects due to the lateral confinement of the ablation cloud. Laser drilling entails the interaction of radiation with ablation products generated from prior laser pulses. Reduced intervals between consecutive laser pulses amplify this impact [57,61]. There have been relatively few investigations into the impact of bursts in laser drilling utilizing ultra-short pulses. An early study on percussion drilling utilizing MHz bursts of aluminum demonstrated benefits regarding drilling duration [156]. Nonetheless, these benefits must be considered within the framework of the repetition rates prevalent in the kHz range during that period. The primary advantage was the increased pulse frequency attained by the burst, leading to a greater volume of ablation per unit of time. The interactions of burst pulses with ablation products in boreholes were examined diagnostically, while the drilling of glass was analyzed by time-resolved imaging [157]. Since the examined material was not metallic, the reported effects are likely somewhat analogous. Despite the apparent shielding effects of radiation within the borehole, the resulting holes had a more uniform geometry with smoother walls. Comparable findings were noted during silicon drilling [158], when MHz burst processing led to a reduction in undesirable side channels and enhanced reproducibility of the hole geometry. This is primarily attributable to re-deposited material, resulting in smoother surfaces. In [159], the drilling of copper and aluminum was examined using double pulses, with the inter-pulse delay varied from -90 ps to 90 ps. The observed decrease in drilling speed while employing double pulses was also ascribed to shielding. This starkly contrasts with the findings presented in [15], which showed significantly better removal rates for cutting copper and silicon using GHz bursts. In [133], an effort was undertaken to replicate these benefits; nevertheless, considerable melting was noted. Thus, as previously indicated, GHz processing appears to exhibit similarities with the use of nanosecond pulse lengths. The highest energy specific volumes attainable with a burst of 200 pulses at an intra-burst repetition rate of 1.76 GHz are nearly comparable when utilizing a laser with a pulse duration of 100 ns for percussive drilling of copper, aluminum, and stainless steel [64,135]. A work on laser drilling of invar using ultra-short MHz and GHz bursts has revealed an equivocal outcome. Variations in sample thickness and intra-burst delay resulted in reported drilling durations that were either marginally shorter or longer than those observed during continuous drilling without bursts [137]. Most research on laser drilling utilizing ultra-short burst pulses indicates a negligible effect on drilling speed, although a heightened production of melt occurs. Regarding the extent to which the unique characteristics of bursts can be advantageous in laser drilling, much research is still required.*3.2. Scribing and Cutting*

In several works, line scribing using bursts have been performed [**22**,**133**,**135**,**151**,**160**] with similar results. The removal rate using bursts is smaller, and the burr formation is increased compared to single pulses. These effects were confirmed by the simulations of Matsumota et al. [**133**]. Using a 10-pulse burst with an intra-burst delay of 16.7 ns (corresponding to an intra-burst repetition rate of 60 MHz) led to a decrease in the ablation depth of about 40% compared to single pulse operation using the same average power in the case of stainless steel [**22**]. The authors have attributed this decrease to plasma shielding, as the small ablation geometry (machining of deep trenches) does not allow a fast dissolution of the plasma between two pulses. In [**160**], a decrease in the removal rate for stainless steel (>15%) and copper (>5%) using bursts with an intra-burst delay of 1 ns (corresponding to an intra-burst repetition rate of 1 GHz) and different energy distributions was reported. Using a similar intra-burst delay of 1.13 ns (corresponding to an intra-burst repetition rate of 0.88 GHz), it was reported by Bonamis et al. that in certain conditions (high fluence and a pulse overlap higher than 70%) the grooves were refilled by molten material [**135**]. The same effect was also observed in line scribing using single pulse operation at a comparably high pulse repetition rate of 4.1 MHz [**161**]. Simulations confirm this phenomenon for GHz bursts [**133**]. It has also been shown that using a pulse burst consisting of 160 pulses with an intra-burst delay of 1.16 ns (corresponding to a frequency of 864 MHz) leads to a larger burr compared to single pulses [**133**].

Principally, no advantages in using ultra-short pulse bursts for line scribing has been reported in the literature so far. Further research is also needed to clarify the influence of pulse bursts in the field of precision cutting.

3.3. Surface Structures

3.3.1. Polishing/Smoothing

In ultra-short pulsed laser machining of some metals like steel and titanium, it is well known that in certain conditions bumpy surfaces can appear [**29,162**]. In most applications, these bumps usually need to be avoided.

As it was found by Bauer et al., those surfaces are created if the surface temperature is higher than about 600 °C before the next laser pulse impinges on the surface in case of stainless steel 1.4301 (AISI 304) [**73**]. Heat accumulation is the limiting factor for the productivity of laser processes where a good surface quality is required. Alternatively, heat accumulation can also be used as an advantage. Lickschat et al. [**105**] and Herrmann et al. [**163**] have shown a smoothing of cavities using pulse bursts. Nyenhuis et al. [**142**] and Michalowski et al. [**143**] have made the same observations and have further developed a smoothing process for stainless steel 1.4301 (AISI 304). Using GHz bursts with an intra-burst delay of 625 ps (corresponding to an intra-burst repetition rate of 1.6 GHz) and a fluence for each individual pulse below the threshold fluence, the heat input into the material can be utilized to create a small melt film that smoothens the bumpy surface (cf. **Figure 13**).

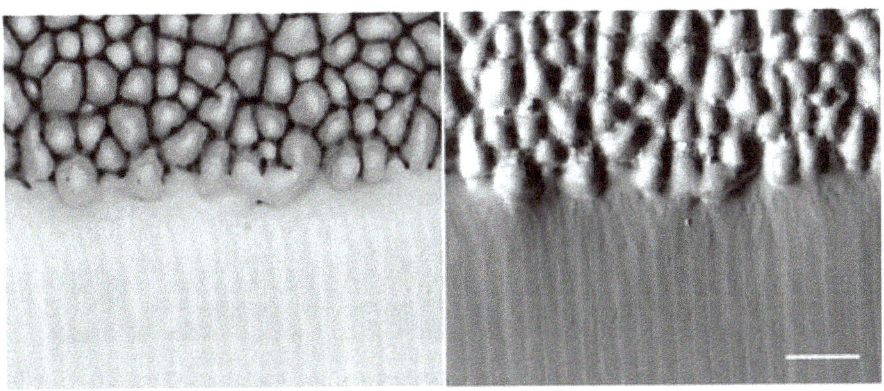

Figure 13. SEM images of a bumpy surface on 1.4301 (AISI 304) steel, where the lower part is polished using GHz bursts. The right-hand image shows the topological view. The length of the scale bar represents 40 µm. Reprinted from [142].

It has been found that high fluences lead to pores in the solidified melt films. Therefore, it is favorable to use more pulses in the burst, i.e., a burst of longer time span at lower fluences of the burst pulses. In comparison to a MHz burst (intra-burst delay of 12.5 ns), the smoothing using GHz bursts is much more energy efficient as the total fluence of the burst needed is about 2/3 smaller using such GHz bursts [**142**]. The low fluences used in the GHz burst processing lead to a very thin melt film which can also be used for polishing of microstructures without damaging them [**142,143**]. Using this kind of polishing process can lead to a surface roughness of Sa=0.13 μm and Sz=3.4 μm [**164**]. Using an intra-burst delay of 200 ps (corresponding to a repetition rate of 5 GHz) Metzner et al. have shown a smoothing on stainless steel surfaces [**21**]. The bumpy surface created during single pulse engraving was smoothed using GHz bursts. The more layers with GHz bursts are applied to the surface, the smoother the surface gets. No bumps are present on the surface anymore after 20 overscans (scan levels), as can be seen in **Figure 14**. After 20 overscans, a surface roughness of Sa= 0.1 ± 0.05 μm and Sz= 1.3 ± 0.3 μm was measured.

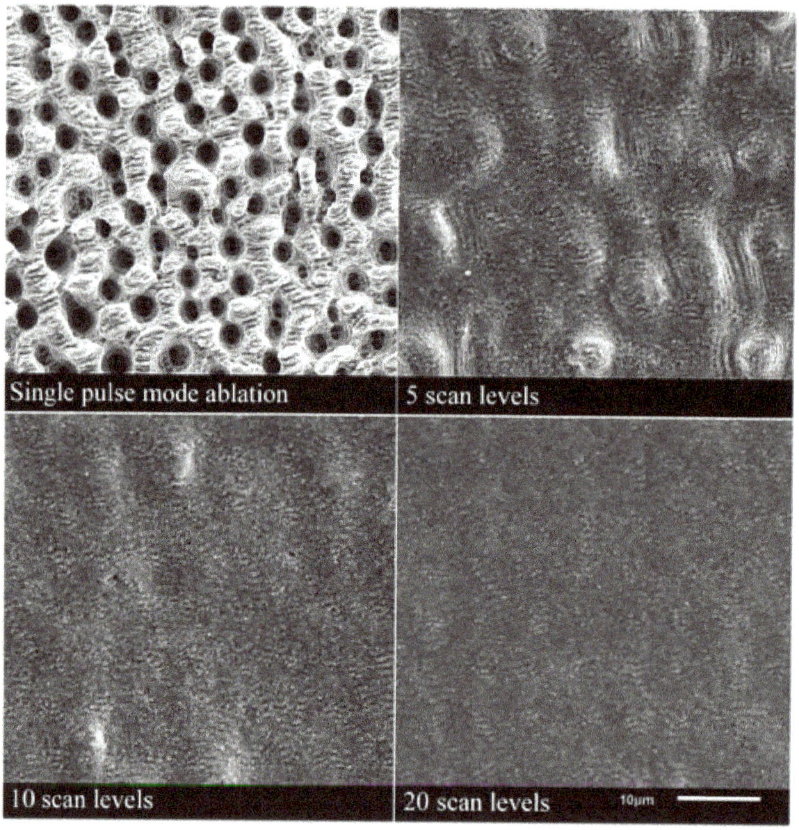

Figure 14. SEM images of surface structures on stainless steel. Top left: Surface structure, ablated in single pulse operation, a scan speed of 1.2 m/s, a pulse duration of 270 fs and 20 overscans (scan levels). The other SEM images visualize the structure after the surface treatment in the GHz burst mode with four pulses per burst, a pulse duration of 270 fs and a variation of the overscans (scan levels) between 5 and 20. Reprinted from [21].

To be able to use the maximum available pulse energy provided by a laser system, engraving processes can be done using MHz bursts instead of single pulse operation, followed by a GHz-polishing step. For such multi-step strategies, surface roughness values of Sa= 0.15 ± 0.02 µm and Sz=1.1 ± 0.1 µm have been measured. Again, after 20 overscans, the surface was found to be smoothed [**21**]. The advantage of processes combining MHz and GHz bursts is the higher productivity, i.e., smaller cycle time during engraving processes.

In contrast, Brenner et al. used MHz bursts with an intra-burst delay tB of 12.5 ns (corresponding to a repetition rate of 80 MHz) at a base repetition rate $frep$ of 2 MHz (inter-burst delay of 500 ns) instead of GHz bursts to polish the surface of conducted hot-working steel (1.2738) [**130**]. The achieved surface roughness amounted to Sa= 0.21 µm and Sz= 3.65 µm. Osbild et al. analyzed the polishing process using MHz bursts in more detail [**128**]. It was found that using a low number of repetitions in the polishing process is favorable to achieve a melt pool depth of only 1 to 3 µm. By doing so, the roughness can be reduced also on small and fine structures.

On a CoCrMo alloy, 3 to 5-pulse bursts with an intra-burst delay of 12.5 ns (corresponding to a repetition rate of 80 MHz) may lead to a much smoother surface compared to single pulse operation [**78**]. The surface roughness for single pulses increases with the structure depth, i.e., with increasing number of overscans, ranging from 300 nm for 10 overscans to 2.7 µm for 399 overscans. Using pulse bursts the smoothing effect can be maintained regardless of the structure depth. In another study, Metzner et al. used bursts with up to 8 pulses and an intra-burst delay of 12.5 ns (corresponding to a repetition rate of 80 MHz) and a repetition rate of 100 kHz (inter-burst delay of 10 µs) to reduce the surface roughness on cobalt and titanium alloys below Sa = 100 nm [**132**]. Depending on the material and the used fluence of an individual pulse in the burst, different numbers of pulses in the burst (in general 6–8 pulses) need to be applied. Using a higher number of pulses in the burst also helps to avoid the formation of nanoscopic cavities on the surface as shown in **Figure 15** for cobalt alloy. The same also can be observed for titanium [**132**].

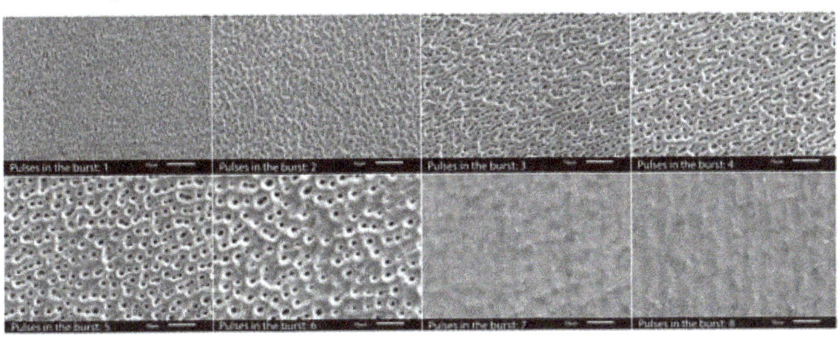

Figure 15. SEM images of the irradiated cobalt alloy surface as a function of the number of pulses in the burst at a fluence of 0.5 J/cm² per pulse. Reproduced under the terms of a Creative Commons Attribution 4.0 International License, (https://creativecommons.org/licenses/by/4.0/ (accessed on 31 May 2021)) [132].

Sassmannshausen et al. recently proposed a method for surface smoothing through continuous surface melting utilizing a laser that emits ultra-short pulses at a repetition rate of 49 MHz, equating to an inter-pulse delay of 20.4 ns. This extreme situation exemplifies classical repetitive processing at a constant repetition rate, while simultaneously being interpretable as an extended burst pulse in the MHz range. This ongoing processing demonstrates the proficiency of extended, precisely calibrated burst pulses. The polishing procedure demonstrated a reduction in surface roughness to approximately 80 nm. Although surface oxidation was also added, this value represents the least ever attained using laser surface treatments [165]. Metzner et al. conducted a detailed investigation comparing the smoothing effects of an MHz burst, a GHz burst, and a biburst (a mixture of MHz and GHz bursts) on stainless steel [131]. The smoothing method demonstrates greater stability and reproducibility when employing GHz bursts rather than MHz bursts. A surface roughness of $Sa < 200$ nm was recorded during processing with the GHz burst under ideal conditions. It is noteworthy that both metals and polymers can be polished using pulse bursts with an intra-burst delay of 400 ps, which corresponds to an intra-burst repetition rate of 2.5 GHz [166].

3.3.2. Coloring

Laser coloring was intensively investigated for ns pulse processing where oxide layers are formed [**167**,**168**,**169**,**170**]. Alternatively, ultra-short pulsed lasers can be used. This is mainly realized by the following approaches: utilizing laser-induced periodic surface structures (LIPSS) as diffraction gratings to produce angle-sensitive colors [**171**,**172**] and creating tempering colors by heating up the surface [**173**], which results in angle-independent colors. Another way is to create micro- and nanostructures on the surface to obtain a certain color on the surface [**171**,**174**]. The coloring strategy using nanostructures was further developed by Guay et al. [**175**]. Plasmonic colors are excited from metallic nanoparticles and nanostructures, which are re-deposited after the laser irradiation leads to angle-independent colors on silver surfaces (cf. **Figure 16**).

Figure 16. Photograph of a colored silver coin (height approx. 1 cm); detailed view on the right-hand side. Reproduced under the terms of a Creative Commons Attribution 4.0 International License (https://creativecommons.org/licenses/by/4.0/ (accessed on 31 May 2021)) [175].

Research demonstrated that employing pulse bursts with an intra-burst delay of 12.8 ns (equating to an intra-burst repetition rate of 82 MHz) yields superior color quality on silver relative to single pulse operation. The authors measured the colors with a Chroma meter (CR-241, Konica Minolta) in the CIELCH color space. Chroma is a metric employed to quantify the colorfulness of a thing. The Chroma is augmented by approximately 50% across the entire color spectrum when utilizing burst pulses, in contrast to colors generated with single pulses [176]. The augmentation of Chroma through bursts can be elucidated by the concurrent formation of LIPSS onto which nanoparticles are redeposited, in contrast to single pulses where LIPSS are absent (cf. Figure 17c). The augmentation in surface area generated by the LIPSS, along with the field enhancement in these recesses, would result in heightened absorption, elucidating the more vivid hues characteristic of the burst process. Furthermore, a more extensive color palette can be generated by pulse bursts. The FlexBurstTM technology developed by Lumentum (formerly Time-Bandwidth Products, Zurich, Switzerland) [177] enables the attenuation of specific pulses within a burst, allowing for adjustments in intra-burst delay in increments of 12.8 ns and modifications to the pulse energy of each individual pulse within the burst. The alteration of the bursts results in an additional enhancement of Chroma (Figure 17a) and an expanded color palette (Figure 17b) [178]. The formulated technique was implemented on the metals: gold, silver, copper, and aluminum.

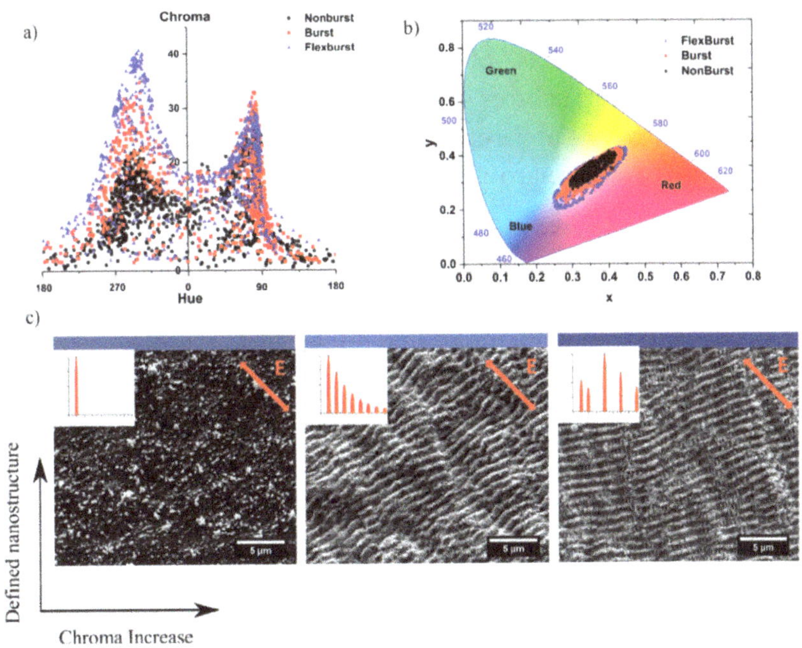

Figure 17. Colors and corresponding topographies on a silver surface. (a) Graph of Chroma versus hue comparing colors obtained after single pulse operation (•), burst (■), and flexburst (▲) coloring. (b) CIE *xy* chromaticity diagram comparing single pulse

operation (•), burst (■), and flexburst (▲) coloring. (c) SEM images of blue surfaces produced using single pulse operation (left), burst (middle), and flexburst (right) coloring. The hue was more or less the same for all squares ($H \approx 295°$), whereas the Chroma values equalled 22.3, 31.2, and 39.44. Nanostructures can be observed on the surfaces for burst and flexburst processing. The relative energy distribution of the burst pulses and the orientation of the electric field are given as insets. Reprinted from [178],

3.3.3. Laser-Induced Periodic Surface Structures (LIPSS)

Laser-induced periodic surface features, commonly referred to as ripples, can be applied to nearly all material surfaces utilizing polarized light. Typically, linearly polarized light is employed to fabricate such structures. Applications for LIPSS are available across multiple domains. Natural surfaces (colored, anti-reflective, hydrophobic, wet and dry adhesion, and friction-reducing surfaces) can be replicated [179], or the melt flow resistance in polypropylene injection molding can be diminished [180]. LIPSS can also be recreated on plastic components [181,182]. The phenomena behind ripple formation is examined in numerous articles [183,184,185,186]. In recent years, numerous studies employing pulse bursts have been conducted utilizing various polarizations of double pulses [187], differing wavelengths in double pulses [188], and alterations in energy distribution [189,190]. The quantity of pulses in the burst has been modified by as much as 32 pulses [191]. The amalgamation of two time-delayed ultra-short pulses exhibiting distinct polarization states can produce many structural kinds, unlike standard LIPSS, which are oriented parallel to the polarization direction in the case of linear polarization [187]. Figure 18 illustrates the alterations in surface structures generated by double pulses with delays of up to 10 ps (equating to an intra-burst repetition rate of 100 GHz) and employing either cross-polarized (XP) or counter-rotating circular-polarized (CP) pulses.

Increasing the number of pulses in the burst from 2 to 32 pulses while maintaining the intra-burst delay of 1.5 ps (corresponding to a repetition rate of 666 GHz) and linear polarization leads to an increase in the spatial period of the low spatial frequency LIPSS on stainless steel, while the depth of the LIPSS remains almost constant [**191**]. It should be mentioned that the spatial period of multiple pulses is always larger than for single pulse operation. Increasing the intra-burst delay from 1.5 ps to 24 ps using double pulses also leads to an increase in the spatial frequency, while the depth of the structures decreases drastically. This decrease is attributed to shielding effects taking place for delays larger than about 6 ps [**191**].

Using cross-polarized light instead of the parallel polarization for all pulses in the burst will lead to similar structures, as shown in **Figure 18** [**192**]. Comparing the wettability of triangular structures produced with a four-pulse burst at an intra-burst delay of 1.5 ps (corresponding to an intra-burst repetition rate of 666 GHz) with normal LIPSS created with a 16-pulse burst and an identical intra-burst delay shows only a difference in the evolution of the water contact angle over time, known as the aging effect, for the two different surface structures. The final water contact angle, which is reached after about eight weeks, was principally the same. It can be concluded that not only does the surface structures have an influence on the wettability but also the surface chemistry [**192**].

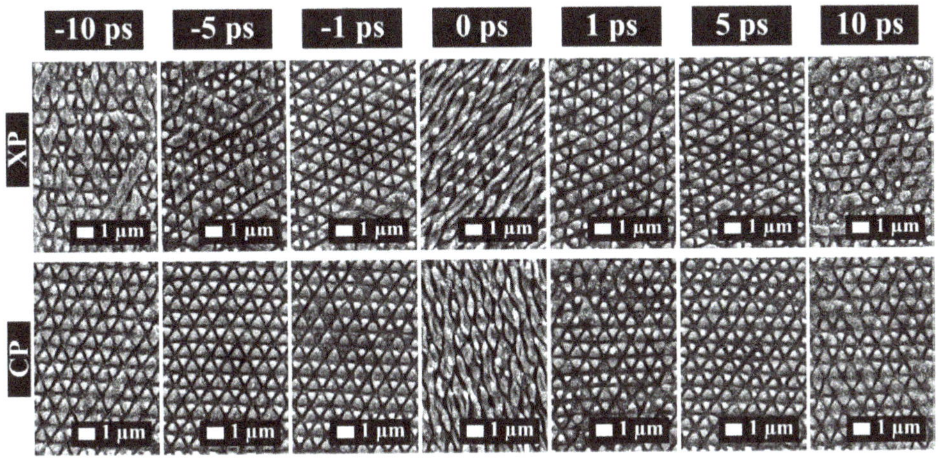

Figure 18. SEM images of the surface structures on stainless steel irradiated with two different polarization configurations: cross-polarized double pulses (XP) and counter-rotating circular-polarized pulses (CP). The rest of the process parameters were fixed: 10 pulses per spot, hatch distance of 1 µm, and a fluence of 0.1 J/cm^2. The delay of the two pulses was varied as indicated on top of the images. Reprinted from [187], copyright 2019, with permission from Elsevier.

Another approach is to combine two cross-polarized pulses with different fluences. The LIPSS' orientation on a titanium substrate can be influenced by the fluence of the delayed pulse and is in good agreement with the direction obtained by the vector sum of the laser fields. For example, using two pulses with polarizations perpendicular to each other with the same pulse energy, the LIPSS are formed at an angle of 45 degrees, while when one pulse has half the energy of the other pulse, LIPSS are formed at an angle of 24.5 degrees [**190**]. A similar setup but with equal polarization states for both pulses and a fixed intra-burst delay of 160 fs (corresponding to a repetition rate of 6.25 THz) was used by Hashida et al. [**189**]. The fluence of the first pulse was always below the threshold fluence, whereas the fluence of the delayed pulse was above the threshold fluence. It was shown that the laser fluence of the first pulse plays an important role for the spacing of the LIPSS. The higher the fluence of the first pulse, the larger the spacing becomes. Comparing the grating spacing produced with double pulses with the one produced with single pulses with the same total energy shows no difference [**189**]. Larger intra-burst delays of 20 ns (corresponding to a repetition rate of 50 MHz) have been used by Wang et al. [**193**] together with a higher number of pulses. Depending on the used fluence of an individual pulse in the burst, either low spatial frequency LIPSS (LSFL) or high spatial frequency LIPSS (HSFL) are formed on the surface. For a fixed total fluence of the burst of 0.25 J/cm^2, the formation of HSFL is enhanced by using three or four pulses in the burst, as the fluence of each pulse in the burst becomes smaller. Increasing the total fluence will lead to microgrooves, nanoholes, and melt formations in the central region of the machined grooves, sometimes referred to as "micro-hills" [**194**].

The use of double pulses with different wavelength (800 nm and 400 nm), fluences, and intra-burst delays and the use of cross-polarization states was investigated by Hashida et al. [**188**]. It was found that the fundamental pulse (800 nm) is responsible for the LIPSS creation and its orientation, while the second harmonic pulse will optimize the uniformity of the LIPSS if intra-burst delays of 0 to 2 ps are used. For delayed pulses (first pulse with 400 nm and the second pulse with 800 nm), the period and the direction of the ripples can be changed [**188**].

Experiments investigating the influence of double pulses onto the LIPSS formation have also been performed on semiconductors and glasses but are not further discussed in this article [**195,196**].

3.4. Milling

The primary advantage of burst mode in milling applications is exemplified by the subsequent analysis: Milling operations are typically executed using galvanometric scanners that provide maximal marking rates of several tens of meters per second [197]. The space between pulses, or pitch, is determined by the laser's repetition rate and the scanner's marking speed. Consequently, the maximum scanner speed constrains the permissible laser repetition rate if a specific minimum pitch value must be maintained. Consequently, pulse energies and peak fluences frequently exceed the optimal amount to maximize the average power of the employed laser system. At this elevated fluence, the energy-specific volume is markedly reduced compared to its maximum value, as demonstrated in Figure 19a, for single pulses with a peak fluence of eight times the optimal value. Partitioning this energy into n sub-pulses, or an n-pulse burst, diminishes the fluence of these sub-pulses and adjusts the energy-specific volume for each pulse inside the burst closer to its optimal value (see Figure 19a). Furthermore, as indicated in Figure 5, the ablation process transitions from the phase explosion to the spallation regime, resulting in enhanced surface quality. The heat accumulation is minimally impacted. Analyses of steel 1.4301 (AISI 304) [14] based on the analytical model in [73] indicate that the temperatures immediately preceding the subsequent burst sequence impacting the surface are consistent between a four-pulse burst and single pulses with quadrupled energy (see Figure 19b).

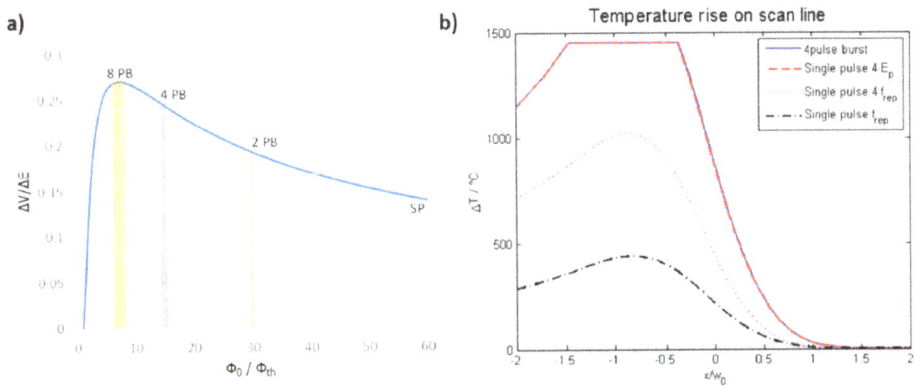

Figure 19. (a) Normalized energy specific volume as a function of the normalized peak fluence and the corresponding energy specific volumes for a total fluence of 60 times the threshold fluence for single pulses, double pulses, and bursts consisting of 4 and 8 pulses. (b) Temperature rise on the scan line just before the next burst sequence impinges on the surface for a 4-pulse burst on steel 1.4301 (AISI 304) with 12 W average power, 500 kHz repetition rate, and a pitch of 8 µm. The temperature rise is plotted for a burst with 4 pulses (blue line), single pulses with 4 times higher pulse energy (red dashed line), single pulses with 4 times higher repetition rate of 2 MHz (magenta dotted line), and single pulses having the same energy at a repetition rate of 500 kHz (black dash-dotted line). The plateau denotes the region where melting happens. The distance x is given in units of the spot radius *w*, and, as the previous pulse has struck the surface at *x*=–0.5·*w*, the maximum temperatures appears around this region. (b) was reprinted from [14].

The machining of stainless steel in single pulse operation is only possible in the low fluence regime, as the hole and spike formation limit the surface quality and therefore the productivity at higher fluences [**14,198**]. Using MHz bursts, the fluence of an individual pulse in the burst can be above the optimal fluence, as the bursts help to avoid this hole formation and therefore are more beneficial [**60**]. As shown in **Section 2.3.**, machining using a triple pulse on copper can be more efficient than using single pulse operation. A milling result using this strategy is shown in **Figure 20**.

It should be mentioned here that, to optimize the productivity in ultra-short pulsed laser milling, not only the ablation process itself must be optimized but also the scan strategy. It has been shown that by using optimized scan speeds, the time for machining a certain geometry, e.g., a line, can be optimized in the case of galvanometer scanners [**197**]. The given values of the energy specific volume are for 100% laser-on time, which is not true for galvanometer-based scanning using the sky-writing mode, which is used to obtain the best machining quality, i.e., no over-engraving near the edges [**199**]. The optimization of the energy specific volume alone makes no sense if the duty cycle of the scanner is near zero in the final application.

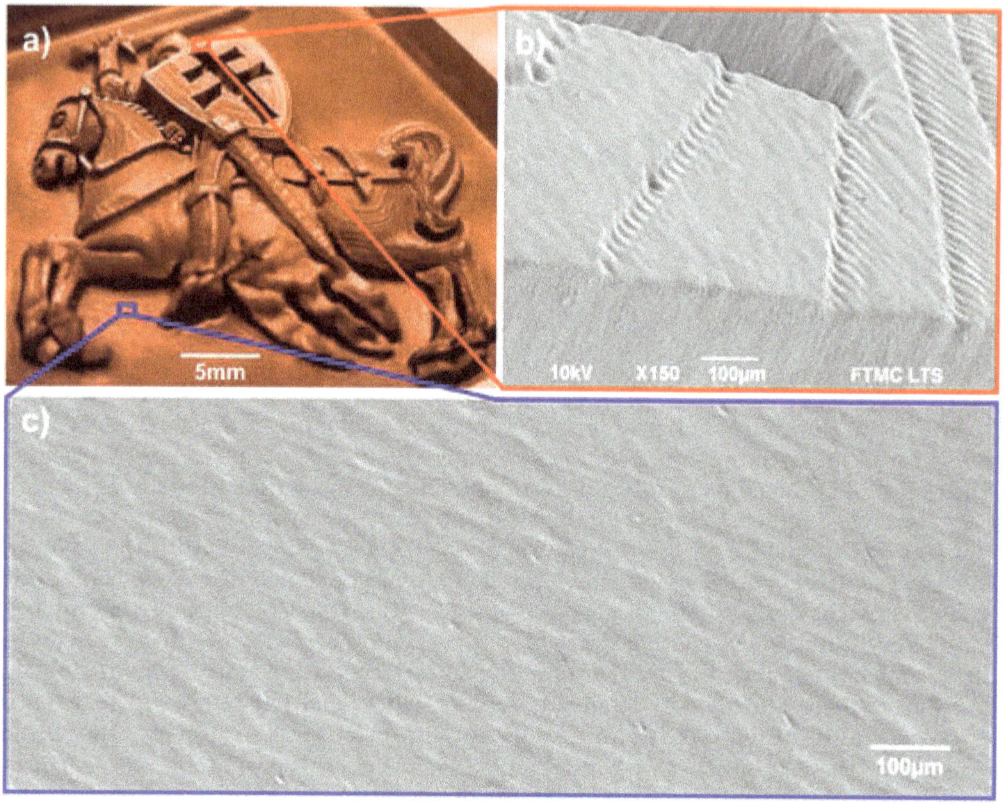

Figure 20. Example of efficient laser milling. (a) Optical image of the coat of arms of Lithuania milled in a copper plate. (b) SEM image of laser-milled surface illustrating layer-by-layer removal. (c) SEM image of the bottom surface of the laser-milled cavity. Laser parameters—triple pulse, laser wavelength of 1030 nm, burst repetition rate 300 kHz, intra-burst repetition rate of 64.5 MHz, and beam scanning speed of 1 m/s. Reproduced under the terms of a Creative Commons Attribution 4.0 International License, (https://creativecommons.org/licenses/by/4.0/ (accessed on 31 May 2021)) [18].

The former considerations are based on the assumption that only the peak fluence of the single pulses in a burst defines the energy specific volume and that the latter does not depend on the number of pulses per burst. This is definitively not the case, as, e.g., shown in [**14,130**] for stainless steel and as discussed in **Section 2.3** and **Section 2.4**, as the number of pulses as well as the intra-burst delay affect the energy specific volume. Further, to compare situations with equal average power single pulse operation, a repetition rate $frep$ should be compared with a n-pulse burst with a repetition rate of $frep/n$. In **Figure 21**, based on the results for stainless steel 1.4301 (AISI 304) machined with 10 ps pulse duration at a wavelength of 1064 nm and a spot radius of 15.5 µm [**14**], the energy specific volumes as well as the removal rates are shown for bursts consisting of eight, four, two, and one pulses at corresponding repetition rates of 200 kHz, 402 kHz, 804 kHz, and 1610 kHz, respectively. Up to an average power of about 7.5 W, the energy specific volume for single pulses at a repetition rate of 1610 kHz exceeds that of the burst with eight pulses, and the corresponding removal rates are higher for single pulses. For higher average powers, the situation changes, and bursts with eight pulses become more efficient, while the removal rate exceeds that of all other presented situations. These effects could eventually be caused by the smoothing effect and the reduction in the surface structures as shown in **Figure 13**, **Figure 14** and **Figure 15**.

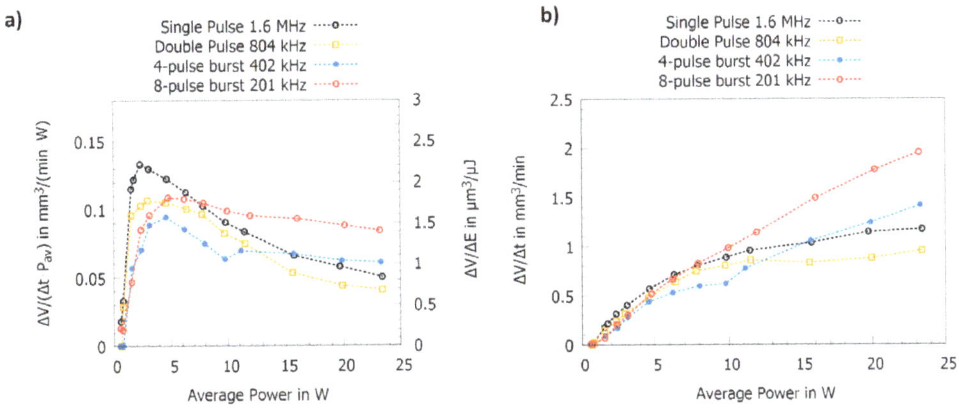

Figure 21. (a) Energy specific volumes and (b) removal rates for steel 1.4301 (AISI 304) machined with pulses of 10 ps pulse duration with a wavelength of 1064 nm and a spot radius of 15.5 µm. To guarantee identical peak fluence of the individual pulses, the laser repetition rate was 1610 kHz for single pulses, 804 kHz for the double pulse burst, 402 kHz for the burst with four pulses, and 200 kHz for the burst with eight pulses. Data extracted from [14].

The situation may change when higher repetition rates are used for single pulses, e.g., by using polygon line scanners, which can now achieve marking speeds of several 100 m/s [**200,201**]. However, the scale-up process is limited either by heat accumulation, especially for steel, and/or shielding effects discussed in the double pulse section, as shown in [**202**].

Therefore, for concrete applications, an optimization of processing strategies with respect to the number of pulses per burst, the laser repetition rate, the spot size, and the machining quality is demanded to find the best-suited parameters. Dual process strategies using different parameter sets, e.g., one for roughing with high energies and high removal rates followed by one with pulse bursts, either in the MHz or even the GHz regime used for finishing/polishing are very promising. For example, a dual process strategy of alternating GHz bursts and conventional processing was presented to achieve a good surface quality of machined pockets while decreasing the processing time by a factor of two, as illustrated in **Figure 22**.

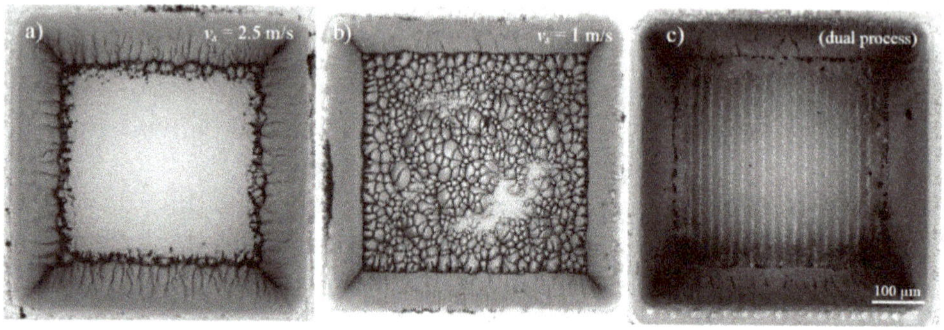

Figure 22. SEM images of milled squares with a scan speed of (a) 2.5 m/s and (b) 1 m/s at a conventional process and 1 m/s for (c) using the dual process strategy. The surface roughness of the inner ground was S_a= 0.23 µm and S_z= 4.1 µm for (a), but was not measurable because of high occurrence of bumps for (b); and S_a= 0.13 µm and S_z= 3.4 µm for (c). Although the scanning speed equaled that of (b), the edge and surface quality was the best and just the line spacing during polishing limited the relatively high S_z value. Reproduced from ref. [164] with permission from Japan Laser Processing Society.

Brenner et al. [**203**] demonstrated another interesting application by combining ablation processing in burst mode followed by a cleaning and an additional polishing process for large-format 3D mold tools. Another new approach of combining MHz and GHz bursts, the bi-burst mode, was recently presented in [**24**,**131**], but, in milling applications for copper and steel, this mode did not reveal higher removal rates compared to single pulses. In the field of laser burst processing, a comparison of literature data regarding roughness and energy specific volumes for different metals is a challenging task due to different study designs. For a comprehensive comparison of data for at least the metals of copper and stainless steel, the reader may refer to [**24**].

To conclude, the combination of different processes taking benefit of the advantages of each milling process to obtain optimum quality and machining time when using MHz and GHz bursts is a widely open field and still demands further research.

3.5. Further Applications

In addition to the micromachining of metals in industrial settings, pulse bursts are employed in various other domains. The subsequent subsections provide a quick introduction to these topics.

3.5.1. Laser-Ablative Space Propulsion

Another possible application of ultra-short laser pulses is the propulsion of either space vessels or space debris. When irradiating material and inducing ablation, the sample experiences a small force in the range of nN or even µN. The ablated material serves as propellant material. This force can be used as thrust to either stabilize and correct the orbits of space vessels, such as satellites, or to decelerate space debris in order to clean the earth's orbit [**204,205,206,207,208,209,210**]. Recently, it was shown that using ultra-short double pulses can give added value to this specific application. When using double pulses, material is pushed back to the initial surface. It was shown that, on the one hand, the measured thrust increases by a factor of two when using a double pulse instead of a single pulse of the same total energy (or using the same average power for processing) [**114**]. On the other hand, since the material is pushed back and can be used as propellant again, the thrust per ablated mass increases by a factor of three, leading to a more efficient propulsion and propellant conservation at the same time [**114**].

3.5.2. Laser-Induced Breakdown Spectroscopy

Due to the ability of ablated material to shield subsequent laser pulses, laser-induced breakdown spectroscopy, a technique for characterizing matter, can benefit from the atomization of the shielding material through the emission of distinctive spectra. LIBS depends on spectra emitted by the vapor and plasma generated following a laser pulse striking a material surface. The occurrence, strength, and ratios of distinct characteristic emission lines are examined to provide information about the material. This characterisation technique originated in the 1960s with the initial demonstration of Q-Switch lasers and was predominantly conducted using lasers with nanosecond pulse durations until the conclusion of the 20th century [**211,212,213**]. In 1998, the first LIBS measurements were reported using ultra-short pulsed lasers [**214**], and it was found that the ablation process itself and the atom excitation is more reproducible for fs than for ns pulses [**215**]. Hence, less energy is needed for excitation, which is crucial for the characterization of specific samples in order to keep the destruction of those limited, e.g., when examining paintings [**216,217,218**] or explosives [**219,220,221**]. Additionally, the sensitivity is higher when using fs lasers [**215**]. Several summarizing papers can be found in the literature regarding the single pulse fs-LIBS technique [**222,223,224**]. Signal amplification of the observed spectra can be achieved through the utilization of double pulses or multiple burst pulses [7,91,223,225,226], as the atomistic and plasma excitation, together with the resultant temperatures, are significantly elevated in comparison to single pulses, as previously described. Material processing scientists may experience ablation suppression with double pulses, whilst spectroscopists anticipate improved excitation of matter resulting from the atomization of the shielding material.

3.5.3. Production of Particle Beams and X-rays

Ultra-short laser pulses can generate streams of particles, including protons, electrons, heavy ions, and X-rays,[227,228,229,230,231,232,233,234,235,236,237] through the excitation of matter and the creation of high-energy plasmas. The spectrum of radiation that can be produced is already vast, although the scope of applications is far broader, encompassing medical uses in radiology and radiotherapy

[227,238,239,240]. Furthermore, the characterization of stuff can be conducted utilizing several investigative techniques reliant on particle beams [233,235,236]. Multiple factors, including produced intensity, spectral quality and purity, together with precisely determined energies of the emitted radiation, are crucial for these applications and are heavily dependent on the accurate excitation of matter. Double pulses, also known as pre-pulses in this particular domain of particle science, provide precise manipulation of the excitation states of heated matter. The pulse delays range from few tens to 100 femtoseconds and extend up to several picoseconds, enabling the tuning of the energy and spectrum of highly charged silver ions [241], the energy and spectrum of generated protons [231,242], and the spectrum of produced X-rays [243].

4. CONCISE SUMMARY

Metal processing utilizing ultra-short burst pulses has emerged as a significant area in laser processing over the past decade and is a topic for continued research. This review article summarizes the physical mechanisms involved in laser burst processing. When developing ablation processes, several factors must be considered, including radiation shielding, material re-deposition, increased absorptance, and heat accumulation, particularly when the intra-burst delay decreases and the pulse count inside bursts rises. While the phenomena associated with double pulses are well comprehended, the mechanisms elucidating the observed removal rates and energy-specific volumes for triple and multi-pulse burst ablation across various metals remain mostly unclear. Consequently, further research in this domain is essential. Material re-deposition can be employed for surface polishing through the application of burst pulses in the MHz range during processing. It facilitates the smoothing of material surfaces with a minimal number of burst pulses due to the re-deposition of material in a liquid condition. Increased pulse counts within a burst result in heat accumulation, which can create stable melt layers on the surface during processing. Increasing the intra-burst repetition rate to several GHz or higher allows for the utilization of heat accumulation effects to either smooth the surface at lower pulse energies or boost ablation at higher pulse energies, rendering it as effective as nanosecond laser processing. The augmentation of ablated volume per energy and duration correlates with an elevated surface roughness, consistent with the effects observed in nanosecond laser ablation. The underlying physical mechanisms facilitate numerous creative applications, such as surface smoothing, plasmonic surface coloration, and an optimal energy distribution to fully harness the average power of ultra-short pulsed lasers. The future is characterized by laser systems capable of producing diverse burst pulse combinations with intra-burst repetition rates in the MHz and GHz ranges, facilitating the execution of integrated operations on the machined workpiece.

AT A GLANCE

Since the early 2000s, laser processing with ultra-short double pulses has garnered significant interest. Over the past decade, pulse bursts including numerous pulses with delays of several tens of nanoseconds or less have been integrated into the field of metal micromachining, facilitating entirely new processing regimes and enhancing both structuring rates and the surface quality of machined samples. Various physical phenomena, including shielding and material re-deposition, have resulted in a revised comprehension of associated machining methods and processing regimes. The outcomes of both experimental and numerical studies are contextualized for various time scales in laser processing. This paper focuses on the essential physical events occurring during burst processing and their corresponding impacts on the machining outcomes of metals within the ultra-short pulse domain, with delays spanning from several hundred femtoseconds to several microseconds. Additionally, a discussion of technical applications derived from these phenomena is presented.

REFERENCES

1. Negel, J.-P.; Voss, A.; Ahmed, M.A.; Bauer, D.; Sutter, D.; Killi, A.; Graf, T. 11 kW average output power from a thin-disk multipass amplifier for ultrashort laser pulses. *Opt. Lett.* **2013**, *38*, 5442–5445. **[Google Scholar]** **[CrossRef]** **[PubMed]**

2. Negel, J.-P.; Loescher, A.; Bauer, D.; Sutter, D.; Killi, A.; Ahmed, M.A.; Graf, T. Second Generation Thin-Disk Multipass Amplifier Delivering Picosecond Pulses with 2 kW of Average Output Power. In *Advanced Solid State Lasers, Proceedings of the Lasers Congress 2016 (ASSL, LSC, LAC), Boston, MA, USA, 30 October–3 November 2016*; The Optical Society of America: Washington, DC, USA, 2016; p. ATu4A.5. **[Google Scholar]**

3. Nubbemeyer, T.; Kaumanns, M.; Ueffing, M.; Gorjan, M.; Alismail, A.; Fattahi, H.; Brons, J.; Pronin, O.; Barros, H.G.; Major, Z.; et al. 1 kW, 200 mJ picosecond thin-disk laser system. *Opt. Lett.* **2017**, *42*, 1381–1384. **[Google Scholar]** **[CrossRef]** **[PubMed]**

4. Gaida, C.; Gebhardt, M.; Heuermann, T.; Stutzki, F.; Jauregui, C.; Limpert, J. Ultrafast thulium fiber laser system emitting more than 1 kW of average power. *Opt. Lett.* **2018**, *43*, 5853–5856. **[Google Scholar]** **[CrossRef]**

5. Schnitzler, C.; Mans, T.G.; Dolkemeyer, J.; Dittmann, P. High Power, High Energy, and High Flexibility: Powerful Ultrafast Lasers Based on InnoSlab Technology. In *High-Power Laser Materials Processing: Applications, Diagnostics, and Systems VIII, Proceedings of the SPIE LASE, 2019, San Francisco, CA, USA, 2–7 February 2019*; International Society for Optics and Photonics: Bellingham, WA, USA, 2019; p. 1091103. **[Google Scholar]**

6. Müller, M.; Aleshire, C.; Klenke, A.; Haddad, E.; Légaré, F.; Tünnermann, A.; Limpert, J. 104 kW coherently combined ultrafast fiber laser. *Opt. Lett.* **2020**, *45*, 3083–3086. **[Google Scholar]** **[CrossRef]** **[PubMed]**

7. Semerok, A.; Dutouquet, C. Ultrashort double pulse laser ablation of metals. *Thin Solid Films* **2004**, *453*, 501–505. **[Google Scholar]** **[CrossRef]**

8. Donnelly, T.; Lunney, J.G.; Amoruso, S.; Bruzzese, R.; Wang, X.; Ni, X. Double pulse ultrafast laser ablation of nickel in vacuum. *J. Appl. Phys.* **2009**, *106*, 013304. **[Google Scholar]** **[CrossRef]** **[Green Version]**

9. Povarnitsyn, M.E.; Itina, T.E.; Khishchenko, K.V.; Levashov, P.R. Suppression of Ablation in Femtosecond Double-Pulse Experiments. *Phys. Rev. Lett.* **2009**, *103*, 195002. **[Google Scholar]** **[CrossRef]** **[PubMed]** **[Green Version]**

10. Povarnitsyn, M.E.; Fokin, V.B.; Levashov, P.R.; Itina, T.E. Molecular dynamics simulation of subpicosecond double-pulse laser ablation of metals. *Phys. Rev. B* **2015**, *92*, 174104. **[Google Scholar]** **[CrossRef]**

11. Hartmann, C.; Gillner, A.; Aydin, U.; Noll, R.; Fehr, T.; Gehlen, C.; Poprawe, R. Investigation on laser micro ablation of metals using ns-multi-pulses. *J. Physics Conf. Ser.* **2007**, *59*, 440–444. **[Google Scholar]** **[CrossRef]** **[Green Version]**

12. Hu, W.; Shin, Y.C.; King, G. Modeling of multi-burst mode pico-second laser ablation for improved material removal rate. *Appl. Phys. A* **2009**, *98*, 407–415. **[Google Scholar]** **[CrossRef]**

13. Hernandez-Rueda, J.; Siegel, J.; Galvan-Sosa, M.; De La Cruz, A.R.; Solis, J. Surface structuring of fused silica with asymmetric femtosecond laser pulse bursts. *J. Opt. Soc. Am. B* **2013**, *30*, 1352–1356. **[Google Scholar]** **[CrossRef]** **[Green Version]**

14. Neuenschwander, B.; Kramer, T.; Lauer, B.; Jaeggi, B. Burst Mode with Ps- and Fs-Pulses: Influence on the Removal Rate, Surface Quality, and Heat Accumulation. In *Laser Applications in Microelectronic and Optoelectronic Manufacturing (LAMOM) XX, Proceedings of the SPIE LASE, San Francisco, CA, USA, 9–12 February 2015*; SPIE: Bellingham, WA, USA, 2015; Volume 9350, p. 93500U. **[Google Scholar]**

15. Kerse, C.; Kalaycıoglu, H.; Elahi, P.; Cetin, B.; Kesim, D.K.; Akcaalan, O.; Yavas, S.; Asik, M.D.; Oktem, B.; Hoogland, H.; et al. Ablation-cooled material removal with ultrafast bursts of pulses. *Nature* **2016**, *537*, 84–88. **[Google Scholar]** **[CrossRef]** **[PubMed]**

16. Jäggi, B.; Förster, D.J.; Weber, R.; Neuenschwander, B. Residual heat during laser ablation of metals with bursts of ultra-short pulses. *Adv. Opt. Technol.* **2018**, *7*, 175–182. **[Google Scholar]** **[CrossRef]**

17. Neuenschwander, B.; Jaeggi, B.; Foerster, D.J.; Kramer, T.; Remund, S. Influence of the burst mode onto the specific removal rate for metals and semiconductors. *J. Laser Appl.* **2019**, *31*, 022203. **[Google Scholar]** **[CrossRef]**

18. Zemaitis, A.; Gecys, P.; Barkauskas, M.; Raciukaitis, G.; Gedvilas, M. Highly-efficient laser ablation of copper by bursts of ultrashort tuneable (fs–ps) pulses. *Sci. Rep.* **2019**, *9*, 1–8. **[Google Scholar]** **[CrossRef]** **[Green Version]**

19. Bornschlegel, B.; Finger, J. In-Situ Analysis of Ultrashort Pulsed Laser Ablation with Pulse Bursts. *JLMN* **2019**, *14*, 88–94. **[Google Scholar]**

20. Gaudiuso, C.; Kämmer, H.; Dreisow, F.; Ancona, A.; Tünnermann, A.; Nolte, S. Ablation of Silicon with Bursts of Femtosecond Laser Pulses. In *Frontiers in Ultrafast Optics: Biomedical, Scientific, and Industrial Applications XVI, Proceedings of the SPIE LASE, San Francisco, CA, USA, 13–18 February 2016*; Heisterkamp, A., Herman, P.R., Meunier, M., Nolte, S., Eds.; SPIE: Bellingham, WA, USA, 2016; p. 974017. **[Google Scholar]**

21. Metzner, D.; Lickschat, P.; Weißmantel, S. High-quality surface treatment using GHz burst mode with tunable ultrashort pulses. *Appl. Surf. Sci.* **2020**, *531*, 147270. **[Google Scholar]** **[CrossRef]**

22. Hodgson, N.; Allegre, H.; Starodoumov, A.; Bettencourt, S. Femtosecond Laser Ablation in Burst Mode as a Function of Pulse Fluence and Intra-Burst Repetition Rate. *JLMN* **2020**, *15*. **[Google Scholar]** **[CrossRef]**

23. Hodgson, N.; Steinkopff, A.; Heming, S.; Allegre, H.; Haloui, H.; Lee, T.S.; Laha, M.; van Nunen, J. Ultrafast Laser Machining: Process Optimization and Applications. In *Laser Applications in Microelectronic and Optoelectronic Manufacturing (LAMOM) XXVI, Proceedings of the SPIE LASE, San Francisco, CA, USA, 6–12 March 2021*; SPIE: Bellingham, WA, USA, 2021; Volume 11673, p. 1167308. **[Google Scholar]**

24. Žemaitis, A.; Gaidys, M.; Gečys, P.; Barkauskas, M.; Gedvilas, M. Femtosecond laser ablation by bibursts in the MHz and GHz pulse repetition rates. *Opt. Express* **2021**, *29*, 7641–7653. **[Google Scholar]** **[CrossRef]**

25. Führa, B.; Russ, S.; Hammers-Weber, P.; Diego-Vallejo, D.; Kahmann, M.; Andreev, A.; Hesse, T. High precision drilling with ultra-short laser pulses. In Proceedings of the Lasers in Manufacturing Conference 2017, Munich, Germany, 26–29 June 2017. **[Google Scholar]**

26. Žemaitis, A.; Gaidys, M.; Brikas, M.; Gečys, P.; Raciukaitis, G.; Gedvilas, M. Advanced laser scanning for highly-efficient ablation and ultrafast surface structuring: Experiment and model. *Sci. Rep.* **2018**, *8*, 17376. **[Google Scholar] [CrossRef]**

27. Jaeggi, B.; Neuenschwander, B.; Meier, T.; Zimmermann, M.; Hennig, G. High Precision Surface Structuring with Ultra-Short Laser Pulses and Synchronized Mechanical Axes. *Phys. Procedia* **2013**, *41*, 319–326. **[Google Scholar] [CrossRef] [Green Version]**

28. Kramer, T.; Remund, S.; Gafner, M.; Zwygart, D.; Neuenschwander, B.; Holtz, R.; Witte, R.; Dury, N. Novel strategy for ultrafast pulsed laser micromachining of rotational symmetric metallic parts. *Procedia CIRP* **2018**, *74*, 611–617. **[Google Scholar] [CrossRef]**

29. Bruening, S.; Hennig, G.; Eifel, S.; Gillner, A. Ultrafast Scan Techniques for 3D-μm Structuring of Metal Surfaces with high repetitive ps-laser pulses. *Phys. Procedia* **2011**, *12*, 105–115. **[Google Scholar] [CrossRef]**

30. Liu, X.; Du, D.; Mourou, G. Laser ablation and micromachining with ultrashort laser pulses. *IEEE J. Quantum Electron.* **1997**, *33*, 1706–1716. **[Google Scholar] [CrossRef]**

31. Shirk, M.D.; Molian, P.A. A review of ultrashort pulsed laser ablation of materials. *J. Laser Appl.* **1998**, *10*, 18–28. **[Google Scholar] [CrossRef]**

32. Sundaram, S.K.; Mazur, E. Inducing and probing non-thermal transitions in semiconductors using femtosecond laser pulses. *Nat. Mater.* **2002**, *1*, 217–224. **[Google Scholar] [CrossRef]**

33. Perry, M.D.; Stuart, B.C.; Banks, P.S.; Feit, M.D.; Yanovsky, V.; Rubenchik, A.M. Ultrashort-pulse laser machining of electric materials. *J. Appl. Phys.* **1999**, *85*, 6803–6810. **[Google Scholar] [CrossRef] [Green Version]**

34. Krüger, J.; Kautek, W. Ultrashort Pulse Laser Interaction with Dielectrics and Polymers. In *Advances in Polymer Science*; Springer Science and Business Media: New York, NY, USA, 2012; Volume 168, pp. 247–290. **[Google Scholar]**

35. Bulgakova, N.M.; Zhukov, V.P.; Collins, A.R.; Rostohar, D.; Derrien, T.J.-Y.; Mocek, T. How to optimize ultrashort pulse laser interaction with glass surfaces in cutting regimes? *Appl. Surf. Sci.* **2015**, *336*, 364–374. **[Google Scholar] [CrossRef]**

36. Samant, A.N.; Dahotre, N.B. Laser machining of structural ceramics—A review. *J. Eur. Ceram. Soc.* **2009**, *29*, 969–993. **[Google Scholar] [CrossRef]**

37. Chichkov, B.N.; Momma, C.; Nolte, S.; von Alvensleben, F.; Tünnermann, A. Femtosecond, picosecond and nanosecond laser ablation of solids. *Appl. Phys. A* **1996**, *63*, 109–115. **[Google Scholar] [CrossRef]**

38. Cheng, J.; Liu, C.; Shang, S.; Liu, D.; Perrie, W.; Dearden, G.; Watkins, K. A review of ultrafast laser materials micromachining. *Opt. Laser Technol.* **2013**, *46*, 88–102. **[Google Scholar] [CrossRef]**

39. Mishra, S.; Yadava, V. Laser Beam MicroMachining (LBMM)—A review. *Opt. Lasers Eng.* **2015**, *73*, 89–122. **[Google Scholar] [CrossRef]**

40. Lei, S.; Zhao, X.; Yu, X.; Hu, A.; Vukelic, S.; Jun, M.B.G.; Joe, H.-E.; Yao, Y.L.; Shin, Y.C. Ultrafast Laser Applications in Man-ufacturing Processes: A State-of-the-Art Review. J. *Manuf. Sci. Eng.* **2020**, *142*, 031005. **[Google Scholar] [CrossRef]**

41. Momma, C.; Nolte, S.; Chichkov, B.; Tunnermann, A.; Von Alvensleben, F. Precise Laser Ablation with Ultra-Short Pulses. In Proceedings of the European Meeting on Lasers and Electro-Optics CLEOE-96, Hamburg, Germany, 8–13 September 1996. **[Google Scholar]**

42. Zhigilei, L. Dynamics of the plume formation and parameters of the ejected clusters in short-pulse laser ablation. *Appl. Phys. A* **2003**, *76*, 339–350. **[Google Scholar] [CrossRef] [Green Version]**

43. Ivanov, D.S.; Zhigilei, L. Combined atomistic-continuum modeling of short-pulse laser melting and disintegration of metal films. *Phys. Rev. B* **2003**, *68*, 064114. **[Google Scholar] [CrossRef] [Green Version]**

44. Povarnitsyn, M.E.; Andreev, N.E.; Apfelbaum, E.M.; Itina, T.E.; Khishchenko, K.; Kostenko, O.F.; Levashov, P.R.; Veysman, M.E. A wide-range model for simulation of pump-probe experiments with metals. *Appl. Surf. Sci.* **2012**, *258*, 9480–9483. **[Google Scholar] [CrossRef]**

45. Wu, C.; He, M. Microscopic mechanisms of laser spallation and ablation of metal targets from large-scale molecular dynamics simulations. *Appl. Phys. A* **2014**, *114*, 11–32. **[Google Scholar] [CrossRef]**

46. Povarnitsyn, M.E.; Fokin, V.B.; Levashov, P.R. Microscopic and macroscopic modeling of femtosecond laser ablation of metals. *Appl. Surf. Sci.* **2015**, *357*, 1150–1156. **[Google Scholar] [CrossRef]**

47. Breitling, D.; Ruf, A.; Dausinger, F. Fundamental Aspects in Machining of Metals with Short and Ultrashort Laser Pulses. In *Photon Processing in Microelectronics and Photonics III—Fundamental Aspects in Machining of Metals with Short and Ultrashort Laser Pulses, Proceedings of the SPIE Lasers and Applications in Science and Engineering, San Jose, CA, USA, 25 January 2004*; SPIE: Bellingham, WA, USA, 2004; p. 49. **[Google Scholar]**

48. Lide, D.R. *CRC Handbook of Chemistry and Physics*; CRC Press: Boca Raton, FL, USA, 2004. **[Google Scholar]**

49. Wright, O.B. Ultrafast nonequilibrium stress generation in gold and silver. *Phys. Rev. B* **1994**, *49*, 9985–9988. **[Google Scholar] [CrossRef]**

50. Hüttner, B.; Rohr, G. On the theory of ps and sub-ps laser pulse interaction with metals I. Surface temperature. *Appl. Surf. Sci.* **1996**, *103*, 269–274. **[Google Scholar] [CrossRef]**

51. Nedialkov, N.N.; Imamova, S.E.; Atanasov, P.A. Ablation of metals by ultrashort laser pulses. *J. Phys. D Appl. Phys.* **2004**, *37*, 638–643. **[Google Scholar] [CrossRef]**

52. Wellershoff, S.-S.; Hohlfeld, J.; Güdde, J.; Matthias, E. The role of electron–phonon coupling in femtosecond laser damage of metals. *Appl. Phys. A* **1999**, *69*, S99–S107. **[Google Scholar]**

53. Lin, Z.; Zhigilei, L.; Celli, V. Electron-phonon coupling and electron heat capacity of metals under conditions of strong electron-phonon nonequilibrium. *Phys. Rev. B* **2008**, *77*, 075133. **[Google Scholar] [CrossRef] [Green Version]**

54. Foumani, A.A.; Förster, D.J.; Ghorbanfekr-Kalashami, H.; Weber, R.; Graf, T.; Niknam, A.R. Atomistic simulation of ultra-short pulsed laser ablation of metals with single and double pulses: An investigation of the re-deposition phenomenon. *Appl. Surf. Sci.* **2021**, *537*, 147775. **[Google Scholar] [CrossRef]**

55. Amoruso, S.; Bruzzese, R.; Pagano, C.; Wang, X. Features of plasma plume evolution and material removal efficiency during femtosecond laser ablation of nickel in high vacuum. *Appl. Phys. A* **2007**, *89*, 1017–1024. **[Google Scholar] [CrossRef] [Green Version]**

56. Düsing, J.F.; Hwang, D.J.; Grigoropoulos, C.; Ostendorf, A.; Kling, R. Optical emission imaging and spectroscopy during femtosecond laser ablation of thin metal films on flexible polymer substrates. In Proceedings of the International Congress on Applications of Lasers & Electro-Optics, Orlando, FL, USA, 2–5 November 2009; p. 888. **[Google Scholar]**

57. Förster, D.J.; Faas, S.; Gröninger, S.; Bauer, F.; Michalowski, A.; Weber, R.; Graf, T. Shielding effects and re-deposition of material during processing of metals with bursts of ultra-short laser pulses. *Appl. Surf. Sci.* **2018**, *440*, 926–931. **[Google Scholar] [CrossRef]**

58. König, J.; Nolte, S.; Tünnermann, A. Plasma evolution during metal ablation with ultrashort laser pulses. *Opt. Express* **2005**, *13*, 10597. **[Google Scholar] [CrossRef]**

59. Neuenschwander, B.; Jaeggi, B.; Schmid, M.; Hennig, G. Surface Structuring with Ultra-short Laser Pulses: Basics, Limitations and Needs for High Throughput. *Phys. Procedia* **2014**, *56*, 1047–1058. **[Google Scholar] [CrossRef] [Green Version]**

60. Kramer, T.; Zhang, Y.; Remund, S.; Jaeggi, B.; Michalowski, A.; Grad, L.; Neuenschwander, B. Increasing the Specific Removal Rate for Ultra Short Pulsed Laser-Micromachining by Using Pulse Bursts. *JLMN* **2017**, *12*, 107–114. **[Google Scholar] [CrossRef]**

61. Jaeggi, B.; Cangueiro, L.; Bruneel, D.; de Campos, J.A.R.; Hairaye, C.; Neuenschwander, B. Micromachining using pulse bursts: Influence of the pulse duration and the number of pulses

in the burst on the specific removal rate. In *Laser Applications in Microelectronic and Optoelectronic Manufacturing (LAMOM) XXIII*; International Society for Optics and Photonics: Bellingham, WA, USA, 2018; Volume 10519, p. 1051905. **[Google Scholar]**

62. Neuenschwander, B.; Jaeggi, B.; Schmid, M.; Rouffiange, V.; Martin, P.-E. Optimization of the Volume Ablation Rate for Metals at Different Laser Pulse-Durations from ps to fs. In Proceedings of the SPIE LASE, San Francisco, CA, USA, 15 February 2012; p. 824307. **[Google Scholar]**

63. Schille, J.; Schneider, L.; Loeschner, U. Process optimization in high-average-power ultrashort pulse laser microfabrication: How laser process parameters influence efficiency, throughput and quality. *Appl. Phys. A* **2015**, *120*, 847–855. **[Google Scholar] [CrossRef]**

64. Domke, M.; Matylitsky, V.; Stroj, S. Surface ablation efficiency and quality of fs lasers in single-pulse mode, fs lasers in burst mode, and ns lasers. *Appl. Surf. Sci.* **2020**, *505*, 144594. **[Google Scholar] [CrossRef]**

65. Raciukaitis, G.; Brikas, M.; Gecys, P.; Voisat, B.; Gedvilas, M. Use of High Repetition Rate and High Power Lasers in Microfabrication: How to Keep the Efficiency High? *JLMN* **2009**, *4*, 186–191. **[Google Scholar] [CrossRef]**

66. Gillner, A.; Hartmann, C.; Dohrn, A. High Quality Micro Machining with Tailored Short and Ultra Short Laser Pulses. In *Proceedings of the Pacific International Conference on Applications of Lasers and Optics*; Laser Institute of America: Orlando, FL, USA, 2008; Volume 2008, p. 685. **[Google Scholar]**

67. Knappe, R.; Haloui, H.; Seifert, A.; Weis, A.; Nebel, A. Scaling ablation rates for picosecond lasers using burst micromachining. In *Laser-Based Micro- and Nanopackaging and Assembly IV, Proceedings of the SPIE LASE, San Francisco, CA, USA, 23–28 January 2010*; SPIE: Bellingham, WA, USA, 2010; Volume 7585, p. 75850H. **[Google Scholar]**

68. Vorobyev, A.Y.; Guo, C. Direct observation of enhanced residual thermal energy coupling to solids in femtosecond laser ablation. *Appl. Phys. Lett.* **2005**, *86*, 011916. **[Google Scholar] [CrossRef]**

69. Vorobyev, A.Y.; Guo, C. Enhanced absorptance of gold following multipulse femtosecond laser ablation. *Phys. Rev. B* **2005**, *72*, 195422. **[Google Scholar] [CrossRef] [Green Version]**

70. Vorobyev, A.Y.; Guo, C. Enhanced energy coupling in femtosecond laser-metal interactions at high intensities. *Opt. Express* **2006**, *14*, 13113–13119. **[Google Scholar] [CrossRef]**

71. Vorobyev, A.; Kuzmichev, V.; Kokody, N.; Kohns, P.; Dai, J.; Guo, C. Residual thermal effects in Al following single ns- and fs-laser pulse ablation. *Appl. Phys. A* **2005**, *82*, 357–362. **[Google Scholar] [CrossRef]**

72. Vorobyev, A.Y.; Guo, C. Thermal response and optical absorptance of metals under femtosecond laser irradiation. *Nat. Sci.* **2011**, *03*, 488–495. **[Google Scholar] [CrossRef] [Green Version]**

73. Bauer, F.; Michalowski, A.; Kiedrowski, T.; Nolte, S. Heat accumulation in ultra-short pulsed scanning laser ablation of metals. *Opt. Express* **2015**, *23*, 1035–1043. **[Google Scholar] [CrossRef]**

74. Förster, D.J.; Weber, R.; Graf, T. Residual heat during ultrashort laser drilling of metals. In Proceedings of the LPM2017—The 18th International Symposium on Laser Precision Microfabrication, Toyama, Japan, 5–8 June 2017. **[Google Scholar]**

75. Bauer, F. GrundlegendeUntersuchungenzumAbtragen von Stahl mitUltrakurzenLaserpulsen. Ph.D. Thesis, Friedrich-Schiller-Universität Jena, Jena, Germany, 2018. **[Google Scholar]**

76. Bornschlegel, B.; Koller, J.; Finger, J. In-Situ Analysis of Heat Accumulation during Ultrashort Pulsed Laser Ablation. *JLMN* **2020**, *15*, 56–62. **[Google Scholar]**

77. Wu, B.; Deng, L.; Liu, P.; Zhang, F.; Duan, J.; Zeng, X. Effects of picosecond laser repetition rate on ablation of Cr12MoV cold work mold steel. *Appl. Surf. Sci.* **2017**, *409*, 403–412. **[Google Scholar] [CrossRef]**

78. Metzner, D.; Lickschat, P.; Weißmantel, S. Influence of heat accumulation during laser micromachining of CoCrMo alloy with ultrashort pulses in burst mode. *Appl. Phys. A* **2020**, *126*, 84. **[Google Scholar] [CrossRef]**

79. Martan, J.; Prokešová, L.; Moskal, D.; de Faria, B.F.; Honner, M.; Lang, V. Heat accumulation temperature measurement in ultrashort pulse laser micromachining. *Int. J. Heat Mass Transf.* **2021**, *168*, 120866. **[Google Scholar] [CrossRef]**

80. Weber, R.; Graf, T.; Berger, P.; Onuseit, V.; Wiedenmann, M.; Freitag, C.; Feuer, A. Heat accumulation during pulsed laser materials processing. *Opt. Express* **2014**, *22*, 11312–11324. **[Google Scholar] [CrossRef] [Green Version]**

81. Weber, R.; Graf, T.; Freitag, C.; Feuer, A.; Kononenko, T.; Konov, V.I. Processing constraints resulting from heat accumulation during pulsed and repetitive laser materials processing. *Opt. Express* **2017**, *25*, 3966–3979. **[Google Scholar] [CrossRef]**

82. Di Niso, F.; Gaudiuso, C.; Sibillano, T.; Mezzapesa, F.P.; Ancona, A.; Lugarà, P.M. Role of heat accumulation on the incubation effect in multi-shot laser ablation of stainless steel at high repetition rates. *Opt. Express* **2014**, *22*, 12200–12210. **[Google Scholar] [CrossRef] [PubMed]**

83. Raciukaitis, G.; Brikas, M.; Gecys, P.; Gedvilas, M. Accumulation Effects in Laser Ablation of Metals with High-Repetition-Rate Lasers. In *High-Power Laser Ablation VII, Proceedings of the SPIE LASE, San Francisco, CA, USA, 23–28 January 2008*; SPIE: Bellingham, WA, USA, 2008. **[Google Scholar]**

84. Faas, S.; Bielke, U.; Weber, R.; Graf, T. Prediction of the surface structures resulting from heat accumulation during processing with picosecond laser pulses at the average power of 420 W. *Appl. Phys. A* **2018**, *124*, 612. **[Google Scholar] [CrossRef] [Green Version]**

85. St-Onge, L.; Sabsabi, M.; Cielo, P. Analysis of solids using laser-induced plasma spectroscopy in double-pulse mode. *Spectrochim. Acta Part B At. Spectrosc.* **1998**, *53*, 407–415. **[Google Scholar] [CrossRef]**

86. Mao, S.; Mao, X.; Greif, R.; Russo, R.E. Influence of preformed shock wave on the development of picosecond laser ablation plasma. *J. Appl. Phys.* **2001**, *89*, 4096–4098. **[Google Scholar] [CrossRef]**

87. Corsi, M.; Cristoforetti, G.; Giuffrida, M.; Hidalgo, M.; Legnaioli, S.; Palleschi, V.; Salvetti, A.; Tognoni, E.; Vallebona, C. Three-dimensional analysis of laser induced plasmas in single and double pulse configuration. *Spectrochim. Acta Part B At. Spectrosc.* **2004**, *59*, 723–735. **[Google Scholar] [CrossRef]**

88. Scuderi, D.; Albert, O.; Moreau, D.; Pronko, P.P.; Etchepare, J. Interaction of a laser-produced plume with a second time delayed femtosecond pulse. *Appl. Phys. Lett.* **2005**, *86*, 071502. **[Google Scholar] [CrossRef]**

89. Hanada, Y.; Sugioka, K.; Miyamoto, I.; Midorikawa, K. Double-pulse irradiation by laser-induced plasma-assisted ablation (LIPAA) and mechanisms study. *Appl. Surf. Sci.* **2005**, *248*, 276–280. **[Google Scholar] [CrossRef]**

90. Le Harzic, R.; Breitling, D.; Sommer, S.; Föhl, C.; König, K.; Dausinger, F.; Audouard, E. Processing of metals by double pulses with short laser pulses. *Appl. Phys. A* **2005**, *81*, 1121–1125. **[Google Scholar] [CrossRef]**

91. Babushok, V.; DeLucia, F.; Gottfried, J.; Munson, C.; Miziolek, A. Double pulse laser ablation and plasma: Laser induced breakdown spectroscopy signal enhancement. *Spectrochim. Acta Part B At. Spectrosc.* **2006**, *61*, 999–1014. **[Google Scholar] [CrossRef]**

92. Suttmann, O.; Wojakowski, B.; Klug, U.; Kling, R.; Ostendorf, A. Picosecond double-pulse ablation in silicon and aluminium with variable delay. *J. Laser Appl.* **2008**, *42*. **[Google Scholar] [CrossRef]**

93. Bogaerts, A.; Chen, Z.; Autrique, D. Double pulse laser ablation and laser induced breakdown spectroscopy: A modeling investigation. *Spectrochim. Acta Part B At. Spectrosc.* **2008**, *63*, 746–754. **[Google Scholar] [CrossRef]**

94. Singha, S.; Hu, Z.; Gordon, R.J. Ablation and plasma emission produced by dual femtosecond laser pulses. *J. Appl. Phys.* **2008**, *104*, 113520. **[Google Scholar] [CrossRef]**

95. Noel, S.; Hermann, J. Reducing nanoparticles in metal ablation plumes produced by two delayed short laser pulses. *Appl. Phys. Lett.* **2009**, *94*, 53120. **[Google Scholar] [CrossRef]**

96. Wojakowski, B.; Suttmann, O.; Klug, U.; Kling, R. Micromachining with Picosecond Double Pulses on Silicon and Aluminium. In *Laser-Based Micro- and Nanopackaging and Assembly III, Proceedings of*

the *SPIE LASE: Lasers and Applications in Science and Engineering, San Jose, CA, USA, 24 January 2009*; SPIE: Bellingham, WA, USA, 2009; p. 72020O. **[Google Scholar]**

97. Roberts, D.; du Plessis, A.; Botha, L. Femtosecond laser ablation of silver foil with single and double pulses. *Appl. Surf. Sci.* **2010**, *256*, 1784–1792. **[Google Scholar] [CrossRef]**

98. Deladurantaye, P.; Cournoyer, A.; Drolet, M.; Desbiens, L.; Lemieux, D.; Briand, M.; Taillon, Y. Material Micromachining Using Bursts of High Repetition Rate Picosecond Pulses from a Fiber Laser Source. In *Fiber Lasers VIII: Technology, Systems, and Applications, Proceedings of the SPIE LASE San Francisco, CA, USA, 22–27 January 2011*; SPIE: Bellingham, WA, USA, 2011; Volume 7914, p. 791404. **[Google Scholar]**

99. Axente, E.; Mihailescu, I.N.; Itina, T.; Hermann, J. Probing electron-phonon coupling in metals via observations of ablation plumes produced by two delayed short laser pulses. *Appl. Phys. Lett.* **2011**, *99*, 081502. **[Google Scholar] [CrossRef] [Green Version]**

100. Sailer, M.; Bauer, F.; Kleiner, J.; Kaiser, M. Scaling of ablation rates. Ablation efficiency and quality aspects of "Burstmode"—Micromachining of metals. In Proceedings of the Lasers in Manufacturing Conference, Munich, Germany, 22–25 June 2015; p. 8. **[Google Scholar]**

101. Hänel, N.; Stolze, M.; Herrmann, T.R.W.; Lhuillier, J.A. Fundamental Investigations of ps-Laser Burst-Mode on Common Metals for an Enhanced Ablation Process. In *Laser-Based Micro- and Nanoprocessing IX, Proceedings of the SPIE LASE San Francisco, CA, USA, 7–12 February 2015*; SPIE: Bellingham, WA, USA, 2015; Volume 9351, p. 93510E. **[Google Scholar]**

102. Kramer, T.; Neuenschwander, B.; Jäggi, B.; Remund, S.; Hunziker, U.; Zürcher, J. Influence of Pulse Bursts on the Specific Removal Rate for Ultra-fast Pulsed Laser Micromachining of Copper. *Phys. Procedia* **2016**, *83*, 123–134. **[Google Scholar] [CrossRef] [Green Version]**

103. Schille, J.; Schneider, L.; Kraft, S.; Hartwig, L.; Loeschner, U. Experimental study on double-pulse laser ablation of steel upon multiple parallel-polarized ultrashort-pulse irradiations. *Appl. Phys. A* **2016**, *122*, 644. **[Google Scholar] [CrossRef]**

104. Finger, J. Puls-zu-Puls-WechselwirkungenbeimUltrakurzpuls-LaserabtragmithohenRepetitionsrat en. Ph.D. Thesis, RWTH Aachen, Aachen, Germany, 2017. **[Google Scholar]**

105. Lickschat, P.; Demba, A.; Weissmantel, S. Ablation of steel using picosecond laser pulses in burst mode. *Appl. Phys. A* **2017**, *123*, 137. **[Google Scholar] [CrossRef]**

106. Mayerhofer, R. Ultrashort-pulsed laser material processing with high repetition rate burst pulses. In *Laser Applications in Microelectronic and Optoelectronic Manufacturing (LAMOM) XXII, Proceedings of the SPIE LASE, 2017, San Francisco, CA, USA, 28 January–2 February 2017*; SPIE: Bellingham, WA, USA, 2017; p. 100910. **[Google Scholar]**

107. Jaeggi, B.; Remund, S.; Zhang, Y.; Kramer, T.; Neuenschwander, B. Optimizing the Specific Removal Rate with the Burst Mode Under Varying Conditions. *JLMN* **2017**, *12*, 107–114. **[Google Scholar]**

108. Rosenfeld, A.; Höhm, S.; Krüger, J.; Bonse, J. Dynamics of Ultrashort Double-Pulse Laser Ablation of Solid Surfaces. In *Encyclopedia of Interfacial Chemistry*; Elsevier: Amsterdam, The Netherlands, 2018; pp. 338–347. **[Google Scholar]**

109. Hashida, M.; Masuno, S.; Furukawa, Y.; Kusaba, M.; Inoue, S.; Sakabe, S.; Sakagami, H.; Tsukamoto, M. Suppression of ablation by double-pulse femtosecond laser irradiation. In *Frontiers in Ultrafast Optics: Biomedical, Scientific, and Industrial Applications XVIII, Proceedings of the SPIE LASE, San Francisco, CA, USA, 27 January–1 February 2018*; SPIE: Bellingham, WA, USA, 2018; p. 105220. **[Google Scholar]**

110. Zhang, K.; Zhang, J.; Jiang, L.; Li, X.; Liu, Y.; Li, B.; Lu, Y. Ablation enhancement of metal in ultrashort double-pulse experiments. *Appl. Phys. Lett.* **2018**, *112*, 261906. **[Google Scholar] [CrossRef]**

111. Takenaka, K.; Tsukamoto, M.; Hashida, M.; Masuno, S.; Sakagami, H.; Kusaba, M.; Sakabe, S.; Inoue, S.; Furukawa, Y.; Asai, S. Ablation suppression of a titanium surface interacting with a two-color double-pulse femtosecond laser beam. *Appl. Surf. Sci.* **2019**, *478*, 882–886. **[Google Scholar] [CrossRef]**

112. Bruening, S.; Du, K.; Gillner, A. Micro processing with ultrafast bursts of pulses. *Procedia CIRP* **2020**, *94*, 856–862. **[Google Scholar] [CrossRef]**

113. Lin, Z.; Ji, L.; Hong, M. Enhancement of femtosecond laser-induced surface ablation via temporal overlapping double-pulse irradiation. *Photon Res.* **2020**, *8*, 271. **[Google Scholar] [CrossRef]**

114. Förster, D.J.; Faas, S.; Weber, R.; Graf, T. Thrust enhancement and propellant conservation for laser propulsion using ultra-short double pulses. *Appl. Surf. Sci.* **2020**, *510*, 145391. **[Google Scholar]** **[CrossRef]**

115. Cheng, C.-W.; Chen, J.-K. Drilling of Copper Using a Dual-Pulse Femtosecond Laser. *Technologies* **2016**, *4*, 7. **[Google Scholar]** **[CrossRef]** **[Green Version]**

116. Förster, G.D.; Lewis, L.J. Numerical study of double-pulse laser ablation of Al. *Phys. Rev. B* **2018**, *97*, 224301. **[Google Scholar]** **[CrossRef]**

117. Roth, J.; Krauß, A.; Lotze, J.; Trebin, H.-R. Simulation of laser ablation in aluminum: The effectivity of double pulses. *Appl. Phys. A* **2014**, *117*, 2207–2216. **[Google Scholar]** **[CrossRef]** **[Green Version]**

118. Povarnitsyn, M.E.; Itina, T.E.; Levashov, P.R.; Khishchenko, K. Simulation of ultrashort double-pulse laser ablation. *Appl. Surf. Sci.* **2011**, *257*, 5168–5171. **[Google Scholar]** **[CrossRef]**

119. Kudryashov, S.I.; Samokhvalov, A.A.; Golubev, Y.D.; Ivanov, D.S.; Garcia, M.E.; Veiko, V.P.; Rethfeld, B.; Mikhailovskii, V.Y. Dynamic all-optical control in ultrashort double-pulse laser ablation. *Appl. Surf. Sci.* **2021**, *537*, 147940. **[Google Scholar]** **[CrossRef]**

120. Spellauge, M.; Winter, J.; Rapp, S.; McDonnell, C.; Sotier, F.; Schmidt, M.; Huber, H.P. Influence of stress confinement, particle shielding and re-deposition on the ultrashort pulse laser ablation of metals revealed by ultrafast time-resolved experiments. *Appl. Surf. Sci.* **2021**, *545*, 148930. **[Google Scholar]** **[CrossRef]**

121. Ackerl, N. Laser Surface Functionalization from Fundamentals to Application. Ph.D. Thesis, ETH Zurich, Zurich, Switzerland, 2020. **[Google Scholar]**

122. Börner, P. Ultra-Short Pulsed Laser Ablation of Diamond. Ph.D. Thesis, ETH Zurich, Zurich, Switzerland, 2019. **[Google Scholar]**

123. Furukawa, Y.; Inoue, S.; Hashida, M. Temporal change in laser penetration length of titanium and platinum for double-pulse ablation measured by a novel ablation method. *J. Laser Appl.* **2021**, *33*, 012023. **[Google Scholar]** **[CrossRef]**

124. Lauer, B.; Jaeggi, B.; Zhang, Y.; Neuenschwander, B. Measurement of the maximum specific removal rate: Unexpected influence of the experimental method and the spot size. In Proceedings of the International Congress on Applications of Lasers & Electro-Optics, Atlanta, GA, USA, 18–22 October 2015; Laser Institute of America: Orlando, FL, USA, 2015; Volume 2015, p. 146. **[Google Scholar]**

125. Kramer, T.; Remund, S.; Jäggi, B.; Schmid, M.; Neuenschwander, B. Ablation dynamics—From absorption to heat accumulation/ultra-fast laser matter interaction. *Adv. Opt. Technol.* **2018**, *7*, 129–144. **[Google Scholar]** **[CrossRef]**

126. Winter, J.; Rapp, S.; Schmidt, M.; Huber, H.P. Ultrafast laser processing of copper: A comparative study of experimental and simulated transient optical properties. *Appl. Surf. Sci.* **2017**, *417*, 2–15. **[Google Scholar]** **[CrossRef]**

127. Neuenschwander, B.; Bucher, G.F.; Nussbaum, C.; Joss, B.; Muralt, M.; Hunziker, U.W.; Schuetz, P. Processing of metals and dielectric materials with ps-laser pulses: Results, strategies, limitations and needs. In *Laser Applications in Microelectronic and Optoelectronic Manufacturing XV, Proceedings of the SPIE LASE, San Francisco, CA, USA, 12 March 2010*; 75840R-SPIE; SPIE: Bellingham, WA, USA, 2010. **[Google Scholar]**

128. Osbild, M.; Brenner, A.; Röther, L.; Finger, J. Ultrashort pulse laser micro polishing of steel—Investigation of the melt pool depth. *Procedia CIRP* **2020**, *94*, 936–941. **[Google Scholar]** **[CrossRef]**

129. Remund, S.; Kramer, T.; Neuenschwander, B.; Jäggi, B. Method for Producing an Implant, and Implant Produced by Said Method. U.S. Patent WO/2020/038925, 27 February 2020. **[Google Scholar]**

130. Brenner, A.; Röther, L.; Osbild, M.; Finger, J. Laser Polishing Using Ultrashort Pulse Laser. In *Laser-Based Micro- and Nanoprocessing XIV, Proceedings of the SPIE LASE, San Francisco, CA, USA, 1–6 February 2020*; SPIE: Bellingham, WA, USA, 2020; Volume 11268, p. 112680. **[Google Scholar]**

131. Metzner, D.; Lickschat, P.; Weißmantel, S. Optimization of the ablation process using ultrashort pulsed laser radiation in different burst modes. *J. Laser Appl.* **2021**, *33*, 012057. **[Google Scholar]** **[CrossRef]**

132. Metzner, D.; Lickschat, P.; Weißmantel, S. Surface treatment on cobalt and titanium alloys using picosecond laser pulses in burst mode. *Appl. Phys. A* **2021**, *127*, 1–9. **[Google Scholar]** **[CrossRef]**

133. Matsumoto, H.; Lin, Z.; Kleinert, J. Ultrafast Laser Ablation of Copper with ~GHz Bursts. In *Laser Applications in Microelectronic and Optoelectronic Manufacturing (LAMOM) XXIII (International Society for Optics and Photonics, Proceedings of the SPIE LASE, San Francisco, CA, USA, 29–31 January 2018*; SPIE: Bellingham, WA, USA, 2018; Volume 10519, p. 1051902. **[Google Scholar]**

134. Povarnitsyn, M.E.; Levashov, P.R.; Knyazev, D.V. Simulation of ultrafast bursts of subpicosecond pulses: In pursuit of efficiency. *Appl. Phys. Lett.* **2018**, *112*, 051603. **[Google Scholar]** **[CrossRef]**

135. Bonamis, G.; Audouard, E.; Hoenninger, C.; Lopez, J.; Mishchik, K.; Mottay, E.; Manek-Hönninger, I. Systematic study of laser ablation with GHz bursts of femtosecond pulses. *Opt. Express* **2020**, *28*, 27702. **[Google Scholar]** **[CrossRef]**

136. Bonamis, G.; Mishchick, K.; Audouard, E.; Hönninger, C.; Mottay, E.; Lopez, J.; Manek-Hönninger, I. High efficiency femtosecond laser ablation with gigahertz level bursts. *J. Laser Appl.* **2019**, *31*, 022205. **[Google Scholar]** **[CrossRef]**

137. Butkus, S.; Jukna, V.; Paipulas, D.; Barkauskas, M.; Sirutkaitis, V. Micromachining of Invar Foils with GHz, MHz and kHz Femtosecond Burst Modes. *Micromachines* **2020**, *11*, 733. **[Google Scholar]** **[CrossRef]** **[PubMed]**

138. Hendow, S.T.; Takahashi, H.; Yamaguchi, M.; Xu, J. Enhanced Ablation Using GHz-Pulsed fs Laser. In *Laser-Based Micro- and Nanoprocessing XIV, Proceedings of the SPIE LASE, San Francisco, CA, USA, 1–6 February 2020*; SPIE: Bellingham, WA, USA, 2020; Volume 11268, p. 1126809. **[Google Scholar]**

139. Hirsiger, T.; Gafner, M.; Remund, S.M.; Chaja, M.W.; Urniezius, A.; Butkus, S.; Neuenschwander, B. Machining Metals and Silicon with GHz Bursts: Surprising Tremendous Reduction of the Specific Removal Rate for Surface Texturing Applications. In *Laser Applications in Microelectronic and Optoelectronic Manufacturing (LAMOM) XXV, Proceedings of the SPIE LASE, San Francisco, CA, USA, 1–6 February 2020*; SPIE: Bellingham, WA, USA, 2020; Volume 11267, p. 112670T. **[Google Scholar]**

140. Cheng, C.-W.; Chen, J.-K. Ultrafast laser ablation of copper by GHz bursts. *Appl. Phys. A* **2020**, *126*, 1–7. **[Google Scholar]** **[CrossRef]**

141. Fehrenbacher, A.; Sailer, M.; Fuehra, B.; Jansen, F.; Tan, C.; Baumbach, S.; Flaig, R.; Eberhardt, C.; Ruebling, S.; Quentin, U.; et al. New generation TruMicro Series 2000: Micromachining Versatility by GHz-burst, Higher Average Power, Flexible Pulse on Demand and Integrated Hollow-Core Fiber Interface. In *Laser-Based Micro- and Nanoprocessing XV, Proceedings of the SPIE LASE, San Francisco, CA, USA, 6–11 March 2021*; SPIE: Bellingham, WA, USA, 2021; Volume 11674, p. 116740U. **[Google Scholar]**

142. Nyenhuis, F.; Michalowski, A.; L'Huillier, J.A. Surface Treatment with GHz-Bursts. In *Laser-Based Micro- and Nanoprocessing XIV, Proceedings of the SPIE LASE, San Francisco, CA, USA, 1–6 February 2020*; SPIE: Bellingham, WA, USA, 2020; Volume 11268, p. 112680B. **[Google Scholar]**

143. Michalowski, A.; Nyenhuis, F.; Kunz, G. Smooth Surfaces by Pulsed Laser Processing with Bursts. *Photonics Views* **2020**, *17*, 42–45. **[Google Scholar]** **[CrossRef]**

144. Obata, K.; Caballero-Lucas, F.; Sugioka, K. Material Processing at GHz Burst Mode by Femtosecond Laser Ablation. *JLMN* **2021**, *16*, 5. **[Google Scholar]**

145. Manninen, M.; Hirvimäki, M.; Poutiainen, I.; Salminen, A. Effect of Pulse Length on Engraving Efficiency in Nanosecond Pulsed Laser Engraving of Stainless Steel. *Met. Mater. Trans. A* **2015**, *46*, 2129–2136. **[Google Scholar]** **[CrossRef]**

146. Mathew, M.M.; Bathe, R.N.; Padmanabham, G.; Padmanaban, R.; Thirumalini, S. A study on the micromachining of molybdenum using nanosecond and femtosecond lasers. *Int. J. Adv. Manuf. Technol.* **2017**, *104*, 3239–3249. **[Google Scholar]** **[CrossRef]**

147. Garnov, S.; Konov, V.; Kononenko, T.; Pashinin, V.; Sinyavsky, M. Microsecond laser material processing at 1.06 μm. *Laser Phys.* **2004**, *14*, 910–915. **[Google Scholar]**

148. Mur, J.; Petkovšek, R. Precision and resolution in laser direct microstructuring with bursts of picosecond pulses. *Appl. Phys. A* **2018**, *124*, 62. **[Google Scholar] [CrossRef]**

149. Gaudiuso, C.; Giannuzzi, G.; Volpe, A.; Lugarà, P.M.; Choquet, I.; Ancona, A. Incubation during laser ablation with bursts of femtosecond pulses with picosecond delays. *Opt. Express* **2018**, *26*, 3801–3813. **[Google Scholar] [CrossRef] [PubMed]**

150. Rong, Y.; Ji, P.; He, M.; Zhang, Y.; Tang, Y. Multiscale Investigation of Femtosecond Laser Pulses Processing Aluminum in Burst Mode. *Nanoscale Microscale Thermophys. Eng.* **2018**, *22*, 324–347. **[Google Scholar] [CrossRef]**

151. Mur, J.; Petkovšek, R. Near-THz bursts of pulses—Governing surface ablation mechanisms for laser material processing. *Appl. Surf. Sci.* **2019**, *478*, 355–360. **[Google Scholar] [CrossRef]**

152. Wang, A.; Das, A.; Grojo, D. Ultrafast Laser Writing Deep inside Silicon with THz-Repetition-Rate Trains of Pulses. *Research* **2020**, *2020*, 1–11. **[Google Scholar] [CrossRef]**

153. Ancona, A.; Gaudiuso, C.; Giannuzzi, G.; Choquet, I.; Lugarà, P.M. Incubation Effect In Burst Mode Fs-Laser Ablation Of Stainless Steel Samples. In *Laser-Based Micro- and Nanoprocessing XII, Proceedings of the SPIE LASE, San Francisco, CA, USA, 27 January–1 February 2020*; SPIE: Bellingham, WA, USA, 2018; Volume 10520, p. 105200A. **[Google Scholar]**

154. Förster, D.J.; Weber, R.; Holder, D.; Graf, T. Estimation of the depth limit for percussion drilling with picosecond laser pulses. *Opt. Express* **2018**, *26*, 11546–11552. **[Google Scholar] [CrossRef] [Green Version]**

155. Holder, D.; Weber, R.; Graf, T.; Onuseit, V.; Brinkmeier, D.; Förster, D.J.; Feuer, A. Analytical model for the depth progress of percussion drilling with ultrashort laser pulses. *Appl. Phys. A* **2021**, *127*, 1–8. **[Google Scholar] [CrossRef]**

156. Lapczyna, M.; Chen, K.; Herman, P.; Tan, H.; Marjoribanks, R. Ultra high repetition rate (133 MHz) laser ablation of aluminum with 1.2-ps pulses. *Appl. Phys. A* **1999**, *69*, S883–S886. **[Google Scholar] [CrossRef]**

157. Esser, D.; Rezaei, S.; Li, J.; Herman, P.R.; Gottmann, J. Time dynamics of burst-train filamentation assisted femtosecond laser machining in glasses. *Opt. Express* **2011**, *19*, 25632–25642. **[Google Scholar] [CrossRef]**

158. Kammer, H.; Dreisow, F.; Tünnermann, A. Analysis of the Hole Shape Evolution in fs-Pulse Percussion Drilling with Bursts. In *Frontiers in Ultrafast Optics: Biomedical, Scientific, and Industrial Applications XVI, Proceedings of the SPIE LASE, San Francisco, CA, USA, 13–18 February 2016*; SPIE: Bellingham, WA, USA, 2016; Volume 9740, p. 974012. **[Google Scholar]**

159. Wang, Q.; Luo, S.; Chen, Z.; Qi, H.; Deng, J.; Hu, Z. Drilling of aluminum and copper films with femtosecond double-pulse laser. *Opt. Laser Technol.* **2016**, *80*, 116–124. **[Google Scholar] [CrossRef]**

160. Balachninaitė, O.; Tamulienė, V.; Eičas, L.; Vaičaitis, V. Laser micromachining of steel and copper using femtosecond laser pulses in GHz burst mode. *Results Phys.* **2021**, *22*, 103847. **[Google Scholar] [CrossRef]**

161. Schille, J.; Loeschner, U.; Ebert, R.; Scully, P.; Goddard, N.; Exner, H. Laser micro processing using a high repetition rate femto second laser. *J. Laser Appl.* **2010**, 1491–1499. **[Google Scholar] [CrossRef]**

162. Tsukamoto, M.; Kayahara, T.; Nakano, H.; Hashida, M.; Katto, M.; Fujita, M.; Tanaka, M.; Abe, N. Microstructures formation on titanium plate by femtosecond laser ablation. *J. Physics Conf. Ser.* **2007**, *59*, 666–669. **[Google Scholar] [CrossRef]**

163. Herrmann, T.; Harth, F.; Henrich, B.; L'Huillier, J.; Hajri, M. How to Improve Efficiency in USP Laser Micromachining. *Laser Tech. J.* **2016**, *13*, 16–19. **[Google Scholar] [CrossRef]**

164. Nyenhuis, F.; Michalowski, A.; L'Huillier, J. Dual Process Strategy to Increase the Usable Power for Laser-Milling. *JLMN* **2020**. **[Google Scholar] [CrossRef]**

165. Sassmannshausen, A.; Brenner, A.; Finger, J. Ultrashort pulse laser polishing by continuous surface melting. *J. Mater. Process. Technol.* **2021**, *293*, 117058. **[Google Scholar]** **[CrossRef]**

166. Kažukauskas, E.; Butkus, S.; Tokarski, P.; Jukna, V.; Barkauskas, M.; Sirutkaitis, V. Micromachining of Transparent Biocompatible Polymers Applied in Medicine Using Bursts of Femtosecond Laser Pulses. *Micromachines* **2020**, *11*, 1093. **[Google Scholar]** **[CrossRef]** **[PubMed]**

167. Antończak, A.J.; Stępak, B.; Kozioł, P.E.; Abramski, K.M. The influence of process parameters on the laser-induced coloring of titanium. *Appl. Phys. A* **2013**, *115*, 1003–1013. **[Google Scholar]** **[CrossRef]** **[Green Version]**

168. Veiko, V.; Odintsova, G.; Ageev, E.; Karlagina, Y.; Loginov, A.; Skuratova, A.; Gorbunova, E. Controlled oxide films formation by nanosecond laser pulses for color marking. *Opt. Express* **2014**, *22*, 24342–24347. **[Google Scholar]** **[CrossRef]**

169. Lecka, K.M.; Wojcik, M.R.; Antonczak, A.J. Laser-Induced Color Marking of Titanium: A Modeling Study of the Interference Effect and the Impact of Protective Coating. *Math. Probl. Eng.* **2017**, *2017*, 1–11. **[Google Scholar]** **[CrossRef]** **[Green Version]**

170. Liu, H.; Lin, W.; Hong, M. Surface coloring by laser irradiation of solid substrates. *APL Photon.* **2019**, *4*, 051101. **[Google Scholar]** **[CrossRef]** **[Green Version]**

171. Vorobyev, A.Y.; Guo, C. Colorizing metals with femtosecond laser pulses. *Appl. Phys. Lett.* **2008**, *92*, 041914. **[Google Scholar]** **[CrossRef]**

172. Dusser, B.; Sagan, S.; Soder, H.; Faure, N.; Colombier, J.-P.; Jourlin, M.; Audouard, E. Controlled nanostructrures formation by ultra fast laser pulses for color marking. *Opt. Express* **2010**, *18*, 2913–2924. **[Google Scholar]** **[CrossRef]** **[Green Version]**

173. Ackerl, N.; Gugger, P.; Warhanek, M.; Gysel, J.; Wegener, K. Ultra-Short Pulsed Laser Marking and Coloration of Metals with Segmented Pixel Parameter Transformation. *JLMN* **2020**, *15*. **[Google Scholar]** **[CrossRef]**

174. Vorobyev, A.Y.; Guo, C. Direct femtosecond laser surface nano/microstructuring and its applications: Direct femtosecond laser surface nano/microstructuring and its applications. *Laser Photonics Rev.* **2013**, 7, 385–407. **[Google Scholar]** **[CrossRef]**

175. Guay, J.-M.; Lesina, A.C.; Cote, G.; Charron, M.; Poitras, D.; Ramunno, L.; Berini, P.; Weck, A. Laser-induced plasmonic colours on metals. *Nat. Commun.* **2017**, *8*, 16095. **[Google Scholar]** **[CrossRef]**

176. Berini, P.; Guay, J.-M.; Lesina, A.C.; Walia, J.; Ramunno, L.; Weck, A.; Krupin, O.; Warren, M. Plasmonic Colours on Bulk Metals: Laser Coloring of Large Areas Exhibiting High Topography. In *Frontiers in Ultrafast Optics: Biomedical, Scientific, and Industrial Applications XVIII, Proceedings of the SPIE LASE, San Francisco, CA, USA, 27 January–1 February 2018*; SPIE: Bellingham, WA, USA, 2018; Volume 10522, p. 105220G. **[Google Scholar]**

177. Lumentum. *FlexBurstTM—Picosecond Micromachining Lasers Offer "Burst Mode" with Full User Control of Intra-Burst Pulse Energy Distribution*; Lumentum: San Jose, CA, USA, 2016. **[Google Scholar]**

178. Guay, J.-M.; Lesina, A.C.; Baxter, J.; Killaire, G.; Ramunno, L.; Berini, P.; Weck, A. Topography Tuning for Plasmonic Color Enhancement via Picosecond Laser Bursts. *Adv. Opt. Mater.* **2018**, 6, 1800189. **[Google Scholar]** **[CrossRef]** **[Green Version]**

179. Stratakis, E.; Bonse, J.; Heitz, J.; Siegel, J.; Tsibidis, G.; Skoulas, E.; Papadopoulos, A.; Mimidis, A.; Joel, A.-C.; Comanns, P.; et al. Laser engineering of biomimetic surfaces. *Mater. Sci. Eng. R Rep.* **2020**, *141*, 100562. **[Google Scholar]** **[CrossRef]**

180. Masato, D.; Sorgato, M.; Batal, A.; Dimov, S.; Lucchetta, G. Thin-wall injection molding of polypropylene using molds with different laser-induced periodic surface structures. *Polym. Eng. Sci.* **2019**, *59*, 1889–1896. **[Google Scholar]** **[CrossRef]**

181. Florian, C.; Kirner, S.V.; Krüger, J.; Bonse, J. Surface functionalization by laser-induced periodic surface structures. *J. Laser Appl.* **2020**, *32*, 022063. **[Google Scholar]** **[CrossRef]**

182. Piccolo, L.; Sorgato, M.; Batal, A.; Dimov, S.; Lucchetta, G.; Masato, D. Functionalization of Plastic Parts by Replication of Variable Pitch Laser-Induced Periodic Surface Structures. *Micromachines* **2020**, *11*, 429. **[Google Scholar]** **[CrossRef]** **[PubMed]** **[Green Version]**

183. Bonse, J.; Höhm, S.; Kirner, S.V.; Rosenfeld, A.; Krüger, J. Laser-Induced Periodic Surface Structures—A Scientific Evergreen. *IEEE J. Sel. Top. Quantum Electron.* **2017**, *23*, 9000615. **[Google Scholar]** **[CrossRef]**

184. Bonse, J.; Graf, S. Maxwell Meets Marangoni—A Review of Theories on Laser-Induced Periodic Surface Structures. *Laser Photonics Rev.* **2020**, *14*, 2000215. **[Google Scholar]** **[CrossRef]**

185. Bonse, J. Quo Vadis LIPSS? Recent and Future Trends on Laser-Induced Periodic Surface Structures. *Nanomaterials* **2020**, *10*, 1950. **[Google Scholar]** **[CrossRef]**

186. Bonse, J.; Kirner, S.V.; Krüger, J. Laser-Induced Periodic Surface Structures (LIPSS). In *Handbook of Laser Micro- and Nano-Engineering*; Sugioka, K., Ed.; Springer International Publishing: New York, NY, USA, 2021; pp. 1–59. **[Google Scholar]**

187. Fraggelakis, F.; Mincuzzi, G.; Lopez, J.; Manek-Hönninger, I.; Kling, R. Controlling 2D laser nano structuring over large area with double femtosecond pulses—ScienceDirect. *Appl. Surf. Sci.* **2019**, *470*, 677–686. **[Google Scholar]** **[CrossRef]**

188. Hashida, M.; Furukawa, Y.; Inoue, S.; Sakabe, S.; Masuno, S.; Kusaba, M.; Sakagami, H.; Tsukamoto, M. Uniform LIPSS on titanium irradiated by two-color double-pulse beam of femtosecond laser. *J. Laser Appl.* **2020**, *32*, 022054. **[Google Scholar]** **[CrossRef]**

189. Hashida, M. Periodic Grating Structures on Metal Self-organized by Double-pulse Irradiation. *J. Laser Micro Nanoeng.* **2014**, *9*, 234–237. **[Google Scholar]** **[CrossRef]**

190. Hashida, M.; Nishii, T.; Miyasaka, Y.; Sakagami, H.; Shimizu, M.; Inoue, S.; Sakabe, S. Orientation of periodic grating structures controlled by double-pulse irradiation. *Appl. Phys. A* **2016**, *122*, 1–5. **[Google Scholar]** **[CrossRef]**

191. Giannuzzi, G.; Gaudiuso, C.; Di Franco, C.; Scamarcio, G.; Lugarà, P.M.; Ancona, A. Large area laser-induced periodic surface structures on steel by bursts of femtosecond pulses with picosecond delays. *Opt. Lasers Eng.* **2019**, *114*, 15–21. **[Google Scholar]** **[CrossRef]**

192. Giannuzzi, G.; Gaudiuso, C.; Cinquino, M.; Di Mundo, R.; Mirenghi, L.; Di Franco, C.; Scamarcio, G.; Lugarà, P.M.; Ancona, A. 1-D and 2-D Surface Structuring of Steel by Bursts of Femtosecond Laser Pulses. In *Laser-Based Micro- and Nanoprocessing XIII, Proceedings of the SPIE LASE, San Francisco, CA, USA, 4 February 2019*; SPIE: Bellingham, WA, USA, 2019; Volume 10906, p. 109060O. **[Google Scholar]**

193. Wang, X.; Li, C.; Ma, C.; Feng, J.; Hong, W.; Zhang, Z. Formation of laser induced periodic structures on stainless steel using multi-burst picosecond pulses. *Opt. Express* **2018**, *26*, 6325–6330. **[Google Scholar]** **[CrossRef]**

194. Fraggelakis, F.; Giannuzzi, G.; Gaudiuso, C.; Manek-Hönninger, I.; Mincuzzi, G.; Ancona, A.; Kling, R. Double- and Multi-Femtosecond Pulses Produced by Birefringent Crystals for the Generation of 2D Laser-Induced Structures on a Stainless Steel Surface. *Materials* **2019**, *12*, 1257. **[Google Scholar]** **[CrossRef]** **[Green Version]**

195. Höhm, S.; Rosenfeld, A.; Krüger, J.; Bonse, J. Area dependence of femtosecond laser-induced periodic surface structures for varying band gap materials after double pulse excitation. *Appl. Surf. Sci.* **2013**, *278*, 7–12. **[Google Scholar]** **[CrossRef]**

196. Höhm, S.; Herzlieb, M.; Rosenfeld, A.; Krüger, J.; Bonse, J. Femtosecond laser-induced periodic surface structures on silicon upon polarization controlled two-color double-pulse irradiation. *Opt. Express* **2015**, *23*, 61–71. **[Google Scholar]** **[CrossRef]**

197. Jaeggi, B.; Neuenschwander, B.; Zimmermann, M.; Zecherle, M.; Boeckler, E.W. Time-Optimized Laser Micro Machining by Using a New High Dynamic and High Precision Galvo Scanner. In *Laser Applications in Microelectronic and Optoelectronic Manufacturing (LAMOM) XXI, Proceedings of the SPIE LASE, San Francisco, CA, USA, 13–18 February 2016*; SPIE: Bellingham, WA, USA, 2016; Volume 9735, p. 973513. **[Google Scholar]**

198. Häfner, T.; Heberle, J.; Holder, D.; Schmidt, M. Adjustment of surface energy on steel surfaces due to CLP generation by picosecond laser processing. In Proceedings of the Lasers in Manufacturing Conference, Munich, Germany, 22–25 June 2015; pp. 1–10. **[Google Scholar]**

199. Jaeggi, B.; Neuenschwander, B.; Hunziker, U.; Zuercher, J.; Meier, T.; Zimmermann, M.; Selbmann, K.H.; Hennig, G. Ultra-High-Precision Surface Structuring by Synchronizing a Galvo Scanner with an Ultra-Short-Pulsed Laser System in MOPA Arrangement. In *Laser Applications in Microelectronic and Optoelectronic Manufacturing (LAMOM) XVII, Proceedings of the SPIE LASE, San Francisco, CA, USA, 21–26 January 2012*; 82430K; SPIE: Bellingham, WA, USA, 2012. **[Google Scholar]**

200. Loeschner, U.; Schille, J.; Streek, A.; Knebel, T.; Hartwig, L.; Hillmann, R.; Endisch, C. High-rate laser microprocessing using a polygon scanner system. *J. Laser Appl.* **2015**, *27*, S29303. **[Google Scholar]** **[CrossRef]**

201. Schille, J.; Schneider, L.; Mauersberger, S.; Szokup, S.; Höhn, S.; Pötschke, J.; Reiß, F.; Leidich, E.; Löschner, U. High-Rate Laser Surface Texturing for Advanced Tribological Functionality. *Lubricants* **2020**, *8*, 33. **[Google Scholar]** **[CrossRef]** **[Green Version]**

202. Jaeggi, B.; Remund, S.; Streubel, R.; Goekce, B.; Barcikowski, S.; Neuenschwander, B. Laser Micromachining of Metals with Ultra-Short Pulses: Factors Limiting the Scale-Up Process. *JLMN* **2017**, *12*, 267–273. **[Google Scholar]**

203. Brenner, A.; Zecherle, M.; Verpoort, S.; Schuster, K.; Schnitzler, C.; Kogel-Hollacher, M.; Reisacher, M.; Nohn, B. Efficient production of design textures on large-format 3D mold tools. *J. Laser Appl.* **2020**, *32*, 012018. **[Google Scholar]** **[CrossRef]** **[Green Version]**

204. Phipps, C.; Luke, J. Diode Laser-Driven Microthrusters: A New Departure for Micropropulsion. *AIAA J.* **2002**, *40*, 310–318. **[Google Scholar]** **[CrossRef]** **[Green Version]**

205. Phipps, C.; Birkan, M.; Bohn, W.; Eckel, H.-A.; Horisawa, H.; Lippert, T.; Michaelis, M.; Rezunkov, Y.; Sasoh, A.; Schall, W.; et al. Review: Laser-Ablation Propulsion. *J. Propuls. Power* **2010**, *26*, 609–637. **[Google Scholar]** **[CrossRef]**

206. Phipps, C.R.; Baker, K.L.; Libby, S.; Liedahl, D.A.; Olivier, S.S.; Pleasance, L.D.; Rubenchik, A.; Trebes, J.E.; George, E.V.; Marcovici, B.; et al. Removing orbital debris with lasers. *Adv. Space Res.* **2012**, *49*, 1283–1300. **[Google Scholar]** **[CrossRef]** **[Green Version]**

207. Scharring, S.; Lorbeer, R.A.; Karg, S.; Pastuschka, L.; Förster, D.J.; Eckel, H.A. The MICROLAS Concept: Precise Thrust Generation in the Micronewton Range by Laser Ablation. In *Technology for Small Satellite Research: Payloads and Subsystem Technologies Small Satellite Applications, Missions, and In-Orbit Experiences*; International Academy of Astronautics: Stockholm, Sweden, 2016; pp. 27–34. **[Google Scholar]**

208. Phipps, C.R. Laser Ablation Propulsion and Its Applications in Space. In *Superconductivity*; Springer Science and Business Media: New York, NY, USA, 2018; Volume 274, pp. 217–246. **[Google Scholar]**

209. Levchenko, I.; Bazaka, K.; Mazouffre, S.; Xu, S. Prospects and physical mechanisms for photonic space propulsion. *Nat. Photon.* **2018**, *12*, 649–657. **[Google Scholar]** **[CrossRef]**

210. Lorbeer, R.-A.; Zwilich, M.; Zabic, M.; Scharring, S.; Eisert, L.; Wilken, J.; Schumacher, D.; Roth, M.; Eckel, H.-A. Experimental verification of high energy laser-generated impulse for remote laser control of space debris. *Sci. Rep.* **2018**, *8*, 1–9. **[Google Scholar]** **[CrossRef]**

211. Cremers, D.A.; Knight, A.K. Laser-Induced Breakdown Spectroscopy. In *Encyclopedia of Analytical Chemistry*; Meyers, R.A., Ed.; John Wiley & Sons: Hoboken, NJ, USA, 2000; p. a5110t. **[Google Scholar]**

212. Song, K.; Lee, Y.-I.; Sneddon, J. Recent developments in instrumentation for laser induced breakdown spectroscopy. *Appl. Spectrosc. Rev.* **2002**, *37*, 89–117. **[Google Scholar]** **[CrossRef]**

213. Miziolek, A.W.; Palleschi, V.; Schechter, I. *Laser-Induced Breakdown Spectroscopy (LIBS): Fundamentals and Applications*; Cambridge University Press: Cambridge, MA, USA, 2006. **[Google Scholar]**

214. Kim, B.-M.; Feit, M.; Rubenchik, A.; Mammini, B.; Da Silva, L. Optical feedback signal for ultrashort laser pulse ablation of tissue. *Appl. Surf. Sci.* **1998**, *127-129*, 857–862. **[Google Scholar]** **[CrossRef]** **[Green Version]**

215. Margetic, V.; Pakulev, A.; Stockhaus, A.; Bolshov, M.; Niemax, K.; Hergenröder, R. A comparison of nanosecond and femtosecond laser-induced plasma spectroscopy of brass samples. *Spectrochim. Acta Part B At. Spectrosc.* **2000**, *55*, 1771–1785. **[Google Scholar]** **[CrossRef]**

216. Burgio, L.; Melessanaki, K.; Doulgeridis, M.; Clark, R.; Anglos, D. Pigment identification in paintings employing laser induced breakdown spectroscopy and Raman microscopy. *Spectrochim. Acta Part B At. Spectrosc.* **2001**, *56*, 905–913. **[Google Scholar] [CrossRef]**

217. Serifaki, K.; Boke, H.; Yalcin, S.; Ipekoglu, B. Characterization of materials used in the execution of historic oil paintings by XRD, SEM-EDS, TGA and LIBS analysis. *Mater. Charact.* **2009**, *60*, 303–311. **[Google Scholar] [CrossRef] [Green Version]**

218. Borba, F.D.S.L.; Cortez, J.; Asfora, V.K.; Pasquini, C.; Pimentel, M.F.; Pessis, A.-M.; Khoury, H.J. Multivariate treatment of LIBS data of prehistoric paintings. *J. Braz. Chem. Soc.* **2012**, *23*, 958–965. **[Google Scholar] [CrossRef]**

219. Dikmelik, Y.; McEnnis, C.; Spicer, J.B. Femtosecond and nanosecond laser-induced breakdown spectroscopy of trinitrotoluene. *Opt. Express* **2008**, *16*, 5332–5337. **[Google Scholar] [CrossRef] [PubMed]**

220. Gottfried, J.L.; De Lucia, F.C.; Munson, C.A.; Miziolek, A.W. Laser-induced breakdown spectroscopy for detection of explosives residues: A review of recent advances, challenges, and future prospects. *Anal. Bioanal. Chem.* **2009**, *395*, 283–300. **[Google Scholar] [CrossRef]**

221. Shaik, A.K.; Epuru, N.R.; Syed, H.; Byram, C.; Soma, V.R. Femtosecond laser induced breakdown spectroscopy based standoff detection of explosives and discrimination using principal component analysis. *Opt. Express* **2018**, *26*, 8069–8083. **[Google Scholar] [CrossRef]**

222. Gurevich, E.L.; Hergenröder, R. Femtosecond Laser-Induced Breakdown Spectroscopy: Physics, Applications, and Perspectives. *Appl. Spectrosc.* **2007**, *61*, A233–A242. **[Google Scholar] [CrossRef]**

223. Malvezzi, A.M. Laser–Matter Interaction in LIBS Experiments. In *Laser-Induced Breakdown Spectroscopy*; Musazzi, S., Perini, U., Eds.; Springer Series in Optical Sciences; Springer: Berlin/Heidelberg, Germany, 2014; Volume 182, pp. 3–29. **[Google Scholar]**

224. Labutin, T.A.; Lednev, V.N.; Ilyin, A.A.; Popov, A. Femtosecond laser-induced breakdown spectroscopy. *J. Anal. At. Spectrom.* **2015**, *31*, 90–118. **[Google Scholar] [CrossRef]**

225. Piñon, V.; Fotakis, C.; Nicolas, G.; Anglos, D. Double pulse laser-induced breakdown spectroscopy with femtosecond laser pulses. *Spectrochim. Acta Part B At. Spectrosc.* **2008**, *63*, 1006–1010. **[Google Scholar] [CrossRef]**

226. Lu, Y.; Zorba, V.; Mao, X.; Zheng, R.; Russo, R.E. UV fs–ns double-pulse laser induced breakdown spectroscopy for high spatial resolution chemical analysis. *J. Anal. At. Spectrom.* **2013**, *28*, 743–748. **[Google Scholar] [CrossRef]**

227. Bulanov, S.V.; Khoroshkov, V.S. Feasibility of using laser ion accelerators in proton therapy. *Plasma Phys. Rep.* **2002**, *28*, 453–456. **[Google Scholar] [CrossRef]**

228. Ditmire, T.; Donnelly, T.; Rubenchik, A.M.; Falcone, R.W.; Perry, M.D. Interaction of intense laser pulses with atomic clusters. *Phys. Rev. A* **1996**, *53*, 3379–3402. **[Google Scholar] [CrossRef] [PubMed] [Green Version]**

229. Malka, V.; Fritzler, S. Electron and proton beams produced by ultra short laser pulses in the relativistic regime. *Laser Part. Beams* **2004**, *22*, 399–405. **[Google Scholar] [CrossRef]**

230. Neely, D.; Foster, P.; Robinson, A.; Lindau, F.; Lundh, O.; Persson, A.; Wahlström, C.-G.; McKenna, P. Enhanced proton beams from ultrathin targets driven by high contrast laser pulses. *Appl. Phys. Lett.* **2006**, *89*, 021502. **[Google Scholar] [CrossRef]**

231. Andreev, A.; Sonobe, R.; Kawata, S.; Miyazaki, S.; Sakai, K.; Miyauchi, K.; Kikuchi, T.; Platonov, K.; Nemoto, K. Effect of a laser prepulse on fast ion generation in the interaction of ultra-short intense laser pulses with a limited-mass foil target. *Plasma Phys. Control. Fusion* **2006**, *48*, 1605–1619. **[Google Scholar] [CrossRef]**

232. Passoni, M.; Bertagna, L.; Zani, A. Target normal sheath acceleration: Theory, comparison with experiments and future perspectives. *New J. Phys.* **2010**, *12*, 045012. **[Google Scholar] [CrossRef]**

233. Daido, H.; Nishiuchi, M.; Pirozhkov, A. Review of laser-driven ion sources and their applications. *Rep. Prog. Phys.* **2012**, *75*, 056401. **[Google Scholar] [CrossRef]**

234. Fourmaux, S.; Buffechoux, S.; Albertazzi, B.; Capelli, D.; Levy, A.; Gnedyuk, S.; Lecherbourg, L.; Lassonde, P.; Payeur, S.; Antici, P.; et al. Investigation of laser-driven proton acceleration using ultra-short, ultra-intense laser pulses. *Phys. Plasmas* **2013**, *20*, 013110. **[Google Scholar] [CrossRef] [Green Version]**

235. Barberio, M.; Veltri, S.; Scisciò, M.; Antici, P. Laser-Accelerated Proton Beams as Diagnostics for Cultural Heritage. *Sci. Rep.* **2017**, *7*, 40415. **[Google Scholar] [CrossRef] [PubMed] [Green Version]**

236. Barberio, M.; Scisciò, M.; Vallières, S.; Cardelli, F.; Chen, S.N.; Famulari, G.; Gangolf, T.; Revet, G.; Schiavi, A.; Senzacqua, M.; et al. Laser-accelerated particle beams for stress testing of materials. *Nat. Commun.* **2018**, *9*, 372. **[Google Scholar] [CrossRef] [PubMed] [Green Version]**

237. Torrisi, L.; Cutroneo, M.; Torrisi, A.; Silipigni, L.; Costa, G.; Rosinski, M.; Badziak, J.; Wołowski, J.; Zaraś-Szydłowska, A.; Parys, P. Protons accelerated in the target normal sheath acceleration regime by a femtosecond laser. *Phys. Rev. Accel. Beams* **2019**, *22*, 021302. **[Google Scholar] [CrossRef] [Green Version]**

238. Bulanov, S.; Esirkepov, T.; Khoroshkov, V.; Kuznetsov, A.; Pegoraro, F. Oncological hadrontherapy with laser ion accelerators. *Phys. Lett. A* **2002**, *299*, 240–247. **[Google Scholar] [CrossRef]**

239. Malka, V.; Faure, J.; Gauduel, Y.A. Ultra-short electron beams based spatio-temporal radiation biology and radiotherapy. *Mutat. Res. Mutat. Res.* **2010**, *704*, 142–151. **[Google Scholar] [CrossRef]**

240. Schardt, D.; Elsaesser, T.; Schulz-Ertner, D. Heavy-ion tumor therapy: Physical and radiobiological benefits. *Rev. Mod. Phys.* **2010**, *82*, 383–425. **[Google Scholar] [CrossRef]**

241. Döppner, T.; Fennel, T.; Diederich, T.; Tiggesbäumker, J.; Meiwes-Broer, K.H. Controlling the Coulomb Explosion of Silver Clusters by Femtosecond Dual-Pulse Laser Excitation. *Phys. Rev. Lett.* **2005**, *94*, 013401. **[Google Scholar] [CrossRef]**

242. Markey, K.; McKenna, P.; Brenner, C.M.; Carroll, D.C.; Günther, M.M.; Harres, K.; Kar, S.; Lancaster, K.; Nürnberg, F.; Quinn, M.N.; et al. Spectral Enhancement in the Double Pulse Regime of Laser Proton Acceleration. *Phys. Rev. Lett.* **2010**, *105*, 195008. **[Google Scholar] [CrossRef] [PubMed] [Green Version]**

243. Uryupina, D.S.; Ivanov, K.A.; Brantov, A.; Savel'Ev, A.B.; Bychenkov, V.Y.; Povarnitsyn, M.E.; Volkov, R.V.; Tikhonchuk, V.T. Femtosecond laser-plasma interaction with prepulse-generated liquid metal microjets. *Phys. Plasmas* **2012**, *19*, 13104. **[Google Scholar] [CrossRef]**

CHAPTER

LASER-INDUCED FORWARD TRANSFER PRINTING AND THE FUTURE OF DIGITAL MANUFACTURING

3

1. INTRODUCTION: AN OVERVIEW

Digital manufacturing encompasses fabrication methods and procedures for producing functioning structures and devices, allowing for design, modification, and final optimization to be conducted digitally. The primary advantage over traditional fabrication is the ability to operate without masks, master samples, or molds, facilitating concurrent design modifications during production. These techniques consistently aim to address critical open concerns, including reduced manufacturing costs, industrial scalability, compatibility with existing methods, broad material compatibility, and enhanced spatial resolution and reproducibility. Researchers and engineers strive to develop a comprehensive technique that addresses these issues; nevertheless, this endeavor remains elusive, as there are typically constraints in at least one of these areas. For instance, although various fabrication techniques demonstrate commendable performance in laboratory settings, a significant challenge arises when these techniques are necessitated in actual industrial environments, where scalability and mass production at minimal costs and maximal speeds present clear objectives for developers. This chapter focuses on a particular laser printing method referred to as laser-induced forward transfer (LIFT). In laser printing through LIFT, a laser source irradiates a donor substrate containing the material of interest. The interaction of radiation with the donor material propels a small quantity of material toward a strategically placed acceptor substrate, facilitating the development of a printed voxel. Sequential voxel printing enables the creation of virtually any configuration. Initially, the technique was primarily applicable to pulsed laser sources utilizing predominantly solid donor materials, which constrained its applicability. However, the current proliferation of dependable and cost-effective laser sources (both pulsed and continuous wave, across various wavelengths), coupled with appropriate material preparation (in solid, paste, or liquid forms), now enables the development of devices for the microelectronics sector, as well as miniaturized assays for microbiology, and it has recently demonstrated promise for tissue engineering applications, among numerous others. This work does not aim to serve as the definitive assessment of the approach, nor does it provide an open discussion on the latest advancements; for those, we refer the reader to relevant review articles. The introduction commences with an examination of review papers concerning recent advancements and sky edge applications that utilize this printing process, as referenced in [1,2,3]. The operational principle for transferring materials in solid, paste, and liquid forms for various applications is examined. Armed with this knowledge, we offer a comprehensive evaluation of the LIFT technique's history, current status, and future prospects, taking into account published patents and patent applications that feature LIFT as the primary technique. Ultimately, through the examination of commercial data within the expansive context of additive manufacturing, we offer commercial projections and potential trends for the LIFT process in applicable scientific domains, including electronics, biology, and medicine, among others.

2. LASER-INDUCED FORWARD TRANSFER (LIFT) TECHNIQUE: AN OVERVIEW

In laser-induced forward transfer (LIFT) printing, the material of interest is typically deposited as a thin film onto a transparent substrate known as the "donor substrate." A laser beam is aimed at the interface between the thin film and the transparent substrate, generating a micro-explosion that ejects the material forward. When an additional "receptor substrate" is situated at an appropriate distance from the donor, the transferred material is securely placed at designated user-defined locations. Traditionally, a pulsed laser is employed, with each pulse generating a printed voxel of material, the volume of which can be as little as a few femtoliters [4]. Recent advancements have introduced methods enabling the printing of materials with multiple laser pulses [5,6], as well as continuous wave (CW) laser sources [7]. The resultant pixel geometry and quantity of transferred material are contingent upon the initial characteristics of the donor material (thickness, viscosity, optical properties), the application of an auxiliary sacrificial absorbing layer (commonly referred to as a dynamic release layer or 'DRL'), the physical and chemical attributes of the receptor substrate (wetting, oxidation, temperature), and the laser parameters (fluence, intensity distribution, wavelength, pulse duration). Figure 1 presents a diagram illustrating the overall arrangement and primary components in a conventional setup for material transfer using LIFT.

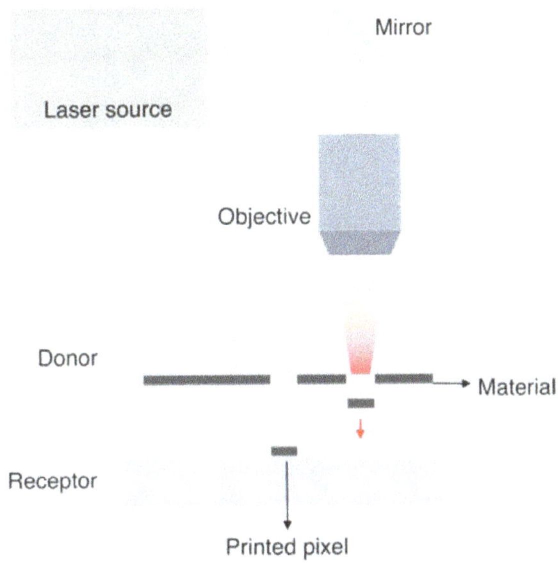

Figure 1. Sketch (not to scale) that includes the main elements of an experimental setup to print materials via LIFT.

To establish a categorization strategy for all LIFT variants, we can utilize the condition of the material to be transferred as a sorting criterion, specifically (i) solid materials, (ii) pastes, or (iii) liquids. We will succinctly examine the operational fundamentals for each scenario and present pertinent examples for each circumstance.

2.1. LIFT from Solids

In this category of applications for the LIFT technique, the initial base material in the donor film consists of a solid layer that typically absorbs incoming radiation and is then ejected toward the receptor substrate [1]. Two primary possibilities may arise based on the modification of the material by laser light; in the first scenario, the solid material undergoes localized melting, resulting in the liquid phase being driven forward [8]. Solidification typically occurs on the receptor substrate as isolated hemispherical droplets, as illustrated in Figure 2A. A second scenario occurs when the material designated for transfer must not experience any phase alteration. This method is appropriate for very sensitive materials that require preservation of crystalline structures post-transfer, or for materials that are particularly susceptible to thermal alterations caused by laser irradiation. An intermediary sacrificial layer is utilized in this instance. This layer will absorb the majority of incoming radiation, resulting in a localized explosion that will propel the substance of interest toward the receptor substrate [9]. The transfer dynamics in this instance are more intricate, since it entails the involvement of an extra substance that will respond differentially to the laser light, as illustrated in Figure 2B.

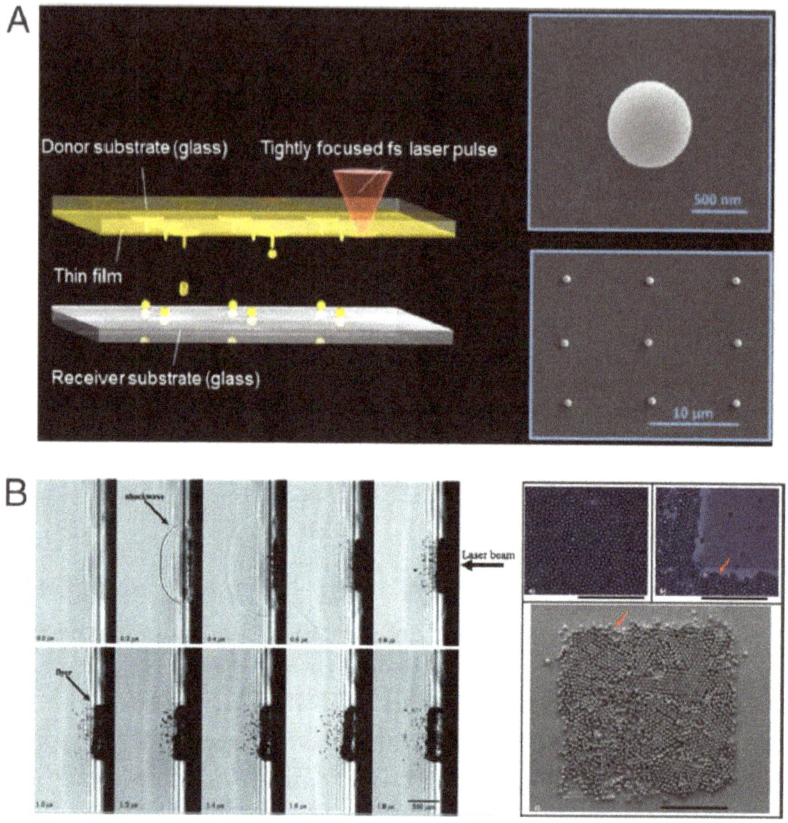

Figure 2. Printing of materials starting from a solid donor film following two different scenarios. (A) Melted material, where the final output corresponds to individual spherical particles. Reprinted under creative commons permission (CC BY-NC-ND 4.0) from [8], copyright 2018, De Gruyter. (B) In this transfer, the use of a sacrificial layer allowed the printing of an array of microspheres that followed the square shape distribution of the incident laser beam.

2.2. LIFT from Pastes

In this specific instance of the LIFT process, often referred to as laser decal transfer (LDT), the donor material consists of an extremely viscous paste. Owing to the characteristics of the material and manufacturing circumstances, the paste is not influenced by deformation from surface tension effects. The geometry of the transferred material usually mirrors the laser intensity distribution, enabling the transfer of voxels with intricate shapes when a standard LIFT configuration is integrated with digital micromirrors, spatial light modulators, or phase masks to establish a specific intensity distribution profile. In the case depicted in Figure 3 [10], the donor material is a silver nanoparticle suspension with a viscosity of 100 Pa.s. Congruent transfers transpire upon attaining the appropriate viscosity. The laser radiation strikes the donor film from above, and the interaction occurs at the interface between the transparent substrate and the paste film. In the illustration, each image represents a frame from a video captured by a high-speed camera at the specified timings in the insets. The entire transfer process constitutes a singular event. When a receiving substrate is positioned nearby, the resultant deposition manifests as a flat cylinder resembling a wafer, with a thickness comparable to that of the donor films.

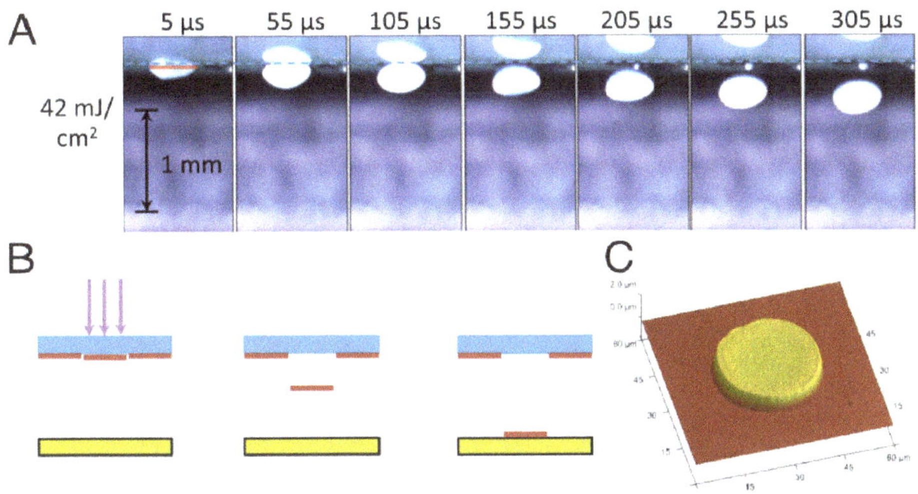

Figure 3. (A) Snapshots of a high-speed video of the transfer via LIFT of a high viscosity silver paste. In the top figure, a single laser pulse with a circular shape comes from the top, pushing downwards a 'coin-like' voxel that flies towards the receptor substrate without breaking. The schematics is included in (B), and a topography acquired via AFM of the final printed material is shown in (C).

2.3. LIFT from Liquids

A multitude of options exists when the material to be printed is applied as a liquid layer on the donor substrate [1,3]. The transfer dynamics resemble the prior examples; however, extra parameters must be considered, especially about energy absorption in the liquid and the material's rheological properties. Typically, upon absorption of the laser by the liquid or the intermediate absorbing layer, a bubble is generated, subsequently leading to the creation of a jet that propels at high velocity towards the receptor substrate strategically positioned near the donor substrate. The liquid jet delicately delivers material that ultimately forms a printed pixel in the configuration of a hemispherical droplet, with dimensions contingent upon the volume of transferred substance and the surface energy of the liquid on the receptor substrate's surface. Figure 4 illustrates instances of LIFT of liquids without an intermediate layer (Figure 4A [11]), with a metallic absorbing layer (Figure 4B [12]), and with an absorbent polymer layer (Figure 4C [13]). Additional comprehensive examples are available in the latest review articles regarding the technique [1,3].

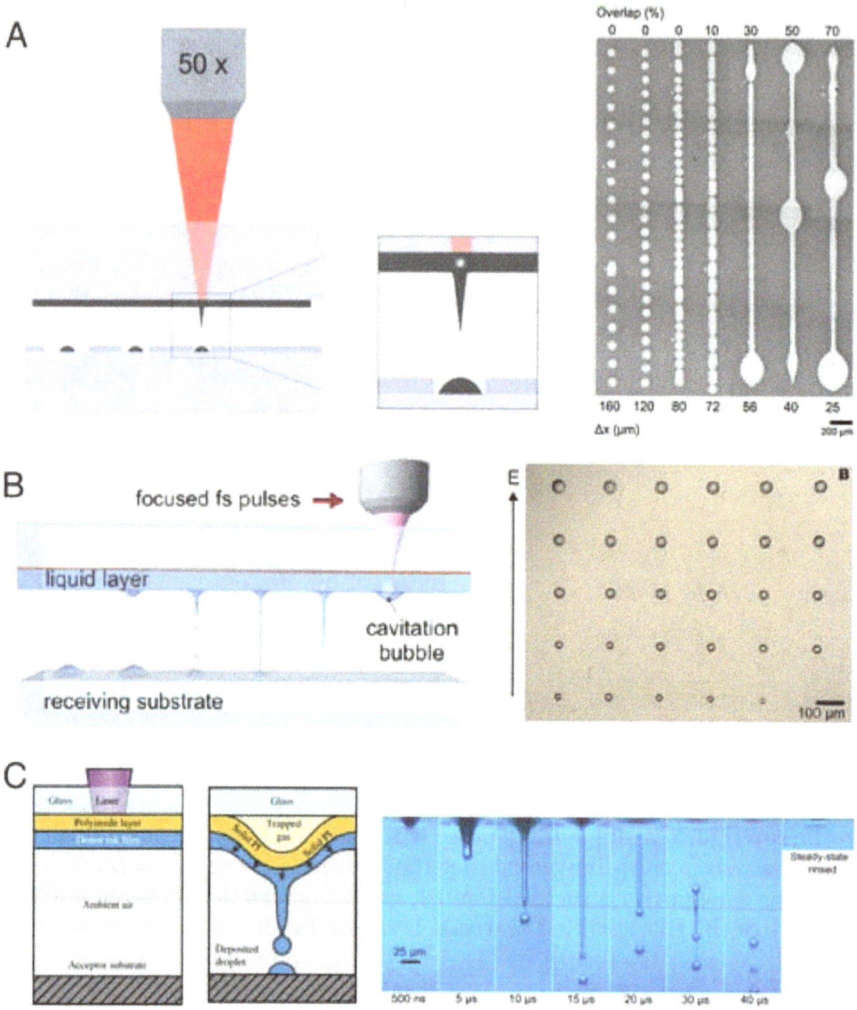

Figure 4. (A) No sacrificial layer (the liquid absorbs the laser radiation). Reprinted from [11], copyright (2016), with permission from Elsevier. (B) Metallic intermediate layer. Reprinted from [12], copyright (2017), with permission from Elsevier. (C) Blister-actuated LIFT based on the use of a polymer absorbing layer. Reprinted/adapted from Ref. [13]

3. LIFT SCIENTIFIC PUBLICATIONS

The initial publication on utilizing a laser beam to drive materials toward a receptor substrate was authored by Levene et al. in 1970 [14]. In that experiment, the donor substrate consisted of a polyethylene typewriter ribbon coated with black ink. Material deposition has been demonstrated to be viable for both forward (as illustrated in Figure 1) and backward configurations (formerly termed 'reverse transfer', wherein the material is expelled in the direction opposite to the laser beam), resulting in continuous printed lines of ink. In 1986, Bohandy et al. [15] introduced the term

LIFT (laser-induced forward transfer) in a highly cited work within the area. Since that time, other studies have been released, totaling 661 peer-reviewed articles, as illustrated in the plot of Figure 5, with a rising number of citations according to the ISI Web of Science database.

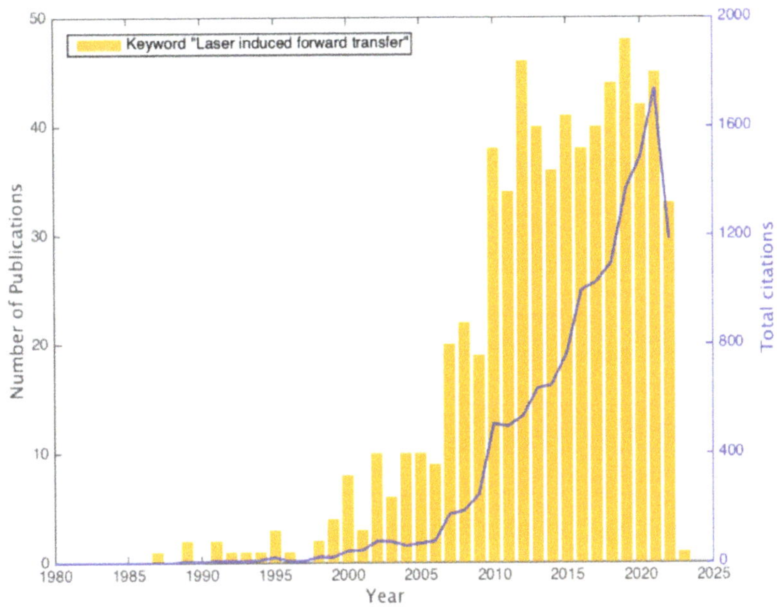

Figure 5. Number of published peer-reviewed articles (grand total of 661) and total number of citations per year. Data obtained via the ISI Web of Science, by sorting the keyword "laser induced forward transfer" as of 11 November 2022.

It has to be noted that, in most cases, when talking about transferring materials via LIFT, the material to be deposited is virtually the same as that disposed in the donor substrate. In the cases where the material is significatively altered compared to the donor material, we have only included studies called reactive LIFT (r-LIFT), in which the material changes its chemical composition mainly due to the action of traveling through a reactive atmosphere or by producing certain structural change due to the laser radiation [16]. It is not the case for the papers regarding matrix-assisted pulsed laser evaporation (MAPLE), which also follows the printing principle of LIFT, but that are not included in the data of **Figure 5**. The MAPLE technique has grown significantly and independently, generating a full string of numerous papers that deserve their own study; for this reason, they are out of scope in the present paper.

Among the publications included in **Figure 5**, the top five journals where the research was published were Applied Surface Science (13%), Proceedings of SPIE (9%), Applied Physics A (7%), Applied Physics Letters (3.6%), and the Journal of Laser Micro Nanoengineering (3.2%). An extended list is included in **Table 1**. Using the classification proposed by the ISI Web of Science, the fields which these publications fall into include applied physics, materials science, optics, physical chemistry, and condensed matter.

Table 1. Journal names and number of papers published on LIFT, according to the ISI Web of Science, by sorting the keyword "laser induced forward transfer" as of 11 November 2022.

Publication Name	Number of Papers	Percentage (%)
Applied Surface Science	84	12.7
Proceedings of the Society of Photo Optical Instrumentation Engineers SPIE	59	8.9
Applied Physics A Materials Science Processing	48	7.2
Applied Physics Letters	24	3.6
Journal of Laser Micro Nanoengineering	21	3.1
Optics Express	14	2.1
Journal of Applied Physics	11	1.6
Conference on Lasers and Electro Optics	9	1.3
Journal of Physics D Applied Physics	9	1.3
Nanomaterials	9	1.3

We have plotted the same data of Figure 5 in a map displayed in Figure 6. The number of publications is sorted by the country where the corresponding author institution was reported. From here, we can see how the most productive region overall is Europe, with notable contributions from North America and Asia.

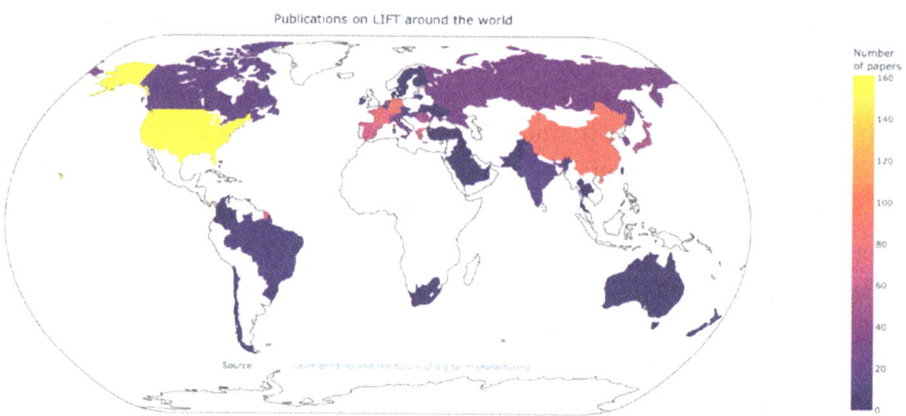

Figure 6. Geographical distribution of the total amount of publications per country on LIFT around the world. Data obtained via the ISI Web of Science by sorting the keyword "laser induced forward transfer" as of 8 June 2022.

Table 2 contains a list with the names of the main funding agencies that made possible most of the contributions of the researchers on the field in these countries, including the number of papers produced and its percentage with respect to the total number of publications (661). This information is also available at the ISI Web of Science database.

Table 2. Main international funding agencies that support the research reflected in the published papers on LIFT, according to the ISI Web of Science, by sorting the keyword "laser induced forward transfer" as of 11 November 2022.

Institution	Country	Number of Papers	Percentage (%)
European Commission	European countries	99	14.9
National Natural Science Foundation of China	China	34	5.1
National Science Foundation	United States	25	3.7
German Research Foundation	Germany	24	3.6
Office of Naval Research	United States	23	3.4
Engineering Physical Sciences Research Council	United Kingdom	21	3.1
UK Research Innovation	United Kingdom	21	3.1
French National Research Agency	France	19	2.8
Spanish Government	Spain	19	2.8
Swiss National Science Foundation	Switzerland	13	1.9

4. INDUSTRIAL PERSPECTIVES

4.1. Patent Applications Landscape

Relevant technological throughputs are often published as patents in order to grant intellectual property rights to the authors. It is important to note that patent applications and the process for their publication follows a procedure that is dramatically different from the one of scientific publications, and technical language is often used in combination with economic and legal terms, in order to provide a solid legal ground for protecting the technology as much as possible. For this reason, we focus our attention on patents that contain the keywords "laser induced forward transfer" from two public patent office databases, considering that the majority of publications on LIFT are located in these two regions according to Figure 6. These patent office databases are the European Patent Office (EPO) and the United States Patent and Trademark Office (USPTO). Based on this premise, we select patents publications that are publicly accessible and published with an assigned patent publication number, regardless of the acceptance or licensing status. In the following data we, therefore, jointly present patents with patent applications. We find 321 patents reported in the EPO and 303 in the USPTO, that, regardless of the publication country, are reported in these databases. Due to the large number of items, the specific lists are included as a Supplementary Material. The following plot in Figure 7 shows the number of patents and patent publications over time, considering the early publication date for the EPO results and the publication date for the USPTO results.

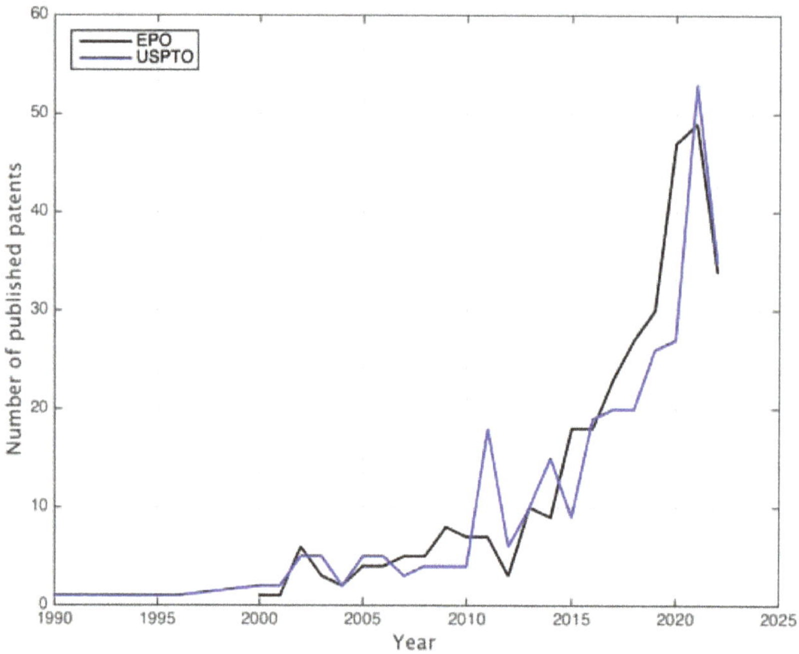

Figure 7. Number of published patent applications reported in the European Patent Office (EPO) and the United States Patent and Trademark Office (USPTO) per year. Data obtained on the respective databases, by sorting the keyword "laser induced forward transfer" as a whole string on 11 November 2022.

4.2. Commercial Forecast and Future Trends

From the current report on "The worldwide market for lasers: market review and forecast 2020" [17] from the European Photonics Industry Consortium (EPIC), that includes projections from 2020 to 2025 on different industries where laser processing of materials is at the center, we can extract an overview for future projections from the economical point of view.

We identify three main areas of revenue where laser printing can find a niche for development into industrial applications, including all the available commercial applications. As a disclosure note, these are just a limited selection of the biggest commercial areas. These areas are additive manufacturing, laser-based paper printing, and medicine. In the case of additive manufacturing, the reported total gross revenue for all applications accounted for 75.3M US dollars in 2022. It is projected a growth of 13.1% over the current decade. The second area of laser-based paper printing accounted for a revenue of around 48.5M US dollars for the year 2022; however, in this case, it is forecast that laser-based technology revenue in this commercial area will decrease over 4% over the current decade, mainly due to a shift toward a paper-less work culture that will mean a decrease in sales of printers in general. Finally, for the area of medicine, the revenue accounted for about 1500M US dollar in 2022, with an average growth estimation of about 7.4% over the current decade. For the LIFT technique, these figures are encouraging, particularly since the characteristic properties of the technique are suitable for the printing of complex materials in all the mentioned areas, as we will discuss in the following.

In the area of additive manufacturing, the selective deposition of materials to fabricate three-dimensional structures has been commercially exploited thanks to important developments in techniques and machines for the printing of polymers, metals, and ceramic-based materials. Commercial 3D printers are one successful example that has been booming in the last decade, and allows us to produce easily accessible methods for functional devices without the need for expensive lab equipment. In these printers, the material of interest is generally deposited in a layer-by-layer fashion, and it is subsequently melted at user-defined locations due to the interaction with an energetic laser beam **[18,19,20]**. At the end, several two-dimensional stacked layers lead to the fabrication of three-dimensional solid structures. The fundamental operation of these 3D printers has been applied for the additive deposition of diverse materials, allowing the fabrication of glass and polymer lab-on-a-chip devices with micrometer resolutions **[21,22]**, as well as the construction of resin boats **[23]** and metallic rocket fuel tanks **[24]** of exceptional quality. The same principles for printing are used with the LIFT technique. There are numerous publications where the printing of solids, pastes, and liquid containing conductive materials have been used for the fabrication of conductive paths **[25,26,27]**, light-emitting diodes **[28]**, solar cells **[29]**, and materials used for microsensors **[30]**. An important advantage over other printing techniques, such as ink-jet printing, relies on the fact that these systems are able to work with a wider viscosity range of inks with nanoparticle suspensions, which opens a wider variety of printable materials suitable for electronic applications, especially in the case of high-viscosity pastes made from particles with particle sizes in the microscale **[1]**.

In the domain of laser-based paper printing, there exists a commercial venture that employed the LIFT process for large-scale graphical design paper printing. The apparatus, designated Lasersonic, could print inks on paper at a rate of 1.3 m^2/min, achieving resolutions of 600 dpi [31]. This application utilized ink disposed of as a thin liquid film, with thickness controlled by cylinders that ensured a virtually unlimited ink supply, facilitating high-speed printing in a roll-to-roll configuration and demonstrating compatibility with contemporary industrial fabrication methods. The identical technique can be utilized for the commercial printing of flexible substrates, predominantly employed in electronic devices, including sensors [32], displays [33], electrical components [34], and solar cells [35], among others. Moreover, recent evidence indicates that the application of conductive inks using LIFT on paper can facilitate the production of economical electronic devices [32]. The LIFT technique is notably significant in the medical field for its role in tissue printing aimed at organ and organelle creation [36,37,38,39,40]. An exemplary commercial success is the business Poietis, which utilizes a 4D Bioprinting technology that integrates a LIFT device into a commercial bioprinting apparatus for single-cell transfer. The LIFT technique's benefits in positioning and controlled volume deposition, coupled with an in situ optical microscope, enable this apparatus to monitor the growth of individual cells and extracellular matrices over time. The company creators own various articles regarding the characterization of the technology for its application in bioprinting [41], along with several patents in which the LIFT technique serves as the fundamental method for transferring diverse biological materials. Despite existing challenges, including unresolved issues related to donor film drying (specifically for liquid donor films), cell aggregation, and possible metal toxicity, advancements in the field indicate that the LIFT technique has proven effective for the transfer of biological materials. Notably, an analysis of the terms "laser induced forward transfer" alongside "organ" and "organelle" in the EPO patent database reveals that, over the past decade, 45 and 12 patents have been filed, respectively, accounting for 17.7% of the total patents illustrated in Figure 7 for the EPO results. This illustrates the potential for exploitation in future technologies and a growing interest from both researchers and companies in the medical and biological sectors.

REFERENCES

1. Serra, P.; Piqué, A. Laser-Induced Forward Transfer: Fundamentals and Applications. *Adv. Mater. Technol.* **2019**, *4*, 1800099. **[Google Scholar] [CrossRef] [Green Version]**
2. Morales, M.; Munoz-Martin, D.; Marquez, A.; Lauzurica, S.; Molpeceres, C. Laser-Induced Forward Transfer Techniques and Applications. In *Advances in Laser Materials Processing: Technology, Research and Applications*; Elsevier: Amsterdam, The Netherlands, 2017; pp. 339–379. ISBN 9780081012529. **[Google Scholar]**
3. Fernández-Pradas, J.M.; Serra, P. Laser-Induced Forward Transfer: A Method for Printing Functional Inks. *Crystals* **2020**, *10*, 651. **[Google Scholar] [CrossRef]**
4. Zenou, M.; Sa'ar, A.; Kotler, Z. Laser Jetting of Femto-Liter Metal Droplets for High Resolution 3D Printed Structures. *Sci. Rep.* **2015**, *5*, 17265. **[Google Scholar] [CrossRef] [PubMed] [Green Version]**
5. Li, Q.; Alloncle, A.P.; Grojo, D.; Delaporte, P. Generating Liquid Nanojets from Copper by Dual Laser Irradiation for Ultra-High Resolution Printing. *Opt. Express* **2017**, *25*, 24164. **[Google Scholar] [CrossRef] [PubMed]**
6. Nakata, Y.; Tsubakimoto, K.; Miyanaga, N.; Narazaki, A.; Shoji, T.; Tsuboi, Y. Laser-Induced Transfer of Noble Metal Nanodots with Femtosecond Laser-Interference Processing. *Nanomaterials* **2021**, *11*, 305. **[Google Scholar] [CrossRef] [PubMed]**

7. Sopeña, P.; González-Torres, S.; Fernández-Pradas, J.M.; Serra, P. Spraying Dynamics in Continuous Wave Laser Printing of Conductive Inks. *Sci. Rep.* **2018**, *8*, 7999. **[Google Scholar] [CrossRef]**

8. Li, Q.; Grojo, D.; Alloncle, A.P.; Chichkov, B.; Delaporte, P. Digital Laser Micro-and Nanoprinting. *Nanophotonics* **2018**, *8*, 27–44. **[Google Scholar] [CrossRef]**

9. Palla-Papavlu, A.; Dinca, V.; Paraico, I.; Moldovan, A.; Shaw-Stewart, J.; Schneider, C.W.; Kovacs, E.; Lippert, T.; Dinescu, M. Microfabrication of Polystyrene Microbead Arrays by Laser Induced Forward Transfer. *J. Appl. Phys.* **2010**, *108*, 033111. **[Google Scholar] [CrossRef]**

10. Mathews, S.A.; Auyeung, R.C.Y.; Kim, H.; Charipar, N.A.; Piqué, A. High-Speed Video Study of Laser-Induced Forward Transfer of Silver Nano-Suspensions. *J. Appl. Phys.* **2013**, *114*, 064910. **[Google Scholar] [CrossRef]**

11. Florian, C.; Caballero-Lucas, F.; Fernández-Pradas, J.M.; Ogier, S.; Winchester, L.; Karnakis, D.; Geremia, R.; Artigas, R.; Serra, P. Printing of Silver Conductive Lines through Laser-Induced Forward Transfer. *Appl. Surf. Sci.* **2016**, *374*, 265–270. **[Google Scholar] [CrossRef]**

12. Florian, C.; Piazza, S.; Diaspro, A.; Serra, P.; Duocastella, M. Direct Laser Printing of Tailored Polymeric Microlenses. *ACS Appl. Mater. Interfaces* **2016**, *8*, 17028–17032. **[Google Scholar] [CrossRef] [PubMed]**

13. Turkoz, E.; Fardel, R.; Arnold, C.B. Advances in Blister-Actuated Laser-Induced Forward Transfer (BA-LIFT). In *Laser Printing of Functional Materials*; Wiley-VCH Verlag GmbH & Co. KGaA: Weinheim, Germany, 2018; pp. 91–121. **[Google Scholar]**

14. Levene, M.L.; Scott, R.D.; Siryj, B.W. *Electro-Optics Handbook (RCA Defense Electronic Products)*; RCA Aerospace Systems Division: Burlington, MA, USA, 1968; Volume 1. **[Google Scholar]**

15. Bohandy, J.; Kim, B.F.; Adrian, F.J. Metal Deposition from a Supported Metal Film Using an Excimer Laser. *J. Appl. Phys.* **1986**, *60*, 1538. **[Google Scholar] [CrossRef]**

16. Piqué, A. The Matrix-Assisted Pulsed Laser Evaporation (MAPLE) Process: Origins and Future Directions. *Appl. Phys. A Mater. Sci. Process.* **2011**, *105*, 517–528. **[Google Scholar] [CrossRef]**

17. *The Worldwide Market for Lasers and Five-Year Forecast (2019–2025)*; Strategies Unlimited: Mountain View, CA, USA, 2020.

18. Duda, T.; Raghavan, L.V. 3D Metal Printing Technology. *IFAC-PapersOnLine* **2016**, *49*, 103–110. **[Google Scholar] [CrossRef]**

19. Yuk, H.; Lu, B.; Lin, S.; Qu, K.; Xu, J.; Luo, J.; Zhao, X. 3D Printing of Conducting Polymers. *Nat. Commun.* **2020**, *11*, 1604. **[Google Scholar] [CrossRef] [Green Version]**

20. Melzer, J.E.; McLeod, E. 3D Nanophotonic Device Fabrication Using Discrete Components. *Nanophotonics* **2020**, *9*, 1373–1390. **[Google Scholar] [CrossRef]**

21. Sanchez Noriega, J.L.; Chartrand, N.A.; Valdoz, J.C.; Cribbs, C.G.; Jacobs, D.A.; Poulson, D.; Viglione, M.S.; Woolley, A.T.; van Ry, P.M.; Christensen, K.A.; et al. Spatially and Optically Tailored 3D Printing for Highly Miniaturized and Integrated Microfluidics. *Nat. Commun.* **2021**, *12*, 5509. **[Google Scholar] [CrossRef] [PubMed]**

22. Mahmood, M.A.; Popescu, A.C. 3D Printing at Micro-Level: Laser-Induced Forward Transfer and Two-Photon Polymerization. *Polymers* **2021**, *13*, 2034. **[Google Scholar] [CrossRef]**

23. The University of Maine. Available online: **https://umaine.edu/news/blog/2019/10/10/umaine-composites-center-receives-three-guinness-world-records-related-to-largest-3d-printer/** (accessed on ٢ November ٢٠٢٢).

24. Relativity Space, Inc. Relativity Space. Available online: **https://www.relativityspace.com** (accessed on ٢ November ٢٠٢٢).

25. Fernández-Pradas, J.M.; Sopeña, P.; González-Torres, S.; Arrese, J.; Cirera, A.; Serra, P. Laser-Induced Forward Transfer for Printed Electronics Applications. *Appl. Phys. A Mater. Sci. Process.* **2018**, *124*, 214. **[Google Scholar] [CrossRef]**

26. Zenou, M.; Kotler, Z. Printing of Metallic 3D Micro-Objects by Laser Induced Forward Transfer. *Opt. Express* **2016**, *24*, 1431. **[Google Scholar] [CrossRef]**

27. Makrygianni, M.; Zacharatos, F.; Andritsos, K.; Theodorakos, I.; Reppas, D.; Oikonomidis, N.; Spandonidis, C.; Zergioti, I. Eco-Friendly Lead-Free Solder Paste Printing via Laser-Induced Forward Transfer for the Assembly of Ultra-Fine Pitch Electronic Components. *Materials* **2021**, *14*, 3353. **[Google Scholar] [CrossRef]**

28. Shaw-Stewart, J.R.H.; Mattle, T.; Lippert, T.K.; Nagel, M.; Nüesch, F.A.; Wokaun, A. The Fabrication of Small Molecule Organic Light-Emitting Diode Pixels by Laser-Induced Forward Transfer. *J. Appl. Phys.* **2013**, *113*, 043104. **[Google Scholar] [CrossRef] [Green Version]**

29. Pozov, S.M.; Andritsos, K.; Theodorakos, I.; Georgiou, E.; Ioakeimidis, A.; Kabla, A.; Melamed, S.; de La Vega, F.; Zergioti, I.; Choulis, S.A. Indium Tin Oxide-Free Inverted Organic Photovoltaics Using Laser-Induced Forward Transfer Silver Nanoparticle Embedded Metal Grids. *ACS Appl. Electron. Mater.* **2022**, *4*, 2689–2698. **[Google Scholar] [CrossRef] [PubMed]**

30. Dinca, V.; Fardel, R.; Shaw-Stewart, J.; di Pietrantonio, F.; Cannatà, D.; Benetti, M.; Verona, E.; Palla-Papavlu, A.; Dinescu, M.; Lippert, T. Laser-Induced Forward Transfer: An Approach to Single-Step Polymer Microsensor Fabrication. *Sens. Lett.* **2010**, *8*, 436–440. **[Google Scholar] [CrossRef]**

31. Hennig, G.; Baldermann, T.; Nussbaum, C.; Rossier, M.; Brockelt, A.; Schuler, L.; Hochstein, G. Lasersonic® LIFT Process for Large Area Digital Printing. *J. Laser Micro Nanoeng.* **2012**, *7*, 299–305. **[Google Scholar] [CrossRef]**

32. Sopeña, P.; Arrese, J.; González-Torres, S.; Fernández-Pradas, J.M.; Cirera, A.; Serra, P. Low-Cost Fabrication of Printed Electronics Devices through Continuous Wave Laser-Induced Forward Transfer. *ACS Appl. Mater. Interfaces* **2017**, *9*, 29412–29417. **[Google Scholar] [CrossRef]**

33. Li, R.Z.; Yang, M.; Guo, L.; Qu, K.; Jian, T.; Yu, Y.; Yan, J. Laser-Induced Forward Transferred Optical Scattering Nanosilica for Transparent Displays. *Nanomaterials* **2022**, *12*, 3674. **[Google Scholar] [CrossRef]**

34. Weidling, A.M.; Turkani, V.S.; Akhavan, V.; Schroder, K.A.; Swisher, S.L. Large-Area Photonic Lift-off Process for Flexible Thin-Film Transistors. *NPJ Flex. Electron.* **2022**, *6*, 14. **[Google Scholar] [CrossRef]**

35. Chen, Y.; Munoz-Martin, D.; Morales, M.; Molpeceres, C.; Sánchez-Cortezon, E.; Murillo-Gutierrez, J. Laser Induced Forward Transfer of High Viscosity Silver Paste for New Metallization Methods in Photovoltaic and Flexible Electronics Industry. *Phys. Procedia* **2016**, *83*, 204–210. **[Google Scholar] [CrossRef] [Green Version]**

36. Murphy, S.V.; Atala, A. 3D Bioprinting of Tissues and Organs. *Nat. Biotechnol.* **2014**, *32*, 773–785. **[Google Scholar] [CrossRef] [PubMed]**

37. Guillotin, B.; Souquet, A.; Catros, S.; Duocastella, M.; Pippenger, B.; Bellance, S.; Bareille, R.; Rémy, M.; Bordenave, L.; Amédée, J.; et al. Laser Assisted Bioprinting of Engineered Tissue with High Cell Density and Microscale Organization. *Biomaterials* **2010**, *31*, 7250–7256. **[Google Scholar] [CrossRef] [PubMed]**

38. Antoshin, A.A.; Churbanov, S.N.; Minaev, N.V.; Zhang, D.; Zhang, Y.; Shpichka, A.I.; Timashev, P.S. LIFT-Bioprinting, Is It Worth It? *Bioprinting* **2019**, *15*, e00052. **[Google Scholar] [CrossRef]**

39. Devillard, R.; Pagès, E.; Correa, M.M.; Kériquel, V.; Rémy, M.; Kalisky, J.Ô.; Ali, M.; Guillotin, B.; Guillemot, F. Cell Patterning by Laser-Assisted Bioprinting. In *Methods in Cell Biology*; Academic Press Inc.: Cambridge, MA, USA, 2014; Volume 119, pp. 159–174. **[Google Scholar]**

40. Kryou, C.; Leva, V.; Chatzipetrou, M.; Zergioti, I. Bioprinting for Liver Transplantation. *Bioengineering* **2019**, *6*, 95. **[Google Scholar] [CrossRef] [Green Version]**

41. Ali, M.; Pages, E.; Ducom, A.; Fontaine, A.; Guillemot, F. Controlling Laser-Induced Jet Formation for Bioprinting Mesenchymal Stem Cells with High Viability and High Resolution. *Biofabrication* **2014**, *6*, 045001. **[Google Scholar] [CrossRef] [PubMed]**

CHAPTER

WORST-CASE X-RAY PHOTON ENERGIES IN ULTRASHORT PULSE LASER PROCESSING

4

1. INTRODUCTION: AN OVERVIEW

Since the 1980s, it has been demonstrated that X-rays can be generated from laser-induced plasmas. Initial examinations were conducted in a vacuum [1]. In the early 2000s, initial research into ultrashort pulse laser materials processing of copper at a repetition rate of 1 kHz in air were undertaken. Thogersen et al. [2] measured a maximum dose rate of X-rays of approximately 10 mSv h-1 at a distance of 13 cm and Bunte et al. [3] determined a maximum dose rate of approximately 50 mSv h-1 at a distance of 10 cm, whereby the specifically measured radiological operational quantity was not explicitly specified in both studies. For steel and tungsten processed at higher repetition rates of 400 kHz, Legall et al. [4] found in 2018 even higher dose equivalent rates $\dot{H}'(0.07)$ of 163 mSv h-1 at a distance of 42 cm exceeding typical radiation protection safety limits, which revealed a potential safety risk for the operating staff of the ultrashort pulse laser (USPL) machine.

Several investigations were performed to analyze the X-ray emission for different laser and process parameters during ultrashort pulse laser machining of various technical materials [5,6,7,8,9,10,11,12,13,14,15,16,17]. Recently, possibly harmful X-ray emission was observed already at laser irradiances below 1013 W cm-2 [18,19]. Especially the burst modes of laser machines were identified as a configuration that can lead to very high dose rates in the ultrashort laser pulse processing of metals [20,21]. During burst mode processing, pulse trains (bursts) with up to 100 sub-pulses per burst and intra-burst pulse repetition rates in the MHz to GHz range are typically generated by current commercial laser systems [22,23]. When using burst mode settings, the pulse energy from a single laser pulse is divided into a predefined number of sub-pulses with corresponding (lower) sub-pulse energies.

In addition to the dose equivalent rates, the spectrum flux of emitted X-ray photons is very significant, as the transmission of X-rays through radiation protection housing is highly contingent upon its spectral distribution. This study seeks to determine an X-ray photon spectrum that exhibits the highest X-ray energy produced during ultrashort pulse laser material processing of steel and tungsten in air, referred to as the "worst-case spectrum." X-ray spectra were recorded during the machining process under various laser and process conditions. The dosage equivalent rates were derived from the spectra.

2. MATERIALS AND METHODS

2.1. Ultrashort Pulse Laser System

The investigation was performed with an ultrashort pulse laser machine emitting an average power of 20 W at a center wavelength of 1030 nm (GL.evo, GFH GmbH, Deggendorf, Germany), equipped with a Pharos-Laser (PH2-20W, Light Conversion, Vilnius, Lithuania). The laser pulse duration of 274 fs was kept constant. The laser beam was focused on the workpiece featuring an $1/e^2$-spot diameter of 33 μm. This value was taken from a configuration file provided by the manufacturer of the laser machine. It was operated here in an industrial setting that regularly does not involve an individual beam diameter evaluation for each laser process. The detailed processing parameters are listed in **Table 1.** Different combinations of laser pulse energy and repetition rate were chosen, whereby the laser pulse energy was varied between 91 μJ and 371 μJ and the pulse repetition rate between 50 kHz and 200 kHz. For each combination of laser pulse energy and repetition rate, single pulse and burst mode experiments were performed.

Table 1. Experimental processing parameters.

Experiment No.	Repetition Rate	Laser Pulse Energy	Average Power	Burst Mode	Peak Irradiance	Material
	in kHz	in μJ	in W		in W cm^{-2}	
1	50	371	18.6	No	3.0×10^{14}	tungsten
2	100	184	18.4	No	1.5×10^{14}	tungsten
3	200	91	18.2	No	7.3×10^{13}	tungsten
4	50	371	18.6	Yes	3.0×10^{14}	tungsten
5	100	184	18.4	Yes	1.5×10^{14}	tungsten
6	200	91	18.2	Yes	7.3×10^{13}	tungsten
7	50	371	18.6	No	3.0×10^{14}	steel
8	100	184	18.4	No	1.5×10^{14}	steel
9	50	371	18.6	Yes	3.0×10^{14}	steel
10	100	184	18.4	Yes	1.5×10^{14}	steel

For the burst mode investigations, the single pulse (with 100% energy) is split into two sub-pulses showing equal energy of 50% of the total single pulse energy. The two sub-pulses have a temporal separation of 200 ps here.

Laser peak irradiances I0 in a range of 7.3 × 1013 W cm−2 to 3.0 × 1014 W cm−2 were calculated under the assumption of a temporal and spatial Gaussian distribution of the laser pulse and an angle of incidence of 0°

$$I_0 = 8\sqrt{\frac{4\ln 2}{\pi}}\frac{Q}{\tau_H \cdot \pi \cdot d^2}, \quad (1)$$

where Q is the pulse energy, H is the pulse duration and d is the laser beam spot diameter (1/e2) [24].

The calculation of the irradiance for the burst mode in Table 1 considers the total energy of the burst (which equals the single pulse energy). Three different repetition rates (50 kHz, 100 kHz, and 200 kHz) and pulse energies (371 µJ, 184 µJ, and 91 µJ) were selected, resulting in an average power of 18.2 W to 18.6 W close to the maximum average power of the laser of 20 W. The experiments were performed with flat 5 mm thick plates of tungsten (99.97% purity) and steel alloy (St37) targets. The target surfaces were prepared by parallel surface grinding achieving an average roughness of 1.6 µm to 4 µm. With a laser beam scanning configuration, 1.5 × 1.5 cm2 squares were treated. The laser processing time was 250 s for each square.

2.2. Spectral X-ray Emission

The assessment of spectral X-ray emission during ultrashort pulse laser processing is complicated by the disparate energy detection ranges of various spectrometers, electromagnetic compatibility (EMC) stipulations, potential pile-up phenomena, and the likelihood of X-ray attenuation by environmental factors. To date, X-ray energy ranging from approximately 2 keV to 30 keV have been documented during ultrashort pulse laser processing in air [4]. Detectors utilizing various sensor materials exhibit reliable measurements across distinct X-ray energy levels. Conventional silicon drift detectors (SDDs) are limited to photon energies of about 15 keV, as the quantum efficiency of 500 µm thick silicon significantly declines at higher energies (see to Figure 1 in [25]). Strüder et al. documented a quantum efficiency consistently above 85% for photon energy ranging from 500 eV to 11 keV for a 450 µm thick SDD [26]. Nevertheless, as the majority of X-ray spectra obtained during laser material interactions encompass a very low photon energy range, studies utilizing an SDD can still yield significant insights. CdTe detectors demonstrate a significantly broader energy detection range compared to SDDs. The X-123 spectrometer (Amptek Inc., Bedford, MA, USA), featuring a sensor thickness of 1000 µm, is specified by the manufacturer to have an optimal energy range of 5 keV to 100 keV. The manufacturer's data indicate a minimum inherent efficiency of 98% for photon energies ranging from 10 keV to 50 keV. Previous investigations indicate that CdTe spectrometers' spectra can be affected by noise and electronic difficulties stemming from electromagnetic fields created by the plasma [14]. Undesired electronic occurrences may arise from laser-target interactions that generate powerful electromagnetic pulses, thereby generating electronic noise in the data collecting system [27]. Furthermore, the spectra for both detector types may be influenced by pile-up [4].

For the reliable measurement of X-ray spectra and to compensate for some disadvantages of the individual spectrometer types, a comparative study using three spectrometers simultaneously was conducted in this work. The spectrometers were placed 60 cm to 70 cm away from the laser interaction point within the housing of the industrial ultrashort pulse laser machining system. The experimental setup with the arrangement of the detectors can be seen in **Figure 1a,b** showing a photograph of the machining process with the optics, the target (with the bright optical plasma emission) and the exhaust system. The measurements were performed in ambient air.

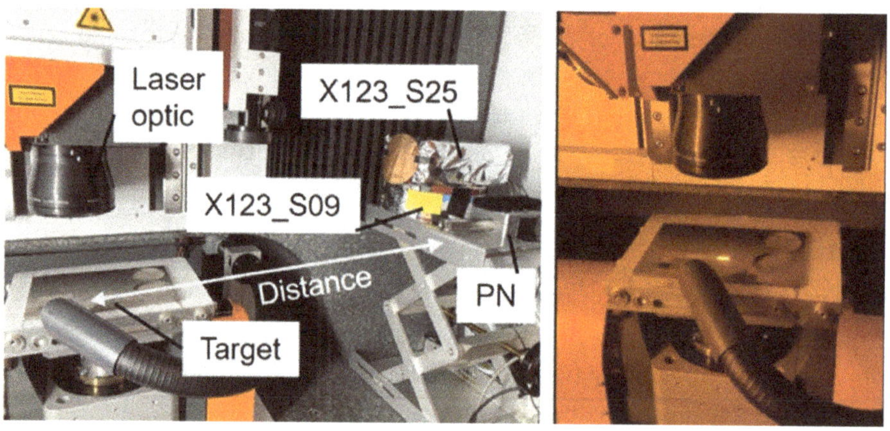

(a) (b)

Figure 1. Experimental setup, (a) Arrangement of the spectrometers, (b) Image during laser processing.

For the measurements of the higher X-ray photon energies (> 13 keV), two CdTe-based spectrometers (Amptek X-123) were used. The first X-123 spectrometer had a sensor size of 3 × 3 × 1 mm3 (X123_S09) and the second had a sensor size of 5 × 5 × 1 mm3 (X123_S25). Both X-123 spectrometers were equipped with a 100 µm Beryllium window. For the lower energy range (< 13 keV), an SDD-based spectrometer (PNDetector with the designation “XRS-30-128-100-BeP Complete” with module type SDD-30-128-BeP) with a sensor area of 30 mm2, a thickness of 450 ± 20 µm and a beryllium window with a thickness of 8 µm was used. This spectrometer is referred to as “PN” in the following.

The X-ray emission investigations showed EMC issues for both CdTe spectrometers (X-123). The SDD (PN) was not affected. To reduce the influence due to the EMC issues, the X-123 spectrometers were encased in special housings. The X123_S09 was placed in a tinplate box and the X123_S25 in an aluminum box. To further suppress electronic noise, aluminum foils were put in front of the measurement window. The thickness of the aluminum foils (different thicknesses for X123_S09 and X123_S25) was calculated from absorption measurements using an X-ray tube with and without aluminum foils (see **Section 3.1**). To evaluate the influence of pile-up effects on the measured spectra, the measurements were performed with and without a 50 μm thick copper foil (see **Section 3.1**).

The placement of the spectrometers with respect to the X-ray emission source and its spatial homogeneity was controlled with a radiographic imaging plate placed at a distance of 60 cm from the laser-induced emission source about 10 cm in front of the three spectrometers. A 2-D Radiography Scanner CR 35 NDT Plus system (DÜRR, Stuttgart, Germany) was used. **Figure 2**a shows the grayscale image of the X-ray field recorded by the 2-D radiography plate along with the positions of the detectors that are marked by green, red, and blue circles, respectively. To compare the X-ray intensities arriving at the three spectrometers, the recorded data contained in the blue and yellow boxes were then binned vertically and then plotted as a function of the horizontal position in **Figure 2**b as blue and yellow curves. The additional green, red, and blue vertical lines mark the individual positions of the three spectrometers, while the corresponding horizontal lines indicate the average grayscale level as a measure of the local X-ray intensity incident to the detector. The spectrometer that received more intensity of laser-generated X-ray intensity (a higher average grayscale value) was the X123_S09. The X123_S25 and the PN received in comparison to X123_S09 about 14.5% and 24% less X-ray intensity during the measurements.

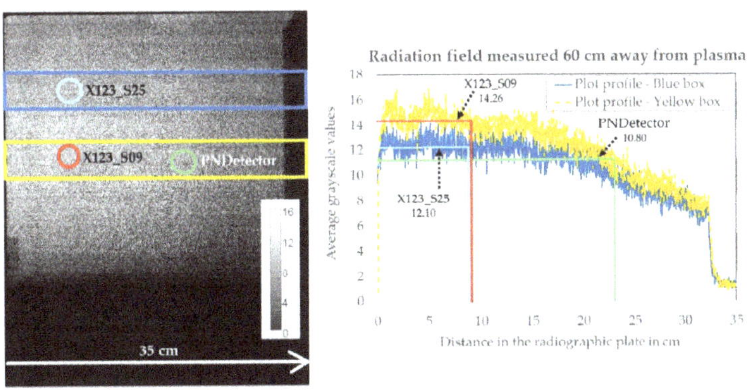

(a) (b)

Figure 2. Grayscale image captured with the radiographic imaging plate (a) placed 10 cm in front of the three spectrometers used in this study. The blue and yellow rectangles are the areas used to calculate the X-ray intensity profile shown in the plot (b). The continuous decrease in grayscale value with an increased distance on the x-axis revealed that the PN spectrometer (PNDetector) received less X-ray intensity in comparison to the two X-123 spectrometers. Additionally, X123_S25 received about 14.5% less radiation than the X123_S09 spectrometer.

3. RESULTS

3.1. Spectrometer Characterization

The spectrometers were characterized by background measurements (Figure 3a) and measurements of an X-ray tube (Figure 3b–d). For a comparison between the different spectrometer properties, the same (known) spectrum of an X-ray tube (Amptek Mini-X Gold (Au) X-ray tube) operated at a current of 10 mA and a voltage of 20 kV was measured by all three devices.

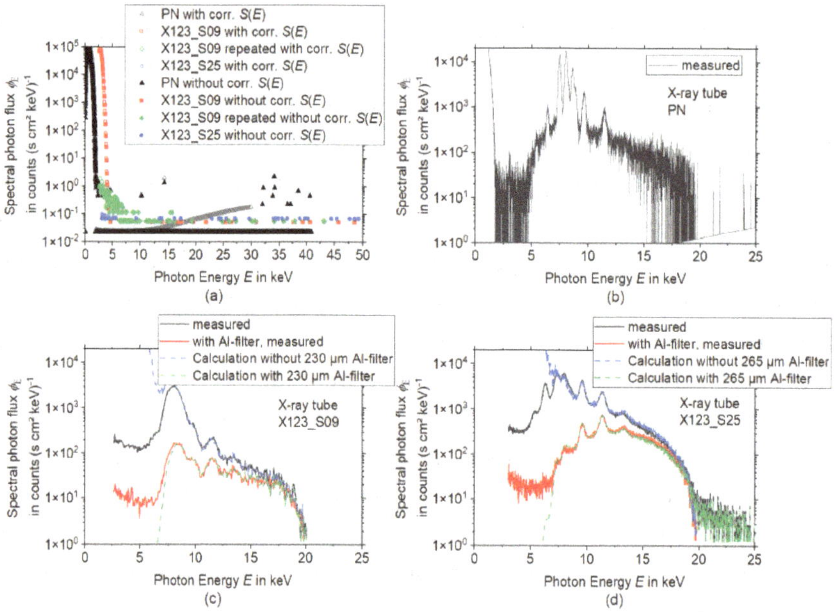

Figure 3. Characterization of the spectrometers without laser material processing, (a) Background measured inside the running USPL machine without laser processing (with and without correction of the spectral sensitivity of the sensor S(E)), (b) X-ray tube spectrum measurement with the PN spectrometer, (c) X-ray tube spectra measured with X123_S09 with and without Al-filter (solid lines) along with two calculated spectra (dashed curves), (d) X-ray tube spectra measured with X123_S25 with and without Al-filter (solid lines) along with two calculated spectra (dashed curves). Note the logarithmic scaling of the ordinate axes.

Every data point in the measured pulse height spectrum X(E) is corrected with the spectral sensitivity of the sensor S(E) and normalized to the area and time with the sensor area A and measuring time interval t and yields the photon flux 0(E):

$$\Phi_0(E) = \frac{X(E)}{A\,t\,S(E)} \quad (2)$$

Typically, in order to determine the photon flux from the pulse height spectrum, an unfolding procedure is required which depends on the inverse response matrix of the detector. Here, due to the low energies and negligible influences due to scattering because of the very low energies, the detector response was estimated by employing only the sensitivity of the detector. Afterward, the spectral photon flux $\Phi_E(E)$ was calculated by the normalization of Φ_0 with the energy channel's width ΔE

$$\Phi_E(E) = \frac{\Phi_0(E)}{\Delta E} \quad (3)$$

To have a comparative situation like during the measurement of the X-ray emission from the laser generated plasma, the X-ray tube was positioned inside the USPL machine on the position, where the laser-induced plasma generation would take place. The X-ray tube measurements were performed when the USPL machine was turned off.

Furthermore, the background signal for all detectors was measured during the USPL machine was running, but no laser material processing took place (**Figure 3a**). This way, it was proven, whether the detection was disturbed by any influence from the USPL machine such as electronic issues, for example, but without the additional influence from the transient plasma generation during the laser processing.

From **Figure 3a,b** one can see that the PN spectrometer has always a high-intensity count range in the low-energy region up to around 2 keV. This noise signal happens due to the high sensitivity of the silicon detector material. This signal also occurred during the background measurement and is independent of the X-ray tube, see **Figure 3a**.

The PN spectrometer shows the highest spectral resolution of the characteristic peaks from the spectra shown in **Figure 3b** and the X123_S25 has a higher resolution than the X123_S09, which can be due to the larger sensor size and the higher number of channels of the X123_S25.

For the PN spectrometer signal (**Figure 3a**) only background noise in the very low count range within the energy region from 2 keV to about 40 keV can be seen. For both Amptek X-123 spectrometers, for energies above 5 keV to 8 keV only background noise with a spectral photon flux Φ_E < 0.1 counts (s cm^2 keV)$^{-1}$ occurs. For the X123_S09 a region with somewhat higher spectral photon flux is found below 8 keV. Repeated measurements showed that this is not a stable signal, and it was not observed during the measurement of the X-ray tube, revealing that the signal is caused by EMC or other electrical issues.

Since all spectral detectors show no significant or stable background within the interesting energy region above 5 keV, a subtraction of the background was not performed. To calculate the influence of the Al foils, which are highly recommended to be used in front of the detector window of an Amptek X-123 spectrometer to avoid EMC issues and spectral artifacts due to pile-up, measurements of the X-ray tube spectrum were performed with and without the aluminum foil. Then the expected spectrum after the transition of aluminum filters with different thicknesses was calculated and the filter thickness was determined with a least-squares-fit. The results can be found in **Figure 3c,d**.

The spectral calculation is based on linear absorption according to Beer's law. Hereby, the measured X-ray flux without filters is reduced by i-filters with their attenuation coefficient from the NIST database [28] μ_i and thickness d_i.

$$(E)=\Phi_0(E)\ e^{-\Sigma_i \mu_i d_i} \quad (4)$$

The procedure was repeated in reverse with the measured spectrum with aluminum foils to calculate the spectrum without foils. For the X123_S09 an aluminum thickness of 230 µm (**Figure 3**c) and for the X123_S25 (**Figure 3**d) an aluminum thickness of 265 µm was determined. For energies higher than 7 keV the calculation and the measurements are in good agreement. The deviations for energies lower than 7 keV are due to absorption in the beryllium window in front of the CdTe sensor volume and electronic noise for the low photon energies.

3.2. Spectroscopy during Ultrashort Pulse Laser Processing

To assess the impact of pile-up effects or EMC on the spectra, measurements were conducted during laser processing both with and without a 50 µm thick copper foil. The spectra following transmission through the copper foil were derived from the measurements taken without the copper foil. This approach enabled a comparison between the measurements obtained with copper foil and the computations conducted. Deviations suggest that the spectra obtained without copper foil remain affected by pile-up in high-energy regions or by electronic noise generated in the spectrometers' acquisition mechanism, attributable to low-frequency electromagnetic fields radiated from the laser-induced plasma.The results for the spectral measurements during the laser processing of tungsten are presented in **Figure 4** and **Figure 5**, while the results during the laser processing of steel are shown in **Figure 6** and **Figure 7** for all three detectors. Except for experiments no. ,1 3, and 7 (see **Table 1**) the measurements without copper foil were performed twice for each set of processing parameters and each detector. In the case of two available measurements, the mean value was used for the evaluation. For experiments no. 3 and 7 a single measurement was utilized for the spectrometers X123_S25 and the PN. For experiment no. 1 a single measurement was used for the spectrometers X123_S09 and the PN.

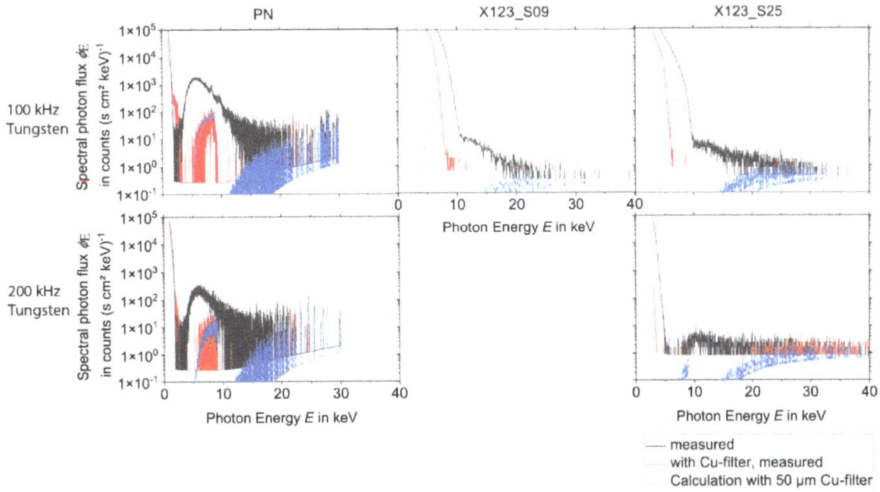

Figure 4. X-ray spectra measured during the laser processing of tungsten without burst mode for the three spectrometers PN, X123_S09, and X123_S25, the measurement with and without Cu-filter was compared with the calculation with copper filter of 50 µm thickness. Note the logarithmic scaling of the ordinate axes.

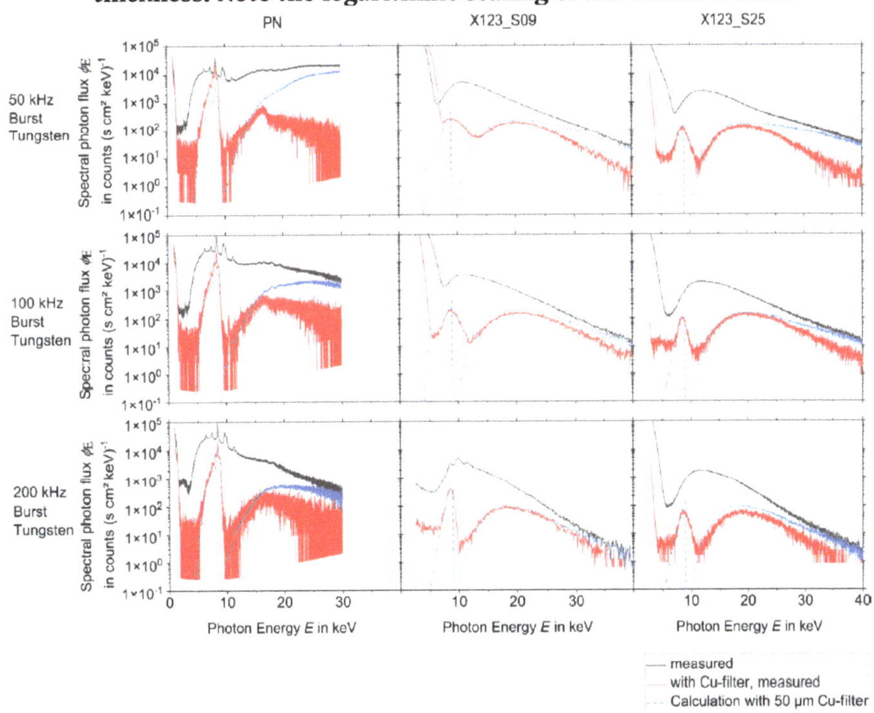

Figure 5. X-ray spectra measured during the laser processing of tungsten with burst mode for the three spectrometers PN, X123_S09, and X123_S25, the measurement with and without Cu-filter was compared with the calculation with copper filter of 50 µm thickness. Note the logarithmic scaling of the ordinate axes.

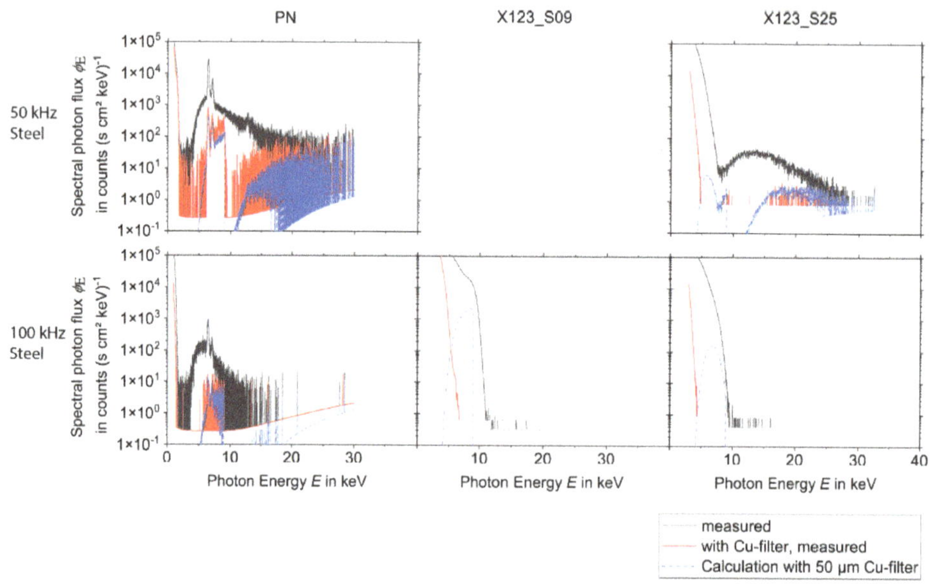

Figure 6. X-ray spectra measured during the laser processing of steel without burst mode for the three spectrometers PN, X123_S09, and X123_S25, the measurement with and without Cu-filter was compared with the calculation with copper filter of 50 µm thickness. Note the logarithmic scaling of the ordinate axes.

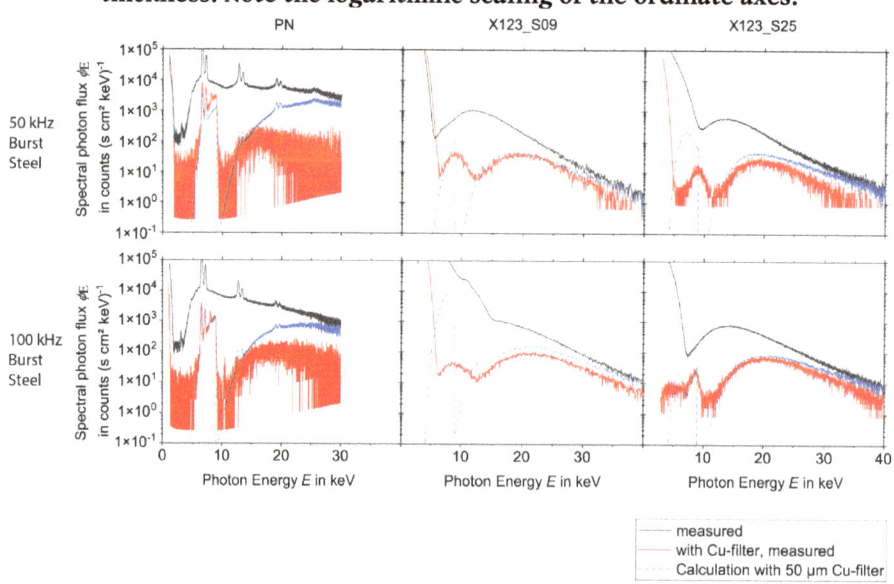

Figure 7. X-ray spectra measured during the processing of steel with burst mode for the three spectrometers PN, X123_S09, and X123_S25, the measurement with and without Cu-filter was compared with the calculation with copper filter of 50 µm thickness. Note the logarithmic scaling of the ordinate axes.

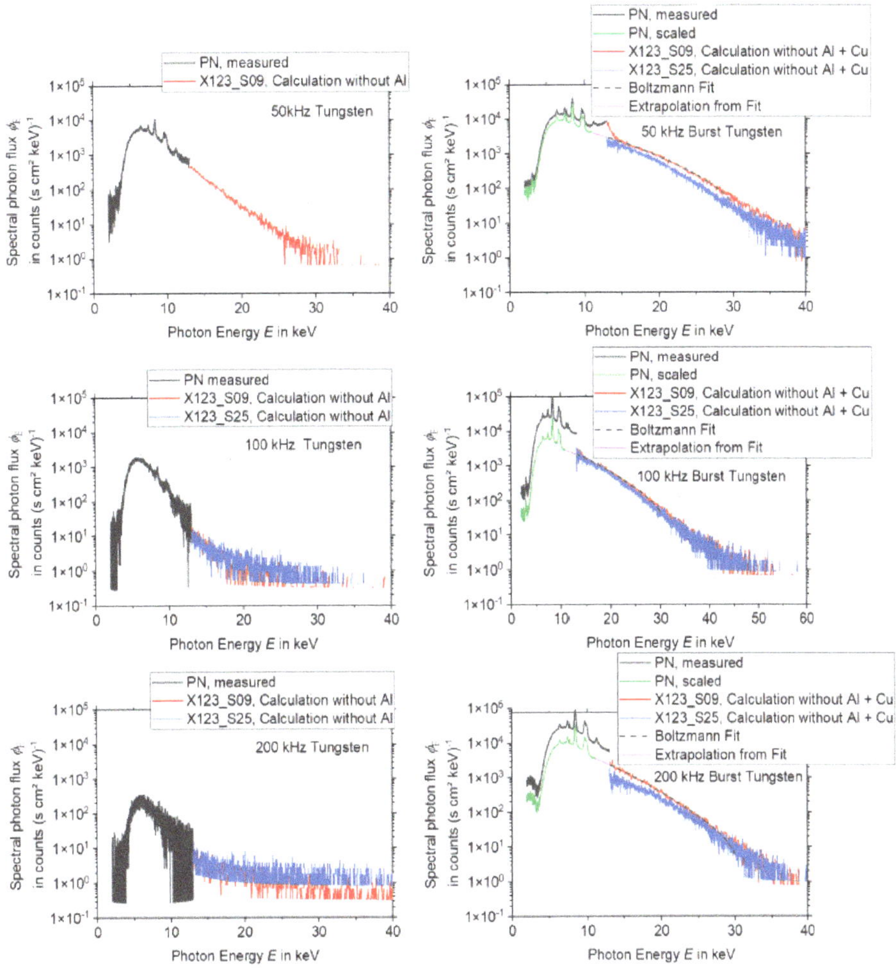

Figure 8. X-ray spectra during the processing of tungsten. Detector type, filter materials, and evaluation constraints are indicated through labels for the different curves. Note the logarithmic scaling of the ordinate axes.

For the PN spectrometer all results, i.e., the comparison of the measurements with copper foil and the corresponding calculations, show a good agreement for X-ray photon energies < 11 keV. However, deviations occur for the measurements with a burst mode for energies > 11 keV, which indicates that pile-up plays a role during the processing with burst mode.

Both X-123 detectors reveal broad spectral peaks with very high intensities for X-ray energies below approximately 9 keV. For the X123_S09, one measurement (100 kHz burst mode, steel) even shows this signal for a photon energy up to 15 keV. It is assumed that these broad peaks are due to EMC issues.

Additionally, also for the X-123 detectors, pile-up can be observed during the laser processing measurements with burst mode. This effect is minimized when the measurement is performed with copper foil. For further evaluation, it follows that the PN spectrometer delivers accurate data (without pile-up) for photon energies < 11 keV. On the other hand, for measurements without burst mode the X-123 spectrometers deliver valid data for photon energies > 13 keV. For energies between 11 and 13 keV, the PN spectrometer data are only valid for processing parameters without burst mode.

As the measurements with burst mode are affected by pile up, the spectra with a copper filter were used. Then the effect of the copper filter was removed by employing Equation (4) and calculating the spectrum without the effects of the copper filter from them.

Due to these findings on the validity of the data in different energy regions, combined spectra with the data from the different detectors were composed for each set of laser processing parameters according to **Table 1**. The combined spectra for the processing of tungsten can be found in **Figure 8**, while the spectra for the processing of steel are depicted in **Figure 9**.

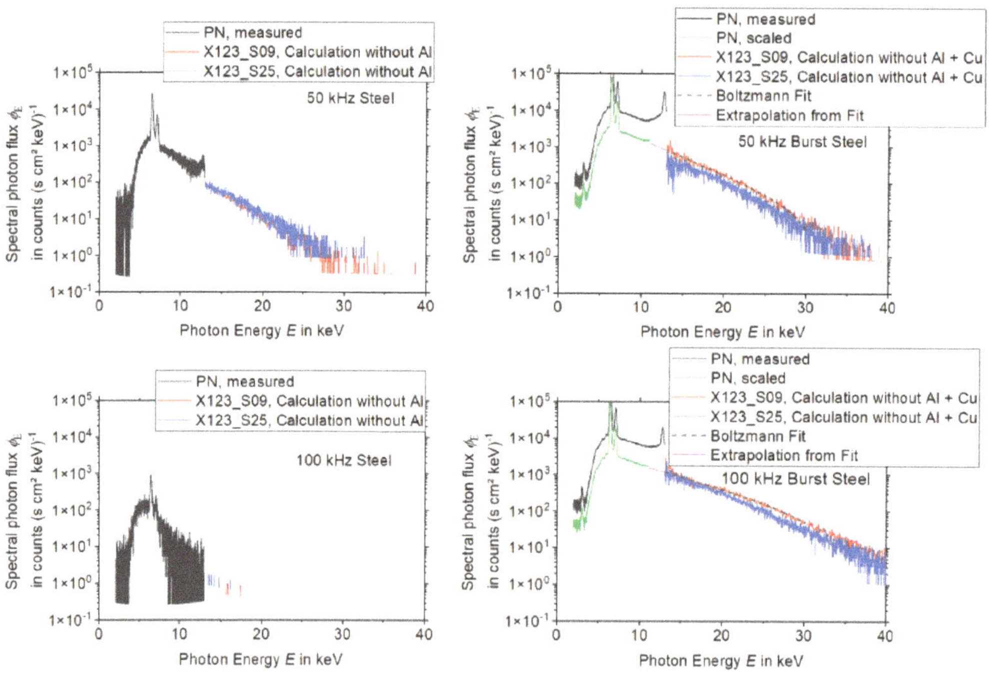

Figure 9. X-ray spectra during the processing of steel. Detector type, filter materials, and evaluation constraints are indicated through labels for the different curves. Note the logarithmic scaling of the ordinate axes.

The data from the PN spectrometer were used for photon energies from 2 keV to 13 keV and the data from the X-123 were utilized for energies > 13 keV. As already mentioned, the PN spectrometer data show a slightly increased photon flux for photon energies between 11 and 13 keV upon processing in the burst mode and these data are probably influenced by pile-up effects. This influence is eliminated by a fitting routine which will be explained later in this chapter. For laser processing without burst mode, i.e., in the normal single pulse mode, the X-123 measurements with aluminum foil were used, whereby the spectra before the transmission of the aluminum foil were calculated by using Equation (4). For burst mode laser irradiation, the X-123 measurements with copper and aluminum foil were selected and the influence of the copper and the aluminum filter was considered also by employing Equation (4).

Taking into account the different energy binning and the different sensor sizes for the calculation of the spectral photon flux, the data of the two X-123 spectrometers are in good agreement. For laser processing parameters without burst mode (i.e., in single pulse mode), an almost continuous transition from the PN spectra to the X-123 spectra can be observed. The PN spectra for laser processing parameters with burst mode have an overall significantly higher photon flux than the spectra from the X-123 detectors. The origin of this deviation is unclear. For an estimation of the uncertainty of the spectra, the curves of the X123_S09 were fitted with a Boltzmann function within the photon energy region of 13–30 keV and extrapolated to the energy region between 11 and 13 keV. In the next step, the PN spectra were scaled down in global amplitude until continuous spectra were achieved together with the extrapolation from the Boltzmann fit and the X-123 spectra. As explained before for the burst mode laser processing, the energy region between 2 and 13 keV is associated with some uncertainty. This uncertainty can be quantified by using two combined spectra. On the one hand, the downscaled PN spectrum together with the Boltzmann extrapolation and the X-123 spectrum as a combined spectrum describes the minimum overall flux. The unchanged PN spectrum together with the X-123 spectrum in a combined spectrum characterizes the maximum overall flux.

The comparison of the spectra recorded with and without the laser burst mode reveals a significant increase in the photon flux for the burst mode processing. A detailed analysis of the estimated dose rates from the spectra will be presented later in **Section 3.3**. An increase in the spectral photon flux can be observed for lower laser repetition rates. This increase can be found for both the processing with and without burst mode. As expected from previous publications, e.g., **[4]** the spectral photon flux emitted during the processing of tungsten is higher when compared to the processing of steel. This is mainly due to the higher atomic number Z of tungsten ($Z = 74$) compared to iron ($Z = 26$) **[4]**. The maximum energies are limited to about 40 keV. The maximum energy of 40 keV was observed during the laser processing of tungsten and steel with the burst mode with 50 and 100 kHz laser pulse repetition rates. For the processing of tungsten with a higher repetition rate (200 kHz), the maximum X-ray energy is approximately 35 keV. For the laser processing without burst mode, i.e., for single pulse mode, a signal for energies up to 40 keV was measured in the case of the 50 kHz repetition rate. However, with a spectral photon flux of around one count (s cm^2 keV)$^{-1}$ the signal is very low. Compared to the burst mode, for the processing without burst mode, the spectral photon flux significantly decreases already for lower energies and reaches the minimum spectral photon flux of about one count (s cm^2 keV)$^{-1}$ for energies between 20 and 30 keV. As a general trend, the maximum X-ray photon energy tends to increase with higher laser pulse energies.

Altogether, the "worst-case" processing scenario of this study is represented by the laser processing of tungsten at a repetition rate of 50 kHz and in burst mode.

3.3. Dose Rate Estimations

The X-ray dose rates ($\dot{H}'(0.07)$, $\dot{H}'(3)$, and $\dot{H}^*(10)$) were calculated from the measured spectra at a distance of 60 cm.

The ambient dose equivalent rate corresponds to the dose equivalent per time at different depths of penetration in the ICRU sphere [28]. Due to the energy-dependent absorption properties of ICRU tissue, three depths of penetration dp were used. The ambient dose equivalent rates can be calculated from the spectral photon flux Φ(E) along with the tissue properties via

$$\dot{H}(d_p) = \int_{E=0}^{E=\infty} \Phi(E)\frac{E_{\mathrm{Photon}}}{\Delta E}\ \frac{e^{-\mu_{en}^{t}(E)\,d_{\mathrm{p}}}\ \mu_t(E)}{\rho^t}dE, \qquad (5)$$

where E is the photon energy, E_{Photon} is the channel energy, ΔE is the bin size from the channel (energy channel's width) and the tissue properties are the X-ray photon energy-dependent energy absorption coefficient μten (E) of Tissue (Soft ICRU-44) and its density ρ^t. [9]. The data for the attenuation coefficients and density of standard tissue were taken from the NIST database [**28**]. Equation (5) is a good approximation for low-energy spectra, where the scattering of the X-ray photons can be neglected. Another method to determine the ambient dose equivalent rates would be to employ conversion coefficients from ICRU report 57 [**29**], for example, used in [**30**] to determine the spectral ambient dose equivalent rate from the spectral photon fluence and then integrate overall available photon energies. Here, both methods have been compared and due to the low energy of the spectra, no significant deviation has been observed.

The results of the dose equivalent rate calculations for the three quantities $\dot{H}'(0.07)$, $\dot{H}'(3)$, $\dot{H}^*(10)$ are presented in **Figure 10**. The dose equivalent rates were calculated for the different parts of the combined spectra shown in **Figure 8** and **Figure 9**. Due to the already discussed large uncertainty for the burst mode processing within the photon energy region between 0 and 13 keV, the dose equivalent rates were calculated for the downscaled PN spectra and the Boltzmann extrapolation as well as for the measured PN spectra. This way, a possible maximum and minimum was estimated for the dose equivalent rates. Furthermore, the dose equivalent rate calculations for the different energy regions of the spectra illustrate how the different quantities $\dot{H}'(0.07)$, $\dot{H}'(3)$ and $\dot{H}^*(10)$ are affected by different energy regions of the laser-generated X-ray spectra. Due to the higher penetration depth of X-rays with higher photon energies, the ambient dose equivalent rate $\dot{H}^*(10)$ depends more on the higher energy region, while the directional dose equivalent rate $\dot{H}'(0.07)$ depends more on the lower energy region of the spectrum. Therefore, the high uncertainty of the energy region between 0 and 13 keV, which is represented by the data from the PN spectrometer, result in larger differences between the maximum and minimum dose rates and, therewith, in high uncertainties for the quantity $\dot{H}'(0.07)$. Compared to that, the uncertainty for the ambient dose equivalent rate $\dot{H}^*(10)$ is smaller. From the "worst-case" spectrum (repetition rate of 50 kHz, burst irradiation mode, tungsten), which was defined due to the highest observed energies, a value of $\dot{H}'(0.07)$ between 2.8 and 4.2 mSv h^{-1}, a value of $\dot{H}'(3)$ between 0.3 and 0.5 mSv h^{-1} and a value of $\dot{H}^*(10)$ between 0.07 and 0.08 mSv h^{-1} were estimated. An even higher maximum dose rate $\dot{H}'(0.07)$ of 7.8 mSv h^{-1} was estimated for the laser processing of tungsten with a repetition rate of 200 kHz in burst mode. However, for these parameters, the deviation between the minimum and maximum dose equivalent rate is significantly larger and, therefore, associated with high uncertainty. If the mean value is calculated from the maximum and minimum dose rate and the uncertainty is calculated by the standard deviation, the maximum dose rate $\dot{H}'(0.07)$ found in this study is 5.3 ± 3.5 mSv h^{-1} at an operator distance of 60 cm.

The values of $\dot{H}'(3)$ and $\dot{H}^*(10)$ are significantly lower, which is due to the lower spectral photon flux within the region of higher energies. Looking at the increase in the dose equivalent rates due to the applied burst irradiation mode compared to the same processing without burst mode, one can see that from $\dot{H}'(0.07)$ to $\dot{H}'(3)$ and to $\dot{H}^*(10)$ the enhancement of the dose equivalent rate increases: for the defined "worst-case" spectrum (repetition rate = 50 kHz, burst mode, tungsten) the value of $\dot{H}'(0.07)$ is increased by a factor of approximately 2, $\dot{H}'(3)$ by a factor of approximately 4 and $\dot{H}^*(10)$ by a factor of approximately 9 when burst mode is applied. This behavior illustrates, that the burst mode laser processing results in a higher spectral photon flux for higher energies and in higher maximum energies. The highest enhancement due to burst mode application could be observed for a laser repetition rate of 200 kHz (tungsten) and a repetition rate of 100 kHz (steel). $\dot{H}'(0.07)$ was increased by a factor of approximately 60 for these parameters. While the absolute dose equivalent rates were rather modest for these parameters in single pulse mode, exceptionally high dose rates were recorded during burst mode laser processing.

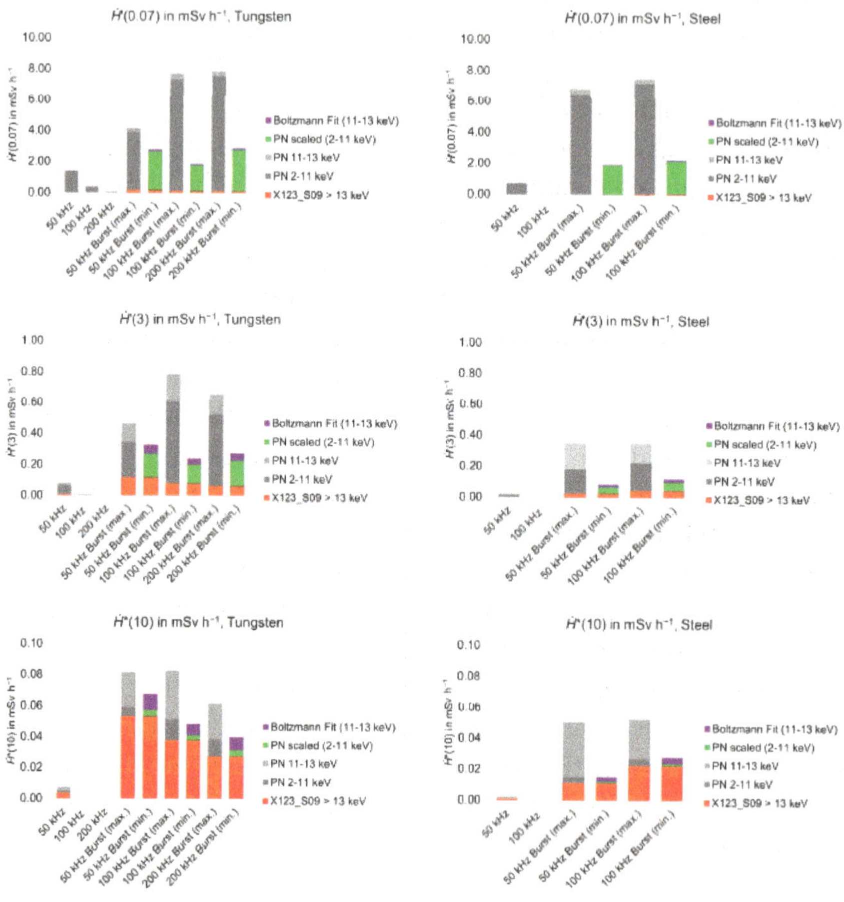

Figure 10. Dose equivalent rate estimations for the quantities Ḣ'(0.07), Ḣ'(3), and Ḣ*(10), as calculated from the spectra presented in Figure 8 and Figure 9. The results for laser processing of tungsten are summarized in the left column, while the results for laser processing of steel are given in the right column.

4. DISCUSSION

The examination of X-ray emission during USPL processing found many challenges in measuring X-ray spectra within the USPL apparatus. The primary challenges include to pile-up effects, electromagnetic incompatibility, a photon energy-dependent diminished sensitivity of the detector materials, and various electronic complications. The mitigation of disruptive influences was accomplished through the synergistic employment of various detector types (CdTe sensor paired with an SDD) and the implementation of shielding, alongside a reduction in pile-up utilizing aluminum and copper foils as filters, necessitating an evaluation routine to exclude the absorption effects caused by the foils from the measured spectra.. The radiation protection operational quantities Ḣ'(0.07), Ḣ'(3), and Ḣ*(10) were calculated from the spectra.

Different combinations for the laser pulse energy (varied between 91 μJ and 371 μJ) and repetition rate (varied between 50 kHz and 200 kHz) were selected for the investigations, whereby the average laser power was kept nearly constant. The average laser power is not a suitable primary parameter for estimating dose equivalent rates. The assessed dose equivalent rates are greatly influenced by the interplay between the laser pulse repetition rate and the chosen pulse energy. Combinations of parameters featuring high laser pulse energy and low repetition rate exhibit markedly elevated dose rates compared to those with high repetition rate and low laser pulse energy, but possessing almost similar average laser power, particularly in single pulse operating scenarios. Moreover, the burst mode laser operation markedly elevates the dosage rates, despite the average laser power remaining constant during its application.

The increase in dose equivalent rate during burst mode processing was already described in the literature. A possible explanation for the significant increase in the X-ray dose rate is the interaction between ultrafast laser radiation and the formed ablation cloud or a high-density plasma **[20,21]**.

Metzner et al. [20] investigated the dependence of the number of pulses per burst and the total fluence on the resulting X-ray dose rate by ablating stainless steel (X100CrMoV-8-1-1) with ultrafast laser pulses emitting a wavelength of 1030 nm and a pulse duration of 0.24 ps in the MHz-(MBM), GHz-(GBM), and the BiBurst (BBM) mode in comparison with conventional ultrafast laser radiation in the single-pulse mode (SPM). Compared to the SPM, producing a maximum dose equivalent rate $\dot{H}'(0.07)$ of approximately 7 × 103 μSv h−1, an increase in the X-ray dose rate by more than a factor of 30 was found in the BBM with two and three sub-pulses in the MHz- and two sub-pulses in the GHz-bursts. For these parameters a dose rate of $\dot{H}'(0.07)$ = 2.5 × 105 μSv h−1 was determined. This is within the range between a factor of 2 and 60 which was determined within this study for the increase in the dose equivalent rate $\dot{H}'(0.07)$ due to burst mode application. Additionally, the absolute values of $\dot{H}'(0.07)$ are comparable with the results presented here for steel and burst mode taking into account the detector distance of 10 cm in [20] and 60 cm in this work.

Schille et al. [21] observed a significant increase in the dose equivalent rate of the emitted X-rays accompanied by pronounced characteristic X-ray emissions for laser bursts irradiating at a 1.0 MHz burst repetition frequency and corresponding lower peak intensity of the intra-burst pulses. The analysis was performed during the processing of technical-grade AISI 304 stainless steel targets. The maximum X-ray emissions were measured to $\dot{H}'(0.07)$ = 32.8 ± 3.6 mSv h−1 (3rd scan) with the two-pulse burst mode. For burst and bi-burst pulses, the second intra-burst pulse was found to significantly enhance the X-ray emission potentially induced by the laser pulse and plasma interaction. The measured X-ray spectra showed an energy region between 0 and 15 keV, which were measured with a SiLi-based detector. This is significantly lower than the maximum energy of 40 keV, which was observed in this study for burst mode processing of tungsten at a repetition rate of 50 kHz and a laser pulse energy of 371 μJ. Until now, maximum X-ray energies up to 30 keV were reported by Legall et al. [4] during USPL processing of tungsten in single-pulse mode. Additionally, higher dose equivalent rates of $\dot{H}'(0.07)$ = 163 mSv h-1 for the processing of tungsten with a repetition rate of 400 kHz and a peak irradiance of 2.6 × 1014 W cm−2 were reported in a distance of 42 cm. The maximum X-ray photon energy must be considered if a safety housing is configured for a USPL machine. Legall et al. [12] calculated the equivalent thickness of different shielding materials for X-ray protection up to photon energies of 60 keV in units of iron. Since the here observed maximum energy of 40 keV is lower, these recommendations should still be valid here. The knowledge about the spectral flux of the emitted X-ray photons can also support the work on new testing concepts for USPL machines. Since the actual testing concept, which is based on the "worst-case" processing scenario, is connected with many difficulties [14] a new testing concept that is based on an alternative X-ray source is highly demanded.

To summarize, a "worst-case" spectrum with X-ray photon energies up to 40 keV was found in this study for the machining of tungsten at a repetition rate of 50 kHz and in burst operation mode. Additionally, equivalent dose rates $\dot{H}'(0.07)$ of the order of 200 mSv h−1 were calculated for the processing of tungsten with 200 kHz repetition rate, burst mode and a peak irradiance of 7.3 × 1013 W cm−2 if a distance between the laser interaction point and detector of 10 cm is assumed without consideration of the absorption in air.

It is important to acknowledge that if the "worst-case" spectrum from USP laser machining is established, it can be replicated using an X-ray tube as a model system to emulate the USP laser machine. This could result in a streamlined testing paradigm for USP laser machines in the future, independent of the volatile USP laser-generated plasma as an X ray source during safety housing evaluations, and aligned with existing standard testing protocols.

5. CONCISE SUMMARY

The investigation focused on ultrashort pulse laser processing of tungsten and steel concerning the production of unwanted secondary X-rays. Spectral X-ray measurements were conducted using complementary CdTe- and SDD-based spectrometers. The radiation protection operational quantities $\dot{H}'(0.07)$, $\dot{H}'(3)$, $\dot{H}^*(10)$ were calculated from the measured X-ray spectra. Regarding the laser processing strategy, the dose equivalent rate of the X-rays increases for lower repetition rates and higher pulse energy if a constant average power is used. The average laser power of ultrashort pulse laser processing alone does not qualify as a suitable parameter for the prediction of X-ray dose rates. Processing with pulse burst mode significantly increases the dose rates and the X-ray photon energies. A maximum dose rate $\dot{H}'(0.07)$ of 5.3 ± 3.5 mSv h−1 at a distance of 60 cm was found for the laser processing of tungsten with a repetition rate of 200 kHz in burst mode and a peak irradiance of 7.3 × 1013 W cm−2. A peak X-ray photon energy of approximately 40 keV was recorded during burst mode processing of tungsten, with a repetition rate of 50 kHz and a peak irradiance of 3 × 1014 W cm−2.

AT A GLANCE

Exceeding a critical laser irradiance of approximately 1013 W cm−2 during ultrashort pulse laser processing may lead to the unintended emission of X-rays. Spectral X-ray emissions were examined during the production of tungsten and steel using three complementary spectrometers (utilizing CdTe and silicon drift detectors) concurrently to identify a worst-case spectral scenario. Consequently, the maximal energies of X-ray photons were ascertained, and the associated dose equivalent rates were computed. A workstation utilizing an ultrashort pulse laser with a pulse duration of 274 fs, a central wavelength of 1030 nm, pulse repetition rates ranging from 50 kHz to 200 kHz, and a Gaussian laser beam focussed to a spot width of 33 μm was utilized in both single pulse and burst laser operation modes. Various combinations of laser pulse energy and repetition rate were employed while maintaining the average laser power near the maximum threshold of 20 W.. Peak irradiances I0 ranging from 7.3 × 1013 W cm^{-2} up to 3.0 × 1014 W cm^{-2} were used. The X-ray dose equivalent rate increases for lower repetition rates and higher pulse energy if a constant average power is used. Laser processing with burst mode significantly increases the dose rates and the X-ray photon energies. A maximum X-ray photon energy of about 40 keV was observed for burst mode processing of tungsten with a repetition rate of 50 kHz and a peak irradiance of 3 × 1014 W cm^{-2}.

REFERENCES

1. Kühlke, D.; Herpers, U.; Von der Linde, D. Soft X-ray emission from subpicosecond laser-produced plasmas. *Appl. Phys. Lett.* **1987**, *50*, 1785–1787. **[Google Scholar] [CrossRef]**

2. Thogersen, J.; Borowiec, A.; Haugen, H.K.; McNeill, F.E.; Stronach, I.M. X-ray emission from femtosecond laser micromachining. *Appl. Phys. A* **2001**, *73*, 361–363. **[Google Scholar] [CrossRef]**

3. Bunte, J.; Barcikowski, S.; Puester, T.; Burmester, T.; Brose, M.; Ludwig, T. Secondary hazards: Particle and X-ray emission. In *Femtosecond Technology for Technical and Medical Applications*; Dausinger, F., Lubatschowski, H., Lichtner, F., Eds.; Springer: Berlin/Heidelberg, Germany, 2004; pp. 309–321. **[Google Scholar]**

4. Legall, H.; Schwanke, C.; Pentzien, S.; Dittmar, G.; Bonse, J.; Krüger, J. X-ray emission as a potential hazard during ultrashort pulse laser material processing. *Appl. Phys. A* **2018**, *124*, 407. **[Google Scholar] [CrossRef] [Green Version]**

5. Behrens, R.; Pullner, B.; Reginatto, M. X-Ray Emission from Materials Processing Lasers. *Radiat. Prot. Dosim.* **2019**, *183*, 361–374. **[Google Scholar] [CrossRef]**

6. Legall, H.; Schwanke, C.; Bonse, J.; Krüger, J. The influence of processing parameters on X ray emission during ultra-short pulse laser machining. *Appl. Phys. A* **2019**, *125*, 570. **[Google Scholar] [CrossRef] [Green Version]**

7. Legall, H.; Schwanke, C.; Bonse, J.; Krüger, J. X-ray emission during ultrashort pulse laser processing. In *Frontiers in Ultrafast Optics: Biomedical, Scientific, and Industrial Applications XIX, Proceedings of the SPIE LASE, San Francisco, CA, USA, 2–5 February 2019*; SPIE: Bellingham, WA, USA, 2019; Volume 10908, p. 1090802. **[Google Scholar]**

8. Weber, R.; Graf, T. X-ray emission during materials processing with ultrashort laser pulses—A serious problem? *Procedia CIRP* **2022**, *111*, 844–849. **[Google Scholar] [CrossRef]**

9. Weber, R.; Giedl-Wagner, R.; Förster, D.J.; Pauli, A.; Graf, T.; Balmer, J.E. Expected X-ray dose rates resulting from industrial ultrafast laser applications. *Appl. Phys. A* **2019**, *125*, 635. **[Google Scholar] [CrossRef] [Green Version]**

10. Legall, H.; Schwanke, C.; Bonse, J.; Krüger, J. X-ray radiation protection aspects during ultrashort laser processing. *J. Laser Appl.* **2020**, *32*, 022004. **[Google Scholar] [CrossRef] [Green Version]**

11. Freitag, C.; Giedl-Wagner, R. X-Ray Protection in an Industrial Production Environment: Industrial implementation of the requirements of the German Radiation Protection Act. *PhotonicsViews* **2020**, *17*, 37–41. **[Google Scholar] [CrossRef]**

12. Legall, H.; Bonse, J.; Krüger, J. Review of X-ray exposure and safety issues arising from ultra-short pulse laser material processing. *J. Radiol. Prot.* **2021**, *41*, R28–R42. **[Google Scholar] [CrossRef]**

13. Mosel, P.; Sankar, P.; Düsing, J.F.; Dittmar, G.; Püster, T.; Jaschke, P.; Vahlbruch, J.W.; Morgner, U.; Kovacev, M. X-ray Dose Rate and Spectral Measurements during Ultrafast Laser Machining Using a Calibrated (High-Sensitivity) Novel X-ray Detector. *Materials* **2021**, *14*, 4397. **[Google Scholar] [CrossRef] [PubMed]**

14. Stolzenberg, U.; Schmitt Rahner, M.; Pullner, B.; Legall, H.; Bonse, J.; Kluge, M.; Ortner, A.; Hoppe, B.; Krüger, J. X-ray Emission Hazards from Ultrashort Pulsed Laser Material Processing in an Industrial Setting. *Materials* **2021**, *14*, 7163. **[Google Scholar] [CrossRef] [PubMed]**

15. Rücker, J.; Fricke, A.; Kahle, M. Unwanted X-ray emission during ultra-short pulse laser material processing and the influence of processing parameters. *Procedia CIRP* **2022**, *111*, 850–854. **[Google Scholar] [CrossRef]**

16. Barkauskas, V.; Plukis, A. Prediction of the irradiation doses from ultrashort laser-solid interactions using different temperature scalings at moderate laser intensities. *J. Radiol. Prot.* **2022**, *42*, 011501. **[Google Scholar] [CrossRef]**

17. Holland, J.; Weber, R.; Sailer, M.; Graf, T. Influence of Pulse Duration on X-ray Emission during Industrial Ultrafast Laser Processing. *Materials* **2022**, *15*, 2257. **[Google Scholar] [CrossRef] [PubMed]**

18. Schille, J.; Kraft, S.; Kattan, D.; Löschner, U. Enhanced X-ray Emissions Arising from High Pulse Repetition Frequency Ultrashort Pulse Laser Materials Processing. *Materials* **2022**, *15*, 2748. **[Google Scholar] [CrossRef] [PubMed]**

19. Mosel, P.; Sankar, P.; Appi, E.; Jusko, C.; Zuber, D.; Kleinert, S.; Düsing, J.; Mapa, J.; Dittmar, G.; Püster, T. Potential hazards and mitigation of X-ray radiation generated by laser-induced plasma from research-grade laser systems. *Opt. Express* **2022**, *30*, 37038–37050. **[Google Scholar] [CrossRef] [PubMed]**

20. Metzner, D.; Olbrich, M.; Lickschat, P.; Horn, A.; Weißmantel, S. X-ray generation by laser ablation using MHz to GHz pulse bursts. *J. Laser Appl.* **2021**, *33*, 032014. **[Google Scholar] [CrossRef]**

21. Schille, J.; Kraft, S.; Pflug, T.; Scholz, C.; Clair, M.; Horn, A.; Loeschner, U. Study on X-ray Emission Using Ultrashort Pulsed Lasers in Materials Processing. *Materials* **2021**, *14*, 4537. **[Google Scholar]** **[CrossRef]**
22. Elahi, P.; Kalaycioğlu, H.; Akçaalan, Ö.; Şenel, Ç.; Ilday, F.Ö. Burst-mode thulium all-fiber laser delivering femtosecond pulses at a 1 GHz intra-burst repetition rate. *Opt. Lett.* **2017**, *42*, 3808–3811. **[Google Scholar]** **[CrossRef]**
23. Bonamis, G.; Mishchick, K.; Audouard, E.; Hönninger, C.; Mottay, E.; Lopez, J.; Manek-Hönninger, I. High efficiency femtosecond laser ablation with gigahertz level bursts. *J. Laser Appl.* **2019**, *31*, 022205. **[Google Scholar]** **[CrossRef]**
24. Shchatsinin, I. Free Clusters and Free Molecules in Strong, Shaped Laser Fields. Ph.D. Thesis, Free University, Berlin, Germany, 3 June 2009. **[Google Scholar]**
25. Becker, J.; Tate, M.W.; Shanks, K.S.; Philipp, H.T.; Weiss, J.T.; Purohit, P.; Chamberlain, D.; Ruff, J.P.C.; Gruner, S.M. Characterization of CdTe sensors with Schottky contacts coupled to charge-integrating pixel array detectors for X-ray science. *J. Instrum.* **2016**, *11*, P12013. **[Google Scholar]** **[CrossRef]** **[Green Version]**
26. Strüder, L.; Niculae, A.; Holl, P.; Soltau, H. Development of the Silicon Drift Detector for Electron Microscopy Applications. *MTO* **2020**, *28*, 46–53. **[Google Scholar]** **[CrossRef]**
27. Cimmino, A.; Horváth, D.; Olšovcová, V.; Stránský, V.; Truneček, R.; Tsinganis, A.; Versaci, R. Radiation Protection at ELI Beamlines: A Unique Laser Driven Accelerator Facility. In Proceedings of the 40th International Conference on High Energy physics—PoS(ICHEP2020), Prague, Czech Republic, 28 July–6 August 2020. **[Google Scholar]**
28. Hubbell, J.H.; Seltzer, S.M. *NIST Standard Reference Database 126*; National Institute of Standards and Technology (NIST): Gaithersburg, MD, USA, 1996.
29. International Commission on Radiation Units and Measurements (ICRU). *Conversion Coefficients for Use in Radiological Protection against External Radiation*; ICRU Report 57; International Commission on Radiation Units and Measurements (ICRU): Bethesda, MD, USA, 1998. **[Google Scholar]**
30. Ankerhold, U.; Behrens, R.; Ambrosi, P. X Ray Spectrometry of Low Energy Photons for Determining Conversion Coefficients from Air Kerma, K_a, to Personal Dose Equivalent, $H_p(10)$, for Radiation Qualities of the ISO Narrow Spectrum Series. *Radiat. Prot. Dosim.* **1999**, *81*, 247–258. **[Google Scholar]** **[CrossRef]**

CHAPTER

AUGMENTED X-RAY EMISSIONS RESULTING FROM HIGH PULSE REPETITION FREQUENCY ULTRASHORT PULSE LASER MATERIAL PROCESSING

5

1. INTRODUCTION: AN OVERVIEW

Ultrashort pulse lasers (USPLs) have emerged as a versatile instrument for high-efficiency and high-precision micromachining, attributable to their superior laser beam characteristics. Furthermore, the prevailing trend towards increased laser power facilitates elevated throughputs and reduced processing durations in contemporary microfabrication, indicating that the USPL will soon transition from laboratory conditions to industrial production. The emission of X-ray photon radiation represents a secondary laser beam danger from ultrashort pulse lasers in materials processing, posing significant health concerns to laser operators. Undesired X-ray photons may be generated when high-intensity laser beams interact with solid surfaces, resulting in the formation of hot, dense plasma that contains electrons with energies of several keV. Since the 1970s, three primary processes have been identified for X-ray radiation fields produced by ultrafast laser-excited electrons: (i) the Bremsstrahlung continuum resulting from free-free transitions of accelerated free electrons, (ii) the recombination continuum from free-bound electron transitions, and (iii) characteristic line emissions stemming from bound-bound transitions of the inner-shell electrons of ionized atoms. The energy distribution of Bremsstrahlung X-ray photons adheres to a Maxwell–Boltzmann distribution, and the emitted X-ray dose rate can be estimated by integrating the complete spectral X-ray photon flux. The X-ray emission dose from a single laser pulse typically varies from a few pico Sievert (pSv) to several nano Sievert (nSv), indicating a relatively minimal risk for laser operators. The dose rates of X-ray emissions and their corresponding hazards and concerns might significantly escalate when utilizing high-average-power ultrashort pulse lasers in materials processing. This reflects the remarkable advancements in the evolution of contemporary USPL systems, achieving kilowatt average power provided by ultrashort pulses at high pulse repetition frequencies (PRF) ranging from ten to hundreds of megahertz (MHz) [1]. In this scenario, millions of ultrashort pulses engage with the substrate surface within a brief period, and the X-ray photons emitted by each individual pulse aggregate to substantial X-ray dosage levels, attaining up to 1 Sv/h [2,3]. The elevated X-ray emissions significantly surpass the permissible limits for both the general population and occupationally exposed individuals as outlined in the German Radiation Protection Act, potentially resulting in severe health detriments to the affected areas of the human body.

X-ray photon emissions have thus far been observed at peak intensities exceeding 10^13 W/cm^2, achievable through ultrashort pulses with a length of 1 ps and a fluence of 10 J/cm^2. This fluence exceeds the optimal fluence for energy-efficient laser processing of metals by a factor of 10 and can be attained by tightly concentrating a laser beam with μJ level optical pulse energy to spot sizes around ten micrometers in the focal plane. A recent research confirmed the generation of damaging X-ray photon dose rates at significantly lower fluence, or intensity, even beneath the legal peak intensity limit [7]. This was accomplished under particular processing conditions where MHz PRF pulses were emitted at a minimal geometrical pulse separation. A comprehensive analysis of the existing literature in this domain discovered over 20 parameters influencing X-ray photon emissions generated by ultrashort laser pulses interacting with solid surfaces (refer to Table 1). These factors interact and can be categorized into three distinct types: (i) laser-related, (ii) process-related, and (iii) material-related parameters of influence. Primarily, regarding the laser, the peak intensity of the laser pulse is posited as the most probable characteristic, as a doubling of the peak intensity results in approximately a tenfold increase in emitted X-ray photon emissions. Consequently, when evaluating pulses with comparable peak intensities, elevated X-ray emission dose rates are observed for extended pulses and increased pulse energies. Another general expectation from the studies thus far is that the X-ray emissions increase practically linearly with rising average laser powers. Moreover, a recent study determined that the PRF significantly influences the emitted X-ray photon flux [7]. This was especially true when MHz PRF pulses, characterized by excellent spatial resolution, overlapped at submicrometer distances within a laser-processed line. The significant rise in X-ray photon emissions was attributed to excellent plasma resonance absorption of the irradiating laser beam. The theory is corroborated by the observation that MHz PRF pulses with minimal time delay can engage with the residual plasma plume generated by prior pulses [8]. Resonance absorption in laser-produced plasma has thus far been documented exclusively for high-intensity laser pulses, where both the polarization of the laser beam and the angle of incidence significantly affect the overall absorption of the incident laser beam [9,10,11]. The paper confirms that even low-intensity pulses can produce significant X-ray emissions when the polarization of the MHz PRF laser beam is positioned parallel to the direction of beam propagation. This indicates that plasma resonance absorption may also be efficient in low-intensity and high pulse repetition frequency ultrashort pulse laser regimes. Moreover, the suction flow velocity in the processing zone is identified as an additional component affecting the X-ray emission dosage rate. Reproducible X-ray emissions can only be observed under controlled and stable suction flow conditions. A third significant conclusion of this work is that X-ray photon emissions can be detected at low peak intensities, reaching up to five times below the legal peak intensity threshold for approval-free laser operation in Germany. This may happen under specific processing conditions, such as when powerful PRF pulses with picosecond durations irradiate at a minimal intra-line distance, the polarization of the laser beam is aligned parallel to its direction of movement, and the suction flow velocity is relatively low.

Table 1. A summary of the influencing factors affecting X-ray photon emissions arising from ultrashort pulse laser materials processing.

Parameter		Impact on X-ray Emission	References
Laser-sided parameters	Peak intensity	- Doubling the peak intensity increases the X-ray dose rate by around ten times	[3,12,13,14]
	Pulse energy	- With higher pulse energies, higher X-ray dose rates tend to be expected, - The released X-ray emission increases non-linearly with higher energies, the effective peak intensity has a stronger effect on X-ray emission	[15,16]
	Average laser power	- The released X-ray emission scales approximately linearly with higher average laser powers, pay attention to the non-linear effects with higher pulse energies, - Strongly increased X-ray dose rates for high-PRF laser processes or in burst mode laser processing, in particular when high-PRF pulses irradiate at small geometrical (micro meter) pulse distances	[3,7,16,17]
	Pulse repetition frequency	- X-ray dose rate increases approximately linearly with higher PRF resulting from the accumulation of X-ray emissions respectively released by the individual pulses. - Caution concerning strong increased X-ray emissions for MHz PRF or burst mode laser processing which results from strong laser pulse with plasma interactions	[3,7,16]

Parameter		Impact on X-ray Emission	References
	Focus spot diameter	- Doubling the focus spot size at constant peak intensities yields 2.5 times higher X-ray dose rates - The spectral distribution and amplitude shifts towards higher X-ray photon energies with increasing focus spot area	[18]
	Wavelength	- Tendency of lower X-ray dose rates for ultrashort pulses of shorter wavelength	[14]
	Polarisation	- Laser radiation polarized parallel to the scan direction increases the dose rate	[11,16,19]
	Pulse duration	- Tendency of higher X-ray dose rates for ultrashort pulses of longer pulse duration	[15,16,20]
Process-sided parameters	Processing regime	- Absorption of the released X-ray emission at the walls of the laser engraved or laser drilled structures reduces the X-ray emission during the ongoing processing - Caution: X-ray dose rate increases as a result of strong laser pulse with plasma/ablation plume interaction, i.e., during stationary laser machining, with deflected beams at high pulse overlap, in laser turning	[7,11,16]
	Scanning direction	- Stronger X-ray emission opposite to the scan direction	[13]
	Intra-line pulse distance	- Pulse irradiations at small geometrical distances cause high surface roughness, in turn lowering the released X-ray dose rate (see surface roughness) - Pulses of small geometrical pulse distances (micro meter) irradiating at MHz PRF amplify the X-ray dose rate by feedback coupling of the pulses with the previously generated laser ablation/plasma plume	[7,11,16]
	Hatch distance	- Larger hatch distances result in higher X-ray dose rates	[11]
	Scan number	- Tendency for lower X-ray dose rates with increasing number of scan crossings	[7]
	Focus position	- Highest X-ray dose rate arises in the focal plane at the position of highest peak intensity - For high-average power laser beams, thermal shift of the optical elements in the beam path can have an effect on the position of highest peak intensity	[2]

Parameter		Impact on X-ray Emission	References
	Cross jet	- A larger volume flow rate of the cross jet induces higher X-ray emissions	[16]
	Fume extraction	- A larger distance between sample and suction nozzle decreases the flow rate in turn increasing the X-ray dose rate	[16]
	Angle of incidence	- Oblique laser beams with increased angle of incident with respect to the plasma flank enhance resonance absorption that tends to cause higher X-ray emissions	[7,16,21]
Material-sided parameters	Material	- Higher X-ray dose rates occur with elements with a higher atomic number - Highest X-ray dose rate was determined on tungsten, the X-ray dose rate is comparably high on steel and stainless steel materials	[2,12,14]
	Suface roughness	- A higher surface roughness leads to lower X-ray dose rates due to the shielding of the X-rays on microscopic substructures, similar to the shielding effect of boreholes or trenches	[11,16]
	Dimension	- Large-area laser processing under similar irradiation conditions and corresponding X-ray dose rates releases a larger X-ray emission dose	

2. EXPERIMENTAL SECTION

Two types of USPL systems were utilized in this study: a UFFL 100 (Active Fiber Systems GmbH, Jena, Germany) and an FX 200 (Edgewave GmbH, Würselen, Germany). This provided a wide range of complementary laser beam parameter settings, i.e., 87 W maximum average laser power, up to 50 MHz PRF, and pulse durations from 340 fs ≤ τ_H ≤ 10 ps (sech^2 pulse shape). Near-infrared laser beams of 1030 nm wavelength were passed through individual beam expanders and were individually expanded and aligned through an f-theta focusing objective of 167 mm focal length. This resulted in a similar laser spot radius w_0 in the focal plane, as determined by $w_{0/UFFL} = w_{0/FX200}$ = 15.0 μm (1/e-^{2}method), for the beam caustic measurements (Micro Spot Monitor, Primes GmbH, Pfungstadt, Germany). In addition, a galvanometer scan system (intelliScan 30, Scanlab GmbH, Puchheim, Germany) was assembled in the beam path for raster-scanning the laser beams in a line pattern across the studied stainless steel (AISI 304) substrate surfaces. The intra-line pulse distance d_X, representing the geometrical distance of the impinging laser pulses within a scanned line, was varied as function of scan speed v_S and PRF f_P according to the following relationship: $v_S = d_X \times f_P$. The distance between the scanned lines in the line pattern was set by the hatch distance d_Y. The polarization state of the linearly polarized laser beams was adjusted by a half-wave retarder plate aligned either parallel ($\vec{E} \parallel \vec{v}_{scan}$) or perpendicular ($\vec{E} \perp \vec{v}_{scan}$) to the laser beam scan direction. In terms of safety precautions, all the experiments carried out in this study were performed inside a laser safety enclosure to protect the operators from direct and scattered laser beams, X-ray radiation, and other secondary harmful effects originating during high PRF ultrashort pulse laser processing.

The maximum pulse energy used in this study was Q_P = 36 µJ. The pulse energy was calculated from the measured average laser power divided by the PRF, taking into account the optical losses in the beam path. The peak power of a laser pulse P_0 was calculated as the pulse energy divided by the pulse duration. Following Equation (1), the peak intensity of the laser pulses I_0 was given by the pulse peak power divided by the focus spot area A_f, considering a pulse-shape-dependent numerical factor of f_S = 0.88 for the sech² ultrashort pulses:

$$I_0 = f_S \frac{P_0}{A_f} = f_S \frac{2\,Q_P}{\tau_H\,\pi\,w_0^2} \tag{1}$$

Accordingly, the maximum peak intensity of the ultrashort pulses was $I_{0/UFFL}$ = × 2.7 10^{13} W/cm² and $I_{0/FX}$ = 10 × 1.5^{13} W/cm² for the UFFL 100 and FX 200 lasers, respectively. Notably, these peak intensity values were only slightly above an intensity of 10 × 1^{13} W/cm², where the X-ray emission dose rate can exceed 1 µSv/h [3,7,12]. In fact, according to the German Radiation Protection Regulations, both the values 1 × 10^{13} W/cm² and 1 µSv/h at a 100 mm distance from the accessible area represent the legal limits for approval-free laser operations [22].

The X-ray emissions arising from the laser beam with AISI 304 substrate interaction were analyzed using two different detectors: a survey meter OD-02 (STEP Sensortechnik und Elektronik Pockau GmbH, Pockau-Lengefeld, Germany) measuring the directional X-ray emission dose $H'(0.07)$ and a SILIX lambda detector (Ingenieurbüro Prof. Dr.-Ing. Günter Dittmar, Aalen, Germany) for monitoring the X-ray emission spectra and X-ray emission dose rate $H''(0.07)$ in the soft X-ray photon energy range between 2 and 20 keV. In each experiment, the detectors were aligned parallel to the laser beam scan direction towards the center of the laser-processed area. The dimensions of the raster-scanned processing field were 20 × 20 mm². The distance between the detectors and the scanning field center point was 200 mm, and the detection angle was 35°. A unidirectional scan regime was applied with a "laser-ON" cycle, with the laser beam moving away from the X-ray detector receiving maximum X-ray emission dose levels [13]. The laser ON/OFF duty cycle was determined to be 46% in oscilloscope measurements; thus, the monitored values were corrected by a factor of 2.17 to discuss process-typical X-ray emissions in the following sections. The background X-ray radiation in the laboratory was determined below 0.1 μSv/h, which was negligibly small compared to the detected laser-generated X-ray dose levels. Furthermore, it should be mentioned that in each case the X-ray emissions were captured from the first scan crossing of the laser beam over the substrates and were obtained after every six seconds. Respectively, the presented X-ray emission dose values represent mean values averaged from at least ten individual X-ray measurements. The standard deviation of the measured values is indicated in the diagrams.

3. RESULTS

3.1. X-ray Emission as a Function of Intra-Line Pulse Distance

First, the X-ray emission dose per pulse $H'(0.07)_{PP}$ was investigated as a function of the intra-line pulse distance and PRF. Therefore, ultrashort pulses with a peak intensity of $I_0 = 10 \times 1.5^{13}$ W/cm^2 were irradiated at f_P = 0.5 MHz and 1.6 MHz. The intra-line pulse distance was varied in the range of 0.15 μm < d_X < 2.0 μm by increasing the scan speed from 75 mm/s to 3200 mm/s, depending on the respective PRF, while the hatch distance was kept constant at d_Y = 20 μm. The X-ray emission dose per pulse, as shown in **Figure 1**, was derived from the averaged SILIX detector readings captured in individual test series on four different days. At f_P = 0.5 MHz, the highest X-ray emission dose per pulse of $H'(0.07)_{PP}$ = 1 pSv was detected at 0.2 μm. For the larger intra-line pulse distances, there was a clear tendency towards lower X-ray emissions. Therefore, at d_X = 2 μm (representing the widest studied distance) the X-ray emission dose per pulse reduced up to two orders of magnitude to $H'(0.07)_{PP}$ = 0.02 pSv, which emphasized the great effect of intra-line pulse distance on laser-induced X-ray emissions.

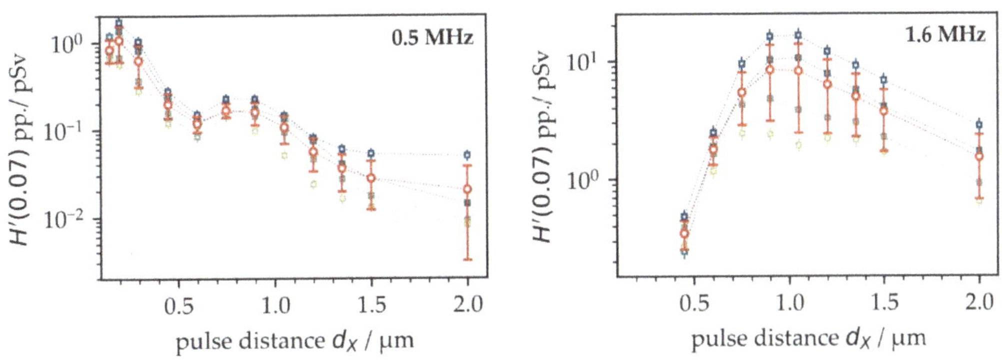

Figure 1. X-ray emission dose per pulse as a function of the intra-line pulse distance at 0.5 MHz (left) and 1.6 MHz (right) PRF obtained on 4 different days. The mean values are highlighted by the red symbols (circles).

At higher PRFs (for example, 1.6 MHz in **Figure 1** (right)), the highest X-ray emissions were found in the range of d_X = 1 μm, steadily decreasing with larger intra-line pulse distances. The highest X-ray emission dose per pulse was assessed at an $H'(0.07)_{PP}$

= 16.0 pSv maximum with about 8.5 pSv average value. In comparison to low PRF pulses, this was more than a ten-fold increase referring to the maximum value and almost a hundred times higher than the X-ray emission dose arising at d_X = 1 µm and 0.5 MHz PRF. This strong increase in the X-ray emissions observed for the high PRF pulses provided evidence of stronger laser beam interaction with plasma during processing, thus, further enhancing X-ray photon generation, which will be discussed more in detail in the following subchapters. However, a significant drop in the X-ray emission dose rate per pulse occurred when 1.6 MHz pulses irradiated at small intra-line pulse distances (d_X < 1 µm), which was in contrast to the X-ray emission characteristics observed at 0.5 MHz. Therefore, a potential explanation can be found in the increased roughness of the high-PRF-processed surfaces, which could behave similar to the X-ray shielding recently observed on kerf walls and boreholes [3,11]. In the particular case of high PRF pulses irradiating at small intra-line pulse distances, self-organizing microscopic surface structures emerged at the substrate surfaces [23]. These rough and protruded surface features potentially screened the laser-induced X-ray photon flux, which in turn caused the lower monitored X-ray emission doses at small intra-line pulse distances seen in **Figure 1** (right). In order to validate the proposed influencing effect of surface roughness on X-ray emission, X-ray photon fluxes arising from the laser processing of four different laser-pretextured substrates were analyzed, **Figure 2.**

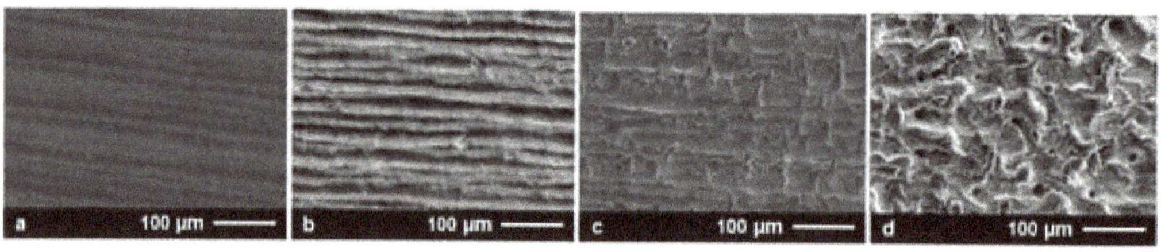

Figure 2. Laser surface textures produced on AISI 304 by raster-scanning of the laser beam across the substrates at 3.0 µm (a), 1.5 µm (b), 0.9 µm (c), and 0.6 µm (d) intra-line pulse distance.

For surface pretexturing, a laser beam of $I_{0/FX}$ = 10 × 1.5^{13} W/cm^2 and f_P = 1.6 MHz was raster-scanned across the AISI 304 substrates at a d_Y = 20 µm hatch distance. The intra-line pulse distances were varied between d_X = 3.0 µm (**Figure 2a**), 1.5 µm (**Figure 2b**), 0.9 µm (**Figure 2c**), and 0.6 µm (**Figure 2d**). In this way, specific surface roughness conditions were provided for the next impinging pulses similar to the ones discussed above in **Figure 1**. In the X-ray measurements, the laser processing conditions were as follows: $I_{0/FX}$ = 10 × 1.5^{13} W/cm^2, f_P = 1.6 MHz, d_X = 0.9 µm, and d_Y = 20 µm.

The highest X-ray emission dose per pulse of $H'(0.07)_{PP}$ = 9.1 pSv was captured for the smooth surface, which was laser-pretextured with d_X = 3.0 µm and S_a = 0.49 µm corresponding area roughness, **Figure 3**. For the d_x = 1.5 µm laser-pretextured surface and resulting S_a = 0.94 µm, the X-ray emission dose per pulse was marginally lower in the range of $H'(0.07)_{pp}$ = 8.2 pSv. These X-ray emission doses were almost equal to the ones measured for the nontextured basic material of similar surface roughness, S_a = 0.44 µm. For the rougher surfaces featured with area roughness measures of S_a = 1.8 µm and 6.8 µm resulting from laser pretexturing at smaller intra-line pulse distances of d_x = 0.9 µm and 0.6 µm, respectively, the detected X-ray emission doses per pulse amounted to $H'(0.07)_{pp}$ = 5.4 pSv and 0.9 pSv, respectively. Remarkably, these dose values were even below the X-ray emissions arising from laser processing of the nontextured steel surface at d_x = 0.6 µm in **Figure 1** (right), which was proof for the screening of laser-induced X-ray photons by rough surface features.

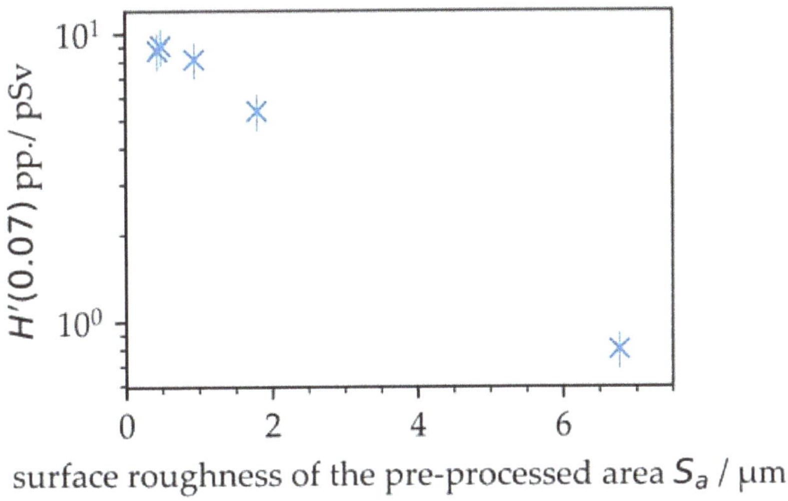

Figure 3. X-ray emission dose per pulse as a function of the area roughness of specifically laser pre-textured AISI ٣٠٤ substrate surfaces.

However, a closer look at **Figure 1** shows a large variance of the X-ray emission dose levels, diverging up to a factor of ten for the measurements carried out on four consecutive days. An influencing effect of the laser beam or, rather, the ambient air conditions on the X-ray emission dose can largely be ruled out, as the laser beam parameters were carefully controlled during the study. Moreover, the experiments were performed in an air-conditioned laboratory room (temperature: 0.5 ± 22.5 °C; humidity: %2 ± %44) in order to provide constant processing conditions.

3.2. Influencing Effect of Dust and Fume Extraction on X-ray Emission

After a detailed inspection of the experimental conditions, both the position of the fume extraction element (FEE) related to the processing area, as well as the suction flow speed v_{fs}, were identified as potentially influencing factors for the monitored different X-ray emission dose levels. For a more precise investigation of this effect, the distance of the FEE from the processing area d_{eff} was varied during the measurements of X-ray photon emission arising from laser surface processing with pulses of $I_{0/FX}$ = 10×1.5^{13} W/cm^2, f_p = 1.6 MHz, and d_X = 0.75 µm. As another control parameter, the suction flow speed was monitored at the center position of the laser-processed field using a resistance anemometer. The measurements showed a continuous reduction in the suction flow speed with increasing distance of the FFE from the processing area, **Figure 4.** At about a 190 mm distance, the suction flow turned from laminar to turbulent, which was indicated by the larger deviation of the measured suction flow speed. From this position, no further reduction in the suction flow speed was detected with larger distances because the cross jet air flow protecting the focusing objective was more predominant within the processing area.

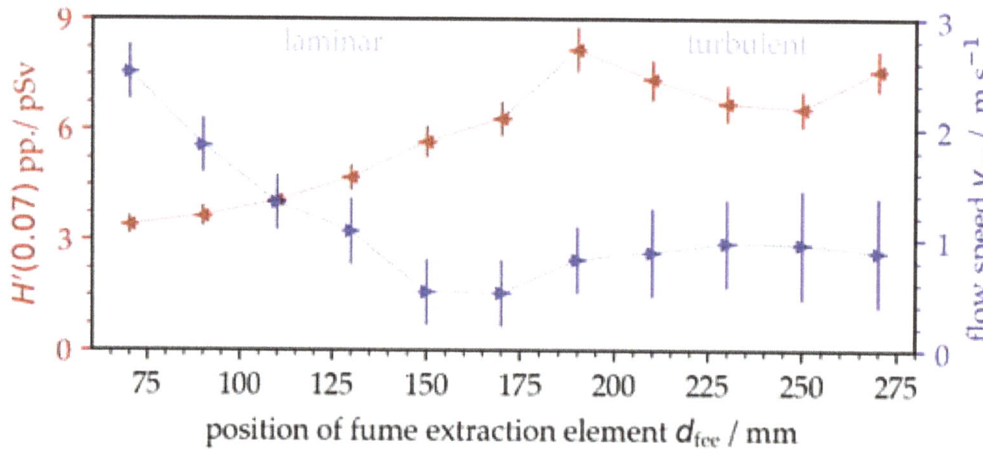

Figure 4. X-ray emission dose per pulse as a function of suction flow speed that was varied by increasing the distance of the fume extraction element from the laser processing area.

The results obtained in this set of experiments validated a strong impact of the surrounding air flow on X-ray photon emission for the first time. In the laminar flow regime, the X-ray emission dose increased almost linearly with lower suction flow speed, while the maximum X-ray photon flux was measured in the turbulent flow regime. Hence, the experiments of Figure 1 were repeated under controlled suction flow conditions. As a result, the X-ray emissions arising from laser processing under stable suction flow speeds (vfs = 0.9 m/s ± 0.3 m/s) showed a considerably reduced standard deviation ranging within the measurement accuracy of the SILIX detector. This was proven for both fP = 0.5 MHz and 1.6 MHz repetitive pulses studied on two different days, Figure 5.

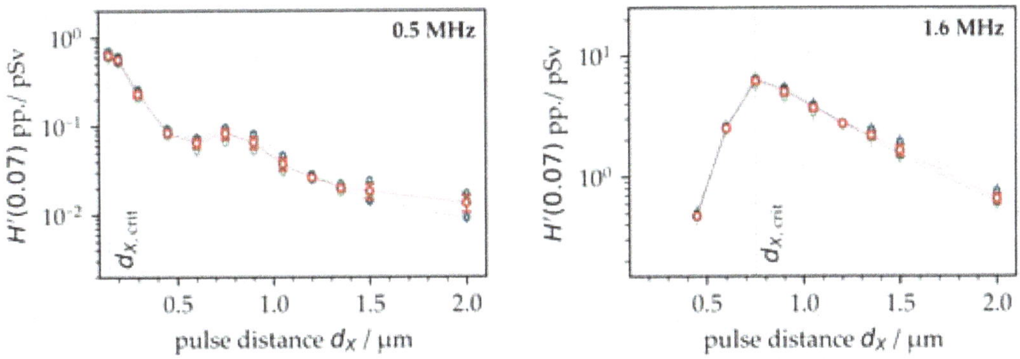

Figure 5. X-ray emission dose per pulse as a function of the intra-line pulse distance at 0.5 MHz (left) and 1.6 MHz (right) PRF obtained on 2 different days under controlled suction flow conditions. The suction flow speed was kept constant at 0.9 m/s ± 0.3 m/s. The mean values are highlighted by the red symbols (circles). The critical intra-line pulse distances dX,crit are indicated.

In **Figure 5**, the critical intra-line pulse distances $d_{X,crit}$ are indicated. These values represent the geometrical pulse distances within a raster-scanned line where the maximum X-ray emission doses were detected. For the low PRF pulses, **Figure 5** (left), the highest X-ray emission was achieved at $d_{X,crit}$ = 0.15 μm, which was similar to the intra-line distance shown in **Figure 1** for the highest X-ray dose level. For high PRF pulses, in contrast, the highest X-ray emission dose was recorded at $d_{X,crit}$ = 0.75 μm, **Figure 5** (right). This intra-line distance was achieved under stable suction flow conditions and was a little smaller than those for uncontrolled laser processing conditions, further underlining the influence of suction flow or, rather, the prevailing dust and fume concentration surrounding the laser processing area on X-ray photon emission.

3.3. X-ray Emission as a Function of PRF and Polarization State

A detailed analysis of the results obtained showed that the critical intra-line pulse distance for the maximum X-ray emission dose was strongly affected by the specifically applied PRF. For example, the critical intra-line distance was $d_{X,crit}$ = 0.2 μm at f_P = 0.5 MHz, enlarging to $d_{X,crit}$ = 0.9 μm at f_P = 1.4 MHz and above, **Figure 6**. Therefore, a potential explanation can be found in the complex interaction of the laser beam with the plasma plume induced by preceding pulses in a pulse train. Moreover, in the particular case of low-intense ultrashort pulses irradiating at MHz PRF, plasma resonance absorption might be the dominant absorption mechanism for the laser beam, which was already evidenced in a previous work [7]. This, in turn, implies optimum plasma characteristics in terms of the degree of ionization, plasma carrier density, dimension and flank angle of the critical plasma density layer, etc., for the most efficient coupling of impinging optical pulse energies with the laser-induced plasma plume. In fact, plasma plume formation and expansion are highly dynamic processes [8]. Therefore, it became evident for the dynamically changing plasma plume that both timing (defined by PRF) and position (defined by intra-line pulse distance) of the next impinging laser pulse had a great effect on optimum laser beam coupling resulting maximum X-ray photon emission. The general trend of increasing X-ray emission dose and enlarging critical intra-line pulse distance with higher PRFs can be seen in **Figure 6** for two different suction flow speeds: v_{fs} = 0.9 m/s and 2.8 m/s. In addition to the X-ray emission dose per pulse, for a better comparison to the literature data, the measured values are also presented in the form of X-ray emission dose rate per irradiated laser power. The higher X-ray dose values and the larger critical intra-line pulse distances were found for the lower suction flow speed, which is in line with the data presented above in **Figure 4.** In this context, it is worth mentioning that the highest X-ray emission doses can be seen in **Figure 6** with about same value of 0.8 mSv/h/W for PRFs ranging between 1.0 MHz and 1.8 MHz. This broad range of MHz PRF for the highest X-ray emissions seems to be contrary to previous results (e.g., **Figure** 5 (right) shows a distinct maximum at 1.6 MHz). The X-ray emission doses of **Figure 6,** however, represent maximum values that were achieved at different intra-line pulse distances.

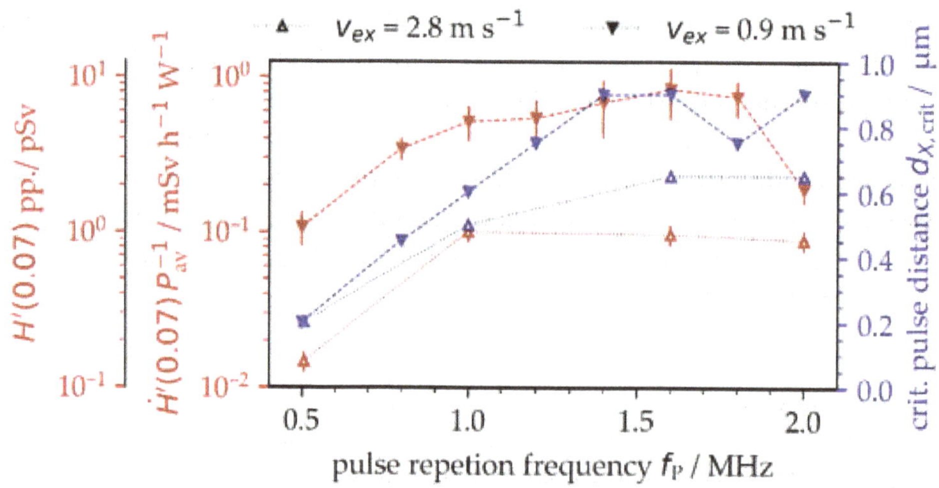

Figure 6. X-ray emission doses per pulse and X-ray emission dose rates related to the applied average laser power as a function of PRF detected at different suction flow speeds. The critical intra-line pulse distances for maximum X-ray emissions are presented.

In addition to PRF and intra-line pulse distance, the polarization state of the laser beam was another enhancing effect of ultrashort pulse laser-induced X-ray emissions, **Figure** 7. This applies in particular to MHz pulses and linear polarization aligned parallel to the beam scanning direction. When comparing with the perpendicularly polarized laser beam, a maximum four-fold increase in X-ray emission was figured out for the f_P = 1.0 MHz repetitive pulses irradiating at a $d_{X,crit}$ = 0.5 µm critical intra-line pulse distance, **Figure** 7 (left, middle). At the higher PRF of 1.6 MHz, the X-ray emission of the parallel-polarized laser beam was about twice the X-ray dose obtained with perpendicular polarization, **Figure** 7 (left, bottom). A difference by the factor of two between the X-ray emission dose levels was also observed for the 0.5 MHz pulses irradiated either parallel or perpendicularly polarized with respect to the laser beam scanning direction, **Figure** 7 (left, top). The maximum X-ray emission dose levels of **Figure** 7, ranging up to $H'(0.07)_{PP,max}$ = 1 pSv, were almost one order of magnitude below the values presented above in **Figure** 5 of $H'(0.07)_{PP,max}$ = 10 pSv. This is mainly due to the fact that different suction flow speeds were applied (v_{fs} = 0.9 m/s in **Figure** 5 vs. 2.8 m/s in **Figure** 7), which further confirms the strong impact of suction flow speed on X-ray photon emission already shown above in **Figure 6.** From these results, it can be postulated that the suction flow speed manipulated the shape of the plasma plume, which, in turn, affected the efficient coupling of the following laser pulses into the plasma under resonant absorption conditions. The proposed influencing effect of laser beam polarization on X-ray emission can also be assessed from the X-ray emission spectra provided in **Figure** 7 (right) showing higher X-ray photon yields and wider X-ray spectra that were further enlarged to higher X-ray photon energies for the parallel-polarized laser beam.

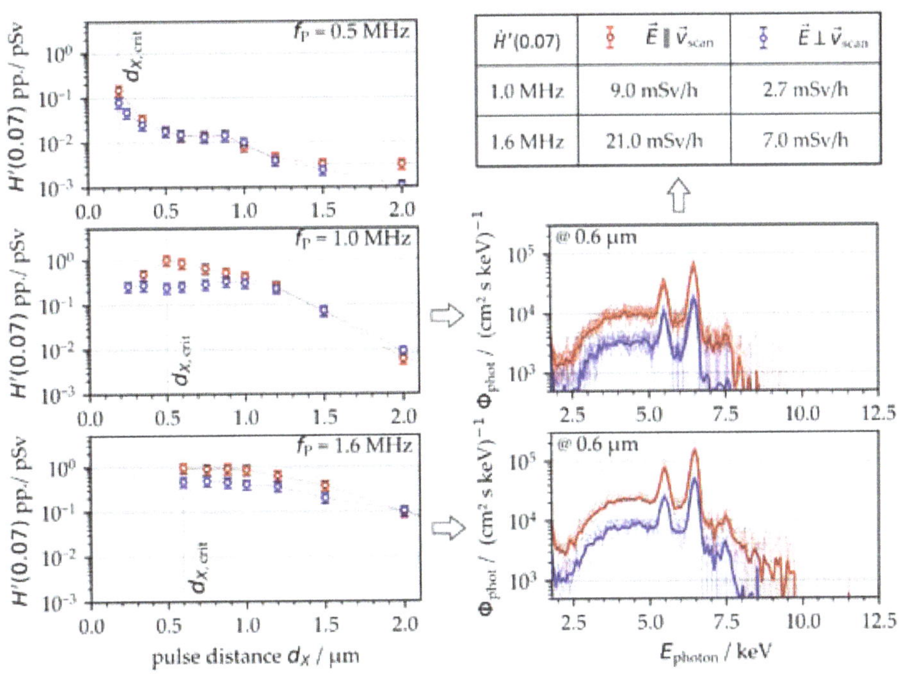

Figure 7. (Left): X-ray emission dose per pulse as a function of pulse distance, PRF, and polarization direction at a 2.8 m/s ± 0.4 m/s suction flow speed; (right): corresponding X-ray photon spectra monitored for 1.0 MHz or 1.6 MHz PRF, 0.6 intra-line pulse distance, 0.9 m/s ± 0.3 m/s suction flow speed, and either parallel (red) or perpendicular (blue) laser beam polarization.

Nonetheless, the significant correlation between X-ray emission spectra and dosage rate with laser beam polarization further suggests that plasma resonance absorption serves as an effective absorption mechanism during the interaction of MHz PRF pulses with well-defined plasma states. The polarization of laser beams has been demonstrated to influence resonance absorption in high-intensity laser-produced plasma. The novelty of our study lies in the demonstrated resonant absorption of subsequent pulses (irradiated with brief time delays at MHz pulse repetition frequency) on the well-defined plasma plume generated by the previous pulses in a pulse train. This was also confirmed by the monitored X-ray photon spectra and corresponding X-ray dose rates increasing disproportionately from $H'(0.07)$ = 2.7 mSv/h to 9.0 mSv/h at 1.0 MHz or, rather, $H'(0.07)$ = 7.0 mSv/h to 21.0 mSv/h at fP = 1.6 MHz when using a parallel- instead of a perpendicular-polarized laser beam.

3.4. Influence of the Peak Intensity on X-ray Emission

In this study, maximum X-ray emission dose rates up to $H'(0.07)$ = 35 mSv/h were produced with pulses of comparably low peak intensity somewhat above 10^{13} W/cm^2. As a rough estimate, the doubling of the peak intensity could induce 10 times higher X-ray emission dose rates, which implied that peak intensity is a significantly influencing variable of X-ray photon emission. Following Equation (1), the peak intensity denotes the maximum optical power related to the laser spot area and can be varied either by the pulse energy, the pulse duration, or the spot size of the irradiating laser beam. For a better understanding of the specific effect of these individual parameters on X-ray emission and further to identify a threshold value for the start of X-ray photon generation, the peak intensity was gradually reduced by (i) increasing the pulse duration between 330 fs $\leq \tau_H \leq$ 10 ps, (ii) reducing the pulse energy by lowering the average laser power to a constant f_P = 0.5 MHz, and (iii) increasing the laser spot area by placing the substrate surface out of the focal plane (defocusing). In the case of defocusing, the effective laser spot radius $w_{86}(z)$ of the laser beam depended on the distance of the substrate surface from the focal plane, hereinafter referenced as defocusing distance z. In order to define the peak intensity of the ultrashort pulses specifically applied in the defocused substrate position, according to Equation (1), the effective laser spot radius was calculated following Equation (2) by taking into account the Rayleigh length of z_R = 0.6 µm and the focus spot radius of $w_{0.86}$ = 15.0 µm

$$w_{86}(z) = w_{0.86}\sqrt{1 + \frac{z^2}{z_R^2}} \quad (2)$$

In Figure 8 (left), the pulse energy showed the greatest effect on the released dose rate when comparing the X-ray emissions from pulses of varying peak intensities.

This was validated by the steepest curve growth (blue, squares) in the $H''(0.07)$ vs. I0 log-log plot. At a fixed peak intensity, the pulses of longer pulse duration induced higher X-ray emission dose levels (green, circles). The results also confirm that pulse duration affected the threshold intensity value, where X-ray photons were detected for the first time. For example, the intensity threshold was figured from I0/UFFL = 1.3 × 1013 W/cm² for 340 fs pulses. In contrast, with the longer 10 ps pulses, the X-ray photon emission started at I0/UFFL = 2.3 × 1012 W/cm². This was about four times below the legal peak intensity limit for approval-free laser operation in Germany. Moreover, the corresponding $H'''(0.07)$ = 10 µSv/h for the applied low-intense 30 W average power laser beam exceeded considerably the permitted 1 µSv/h maximum X-ray emission dose rate.

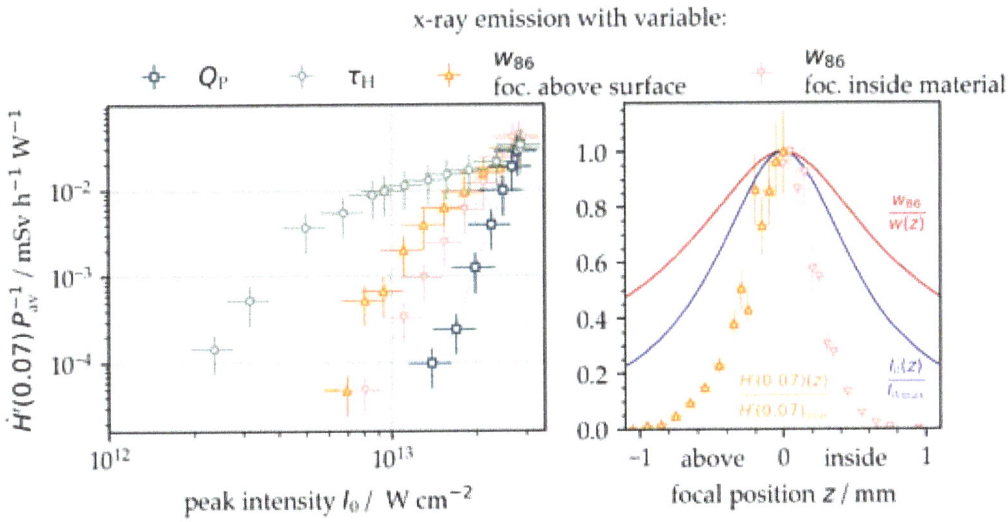

Figure 8. (Left): X-ray emission dose rate as function of pulse energy, pulse duration, and laser spot size; (Right): effect of laser beam defocusing on laser spot radius, peak intensity, and X-ray emission dose rate obtained at 0.5 MHz PRF, 1.0 µm intra-line pulse distance, and 2.8 m/s ± 0.4/s suction flow speed.

The laser beam's spot size at the substrate surface was adjusted using defocusing. For bigger spot sizes with focus points above the laser beam rather than within it, the substrate surface exhibited X-ray emission dose rates up to three times greater. A plausible explanation can be attributed to the heel effect, which involves the screening of X-ray photons produced within the craters on the uneven and elevated surface features. Indeed, positioning the focus within the material results in the outer regions of the laser beam being obstructed by the original surface, so diminishing the effective peak intensity of the laser beam penetrating the deeper sections being treated. Figure 8 (right) illustrates the primary impacts of laser beam defocusing on X-ray emissions, revealing increased laser spot radii, diminished peak intensity, and reduced X-ray emissions as the distance from the laser focus plane to the substrate surface increases.

4. CONCISE SUMMARY

The generation of unwanted X-ray photon radiation during ultrashort pulse laser materials processing is a secondary laser beam danger and can pose significant health hazards to laser operators. The production of detrimental X-ray dose levels is influenced by various laser characteristics and processing conditions. Over 20 significant elements related to the laser, process, and material aspects have been found as substantially impacting the X-ray emission dose, with individual parameters also exerting direct influence on one another. This article identifies peak intensity as the primary variable, while also highlighting the PRF, intra-line pulse distance, polarization direction, and suction flow speed as additional factors that augment laser-induced X-ray emissions. The investigation yielded the following principal findings:X-ray emission dose levels could increase up to ten-fold when ultrashort laser pulses irradiate at MHz PRF and submicrometer intra-line pulse distances:

- A critical intra-line distance depending on PRF existed for maximum X-ray emissions;
- Higher X-ray emissions could be released from smoother surfaces;
- The suction flow conditions in the laser processing area affected X-ray photon emissions, where higher suction flow speeds induced lower X-ray dose levels;
- A laser beam parallel-polarized to the beam moving direction induced higher X-ray emissions, which further confirmed plasma resonance absorption as an efficient laser beam absorption mechanism in a low-intense ultrashort pulse laser regime;
- Harmful X-ray emission dose rates exceeding the legal limit of 1 μSv/h could be generated with peak intensities below 1 × 1013 W/cm, and, finally;
- The peak intensity threshold value for X-ray emissions decreased with larger laser spot sizes and longer pulse durations.

These summing comments offer novel insights into the fundamental mechanics of Xray photon emissions generated by ultrafast laser processing in materials.A comprehensive understanding of the interrelationship among the primary contributing factors is crucial for devising appropriate Xray protection techniques, ensuring safe and riskfree operation of strong ultrafast laser systems in industrial and academic research applications.Moreover, the documented Xray emission dose rates seen in industrially pertinent USPL procedures can serve as a foundation for the appropriate design and sizing of X-ray radiation shielding enclosures.

AT A GLANCE

The current trend in the advancement of potent ultrashort pulse lasers has garnered heightened interest for their application in extensive surface engineering and contemporary microfabrication. Recent reports indicate the generation of undesirable X-ray photon radiation even under industrially pertinent laser irradiation conditions, posing significant health concerns to laser operators. Currently, over twenty influential parameters have been recognized that significantly impact X-ray photon emission generated by ultrashort pulse laser operations. The study on augmented X-ray emission resulting from high pulse repetition frequency ultrashort pulse laser processing offers novel insights into the interconnection of the most significant parameters. The example of AISI 304 substrates demonstrates that X-ray photon production can significantly surpass the permissible dosage rate limit when ultrashort laser pulses with peak intensities below 1×10^{13} W/cm^2 are irradiated at a 0.5 MHz pulse repetition frequency. The threshold value for peak intensity of X-ray emissions diminishes with increased laser spot diameters and extended pulse durations. A significant finding of this study is that the suction flow conditions in the laser processing region can influence the emitted X-ray dosage rate. The results obtained endorse the formulation of efficient X-ray protection measures for secure and hazard-free ultrashort pulse laser use in both industrial and academic research contexts.

REFERENCES

1. Schille, J.; Löschner, U. Ultrashort pulse lasers in high-rate laser micro processing—Quo vadis? *Adv. Opt. Technol.* **2021**, *10*, 233–237. **[Google Scholar] [CrossRef]**

2. Mosel, P.; Sankar, P.; Düsing, J.F.; Dittmar, G.; Püster, T.; Jäschke, P.; Vahlbruch, J.-W.; Morgner, U.; Kovacev, M. X-ray Dose Rate and Spectral Measurements during Ultrafast Laser Machining Using a Calibrated (High-Sensitivity) Novel X-ray Detector. *Materials* **2021**, *14*, 4397. **[Google Scholar] [CrossRef] [PubMed]**

3. Weber, R.; Giedl-Wagner, R.; Förster, D.J.; Pauli, A.; Graf, T.; Balmer, J.E. Expected X-ray dose rates resulting from industrial ultrafast laser applications. *Appl. Phys. A* **2019**, *125*, 635. **[Google Scholar]**

4. Raciukaitis, G.; Brikas, M.; Gečys, P.; Voisiat, B.; Gedvilas, M. Use of high repetition rate and high power lasers in microfabrication: How to keep the efficiency high? *J. Laser Micro Nanoeng.* **2008**, *4*, 186–191. **[Google Scholar] [CrossRef]**

5. Neuenschwander, B.; Bucher, G.F.; Nussbaum, C.; Joss, B.; Muralt, M.; Hunziker, U.W.; Schuetz, P. Processing of metals and dielectric materials with ps laser pulses: Results, strategies, limitations and needs. In Proceedings of the Laser Applications in Microelectronic and Optoelectronic Manufacturing XV, San Francisco, CA, USA, 23–28 January 2010. **[Google Scholar]**

6. Schille, J.; Schneider, L.; Löschner, U. Process optimization in high-average-power ultrashort pulse laser microfabrication: How laser process parameters influence efficiency, throughput and quality. *Appl. Phys. A* **2015**, *120*, 847–855. **[Google Scholar] [CrossRef]**

7. Schille, J.; Kraft, S.; Pflug, T.; Scholz, C.; Clair, M.; Horn, A.; Löschner, U. Study on X-ray Emission Using Ultrashort Pulsed Lasers in Material Processing. *Materials* **2021**, *14*, 4537. **[Google Scholar] [CrossRef] [PubMed]**

8. Kraft, S.; Schille, J.; Mauersberger, S.; Schneider, L.; Löschner, U. Pump-probe imaging for process control and optimization in high-speed laser micro machining. In Proceedings of the Laser-based Micro-and Nanoprocessing XIV, San Francisco, CA, USA, 1–6 February 2020. **[Google Scholar]**
9. Ballmer, J.E.; Donaldson, T.P. Resonance Absorption of 1.06-μm Laser Radiation in Laser-Generated Plasma. *Phys. Rev. Lett.* **1977**, *39*, 17. **[Google Scholar] [CrossRef]**
10. Teubner, U.; Bergmann, J.; van Wonterghem, B.; Schaefer, F.P.; Sauerbrey, R. Angle-dependent x-ray emission and resonance absorption in a laser-produced plasma generated by a high intensity ultrashort pulse. *Phys. Rev. Lett.* **1993**, *70*, 794. **[Google Scholar] [CrossRef] [PubMed]**
11. Legall, H.; Schwanke, C.; Bonse, J.; Krüger, J. The influence of processing parameters on X-ray emission during ultra-short pulse laser machining. *Appl. Phys. A* **2019**, *125*, 570. **[Google Scholar]**
12. Legall, H.; Schwanke, C.; Pentzien, S.; Dittmar, G.; Bonse, J.; Krüger, J. X-ray emission as a potential hazard during ultrashort pulse laser material processing. *Appl. Phys. A* **2018**, *124*, 407. **[Google Scholar] [CrossRef] [Green Version]**
13. Freitag, C.; Giedl-Wagner, R. X-Ray Protection in an Industrial Production Environment. *Photonics Views* **2019**, *17*, 37. **[Google Scholar] [CrossRef]**
14. Legall, H.; Bonse, J.; Krüger, J. Review of x-ray exposure and safety issues arising from ultra-short pulse laser material processing. *J. Radiol. Prot.* **2021**, *41*, R28. **[Google Scholar] [CrossRef] [PubMed]**
15. Chichkov, B.N.; Momma, C.; Tünnermann, A.; Meyer, S.; Menzel, T.; Wellegehausen, B. Hard-x-ray radiation from short-pulse laser-produced plasmas. *Appl. Phys. Lett.* **1996**, *68*, 2804. **[Google Scholar] [CrossRef]**
16. Kattan, D. Untersuchung der Röntgenemissionen durch laserinduzierte Plasmen in der Materialbearbeitung mit Ultrakurzpulslasern an Luft (written in German). Master's Thesis, Heinrich Heine University, Düsseldorf, Germany, 18 August 2021. **[Google Scholar]**
17. Metzner, D.; Olbrich, M.; Lickschat, P.; Horn, A.; Weißmantel, S. X-ray generation by laser ablation using MHz to GHz pulse bursts. *J. Laser Appl.* **2021**, *33*, 032014. **[Google Scholar] [CrossRef]**
18. Dittmar, G. Arbeitsblatt 2: Zunahme der laserinduzierten ionisierenden Strahlung (LIS) bei größer werdender Fokusfläche trotz gleicher Bestrahlungsstärke. Available online: **https://www.researchgate.net/publication/339537884**
19. Sauerbrey, R.; Fure, J.; Le Blanc, S.P. Reflectivity of laser-produced plasmas generated by a high intensity ultrashort pulse. *Phys. Plasmas* **1994**, *1*, 5. **[Google Scholar] [CrossRef]**
20. Holland, J.; Weber, R.; Sailer, M.; Graf, T. Influence of Pulse Duration on X-ray Emission during Industrial Ultrafast Laser Processing. *Materials* **2022**, *15*, 2257. **[Google Scholar] [CrossRef] [PubMed]**
21. Cerchez, M.; Jung, R.; Osterholz, J.; Toncian, T.; Willi, O.; Mulser, P.; Ruhl, H. Absorption of ultrashort laser pulses in strongly overdense targets. *Phys. Rev. Lett.* **2008**, *100*, 245001. **[Google Scholar] [CrossRef] [PubMed] [Green Version]**
22. German Federal Office for Radiation Protection. Limit Values in Radiation Protection. Available online: **https://www.bfs.de/EN/topics/ion/radiation-protection/limit-values/limit-values_node.html** (accessed on 1 February 2022).
23. Schille, J.; Schneider, L.; Mauersberger, S.; Szokup, S.; Höhn, S.; Pötschke, J.; Reiss, F.; Leidich, E.; Löschner, U. High-Rate Laser Surface Texturing for Advanced Tribological Functionality. *Lubricants* **2020**, *8*, 33. **[Google Scholar] [CrossRef] [Green Version]**

CHAPTER 6

HAZARDS OF X-RAY EMISSION FROM ULTRASHORT PULSED LASER MATERIAL PROCESSING IN AN INDUSTRIAL ENVIRONMENT

1. INTRODUCTION: AN OVERVIEW

Utilizing ultrashort laser pulses in air for material processing offers numerous advantages, including lateral and vertical precision of surface contours at the nanoscale scale and great reproducibility of laser-generated structures [1].Owing to advancements in the laser industry, average powers in the kilowatt range with pulse repetition rates surpassing the megahertz threshold are presently accessible [2].Machining utilizing highintensity laser pulses can produce a nearsurface electron plasma through material absorption and ionization, followed by plasma heating from the laser pulse, ultimately resulting in interactions between "hot" plasma electrons and the material, which generate continuous and characteristic X-ray emissions.

The fact that ultrashort sub-ps pulsed laser material interaction can lead to X-ray radiation has been well established since the late 1980s [3,4,5,6,7,8]. Twenty years ago, X-ray emission was reported as an unwanted secondary effect during femtosecond laser micromachining of copper in air using kHz repetition rates [9,10]. The authors registered X-ray dose rates requiring radiation protection measures. The amount of this X-ray radiation is determined by the laser parameters (pulse duration, peak intensity, pulse energy, wavelength, and polarization), the workpiece (atomic number and surface preparation), and the laser process management (scanning or stationary regime, laser turning, etc.). The use of laser peak intensities above 1013 W/cm2 in combination with laser pulse repetition rates in the few 100 kHz range can already lead to X-ray dose rates clearly exceeding the permitted limits for members of the public. Tungsten and steel exhibit notable X-ray emission[11,12,13,14,15,16,17,18,19]. In [11], it was shown that materials like aluminum and glass have much reduced X-ray emissions. The measured radiation rate for an aluminum target was roughly two orders of magnitude inferior to the dose rates of steel and tungsten. An increase in the measured dosage rate with an ascending atomic number Z was noticed as a general tendency at a constant intensity. Under typical ultrashort pulse laser machining conditions in the intensity range of 1013 to 1015 W/cm2, collision-less resonance absorption appears to be the driving X-ray generation mechanism [17].

Recent experimental evidence demonstrates that substantially elevated X-ray dose rates can be produced during USP-laser machining by employing a laser burst mode rather of a single-pulse mode, while maintaining the same total laser intensity. The enhancement in X-ray production was ascribed to the interaction between ultrafast laser light and an ablation cloud or high-density plasma [19,20]. This particular regime was not examined in the context of this article. Given the potential production of ionizing radiation during material processing with a USP-laser machine, it is imperative to adopt radiation monitoring and protective precautions for personnel using these devices. One such measure involves dosimetry assessments conducted outside the protective enclosure or structural radiation shielding during the operational authorization process for USP-laser devices. To guarantee conservativeness during the radiation protection examination, the material processing scenario with the most radiation exposure and the highest photon energies of laser-induced X-ray radiation must generally be utilized. Nonetheless, the extensive array of pertinent laser and processing parameters renders the identification and development of a worst-case scenario laborious and impossible during routine safety inspections. However, to provide uniform and reproducible inspection outcomes, a reference condition must be established. In the context of USP-laser machines, it is particularly crucial that laser-plasma interactions, along with the resultant energy spectra and intensities of X-ray emission, are significantly influenced by several operational parameters, including laser specifications, the manufacturing process and its geometry, as well as the previously referenced laser-irradiated material. The reference case utilized herein pertains to the laser surface processing of a metallic specimen, specifically steel or tungsten. A bidirectional arrangement is employed for the measurements. In this procedure, grooves of a certain length (e.g., 15 mm) are machined into the target material. Upon completion of one groove, the material is displaced by one focus diameter perpendicularly to the groove, and a subsequent groove is created in the opposite direction. This procedure is reiterated until a surface scan region of, for instance, 15×15 mm^2 has been encompassed. This entails the ablation of a singular "process layer". To extend the period of laser processing, one may either enlarge the dimensions of an individual layer or vertically superimpose numerous layers. Nonetheless, as evidenced in [13], the processing of several ablated layers might alter X-ray emission properties due to the altered surface topography following the machining of each layer. This reference material processing scenario benefits from a well-defined radiation field direction.

The maximum dose rates of laser-induced radiation are recorded in the direction opposite to the laser's movement, namely parallel to the grooves created by laser ablation, due to intrinsic absorption in the target material [11,13,14]. A reference scenario utilized in radiation protection inspections must be conservative regarding the ambient dose rate of X-ray emissions and the maximum photon energy of the X-ray spectrum. This guarantees that the transmission of ionizing radiation is not undervalued during the radiation protection assessment of the USP-laser machine's protective enclosure. The literature [11] indicates that the laser peak intensity is a critical parameter for producing laser-induced X-ray radiation with substantial ambient dose rates. Augmenting the laser peak intensity by the enhancement of laser pulse energy would result in an elevation of the ambient dose rate of the produced laser-induced X-ray radiation. Previous investigations have demonstrated that the radiated dose rate is significantly influenced by the individual material of the workpiece [11,14]. The maximum dosage rates have been recorded when utilizing either steel or tungsten as targets. Consequently, it can be inferred that, during radiation protection evaluations of a USP-laser apparatus, the optimal strategy would involve utilizing either of these two materials and maximizing the available laser pulse energy for a specified pulse time. An additional crucial criterion of the reference scenario is a temporally stable X-ray emission with a nearly constant dose rate during the radiation protection examination. Furthermore, the geometry of the complete radiation field must be thoroughly understood and possess a substantial area of uniform X-ray emission. Both criteria guarantee dependable radiation protection assessments throughout the entire region of the protective enclosure.

The criteria for assessing radiation protection inspection methods for USP-laser equipment in authentic industrial environments were satisfied at SCHOTT AG. A USP laser with a pulse duration of 1 ps, a wavelength of 1030 nm, and an average power of 220 W, encased in an aluminum protective casing, was utilized for the tests. The theoretically achievable peak intensity of 3.3 × 1015 W/cm2 is comparatively high and normally used for the processing of transparent glasses making use of nonlinear absorption mechanisms. So far, only Schille et al. have employed higher laser peak intensities of up to 5.2 × 1016 W/cm2 in a recent study [19]. The aluminum protection housing at the USP-laser machine at SCHOTT AG was an interesting test subject as the protective effect is expected to be much smaller than for steel enclosures. Finally, the machine incorporates several noteworthy safety features, which will be elaborated upon thereafter. The subsequent section summarizes the experimental setup, difficulties encountered during measurements, and the results obtained in an industrial context at SCHOTT AG using two distinct experimental configurations. The limits on the positioning of the X-ray measurement apparatus, dictated by safety measures within the protective enclosure, significantly affected the measurements. Due to these difficulties, and despite experience with laser-induced X-ray radiation emission, the dose rates recorded at SCHOTT AG are inferior to those measured by other groups at comparable laser peak intensities.

2. MATERIALS AND METHODS

2.1. USP-Laser Machine

The measurements were conducted using a walk-in microSHAPETM machine (3D-Micromac, Chemnitz, Germany), equipped with an XHE 200 W laser system (Amphos, Aachen, Germany). Figure 1 displays an image of the system's innards.

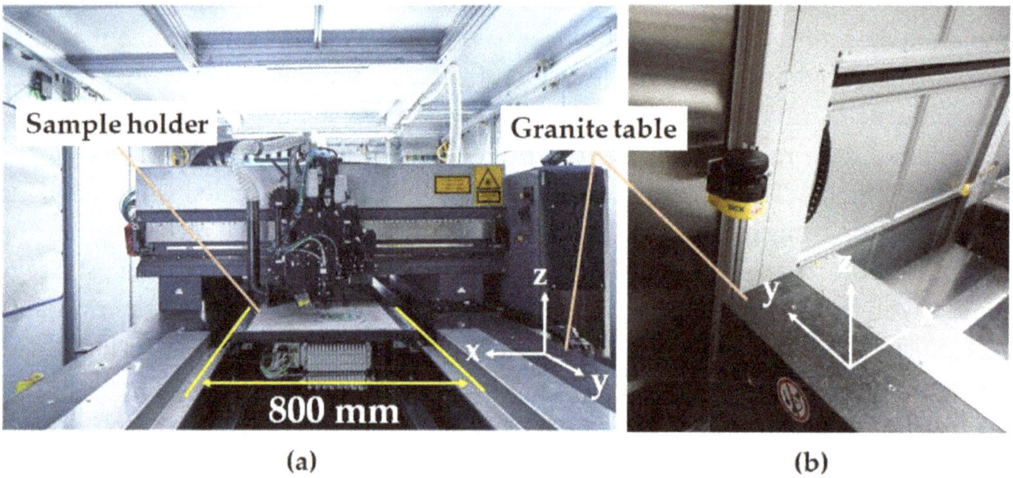

Figure 1. (a) Interior of the microSHAPETM USP laser system at SCHOTT AG. The system is accessible and equipped with several safety systems that are intended to prevent staff from entering the machine during laser operation. One of these safety systems is the Sick laser scanner device (b) installed in the wall in front of the operating machine system. The monitored area covers the region between 10 cm and 30 cm above the granite table.

The laser was used with fixed optics (no laser scanner was employed) that focused the laser beam, using standard processing laser parameters, onto the workpiece, creating a spot diameter of about 10 μm. The focus diameter was calculated by the control system of the USP-laser machine from manufacturer information of the laser beam diameter and the different focal lengths of lenses in the focusing optics system. Since the movement of the laser spot and the optics (x- and z-direction) on the workpiece had to be carried out in relation to the sample holder station (y-direction) by translating the sample, the feed speed rate had to be set considerably low at 100 mm/s. The sample holder has a Y-stage assembly and provides a focal plane with a working area of up to 0.5 m2 (see Figure 1a).

The industrial USP-laser machine utilized in this study was configured as a walk-in model, incorporating multiple safety features to prevent unauthorized access during laser operation. Figure 1b illustrates a laser safety scanner manufactured by Sick (Düsseldorf, Germany) [21] positioned at the entrance of the microSHAPETM USP-laser apparatus. This gadget employs the time-of-flight principle (ToF) to detect objects and individuals within the room, which was very pertinent for establishing the positioning of the X-ray measurement instruments for this study.

2.2. Sample Workpiece and Processing Parameters

A surface process on a tungsten plate was utilized as a reference situation, as previously stated. The machined regions were 15 × 15 mm^2 squares (Figure 2a), subjected to laser beam scanning in a bidirectional manner (Figure 2b) n times on the tungsten plate (Figure 2c), without altering the distance between the optics and the sample surface during the measurements. The duration of the laser pulse was 1 picosecond. Three distinct configurations of laser pulse energy and repetition rate were evaluated. The configurations, along with the computed peak intensity and laser spot distance (in the y-direction), are consolidated in Table 1. Table 1 additionally includes the measurement parameters obtained from two other experimental setups utilizing an HR50 laser (Coherent, Santa Clara, CA, USA) and a GL.evo (GFH GmbH, Deggendorf, Germany) for comparative analysis. The separation between two contiguous lines equated to one focus diameter in the x-direction. Under these configurations, two contiguous processing locations were situated around 0.250 μm to 2 μm apart along a linear axis. The significant overlap between two contiguous machining areas resulted in the formation of deep holes on the material surface, leading to substantial material removal with each laser scan.

(a)

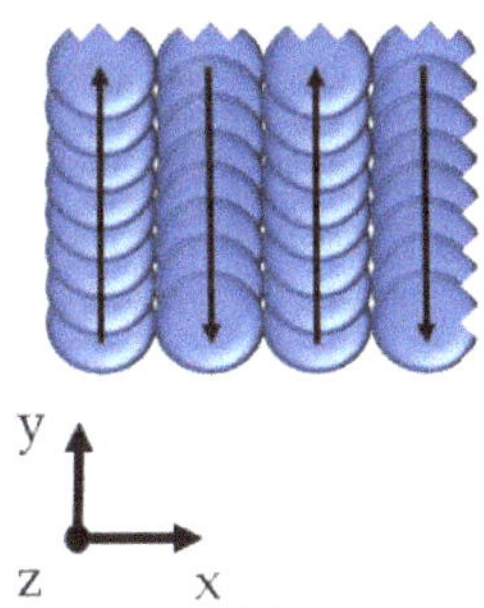

(b)

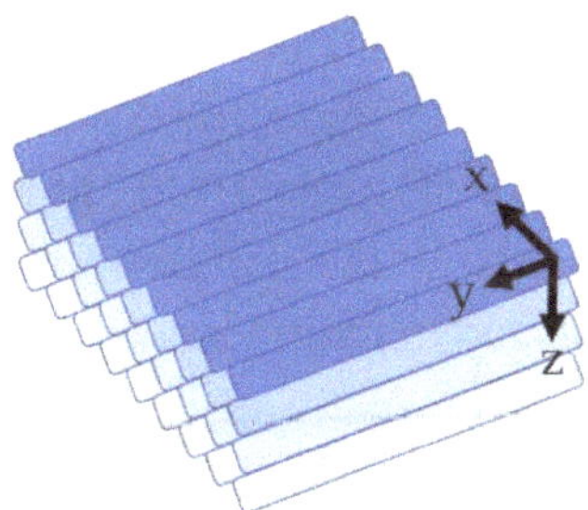

(c)

Figure 2. (a) The 15 × 15 mm2 scan areas after USP-laser processing with the microSHAPETM machine at peak intensities of 1 × 1015 W/cm2 and above (see Table 1). The visibly different depths of the processed areas are caused by different numbers of processing steps and variations of the laser pulse energy. The square in the top left corner was processed at PTB with a much smaller peak intensity of 3.0 × 1014 W/cm2. (b) Schematic of the 2D bidirectional laser spot movement with respect to the coordinate system as given in Figure 1. In (c), multiple processing layers in the z-direction are indicated.

Table 1. Summary of laser processing parameters and corresponding measured dose rates for different USP-laser machines and their respective studied configurations. Pulse durations were 1 ps and 0.274 ps, and laser focus spot diameters were 12 μm and

33 μm (both manufacturer information) for the setups operated at SCHOTT AG and at PTB, respectively. The wavelengths of all lasers listed here were approximately 1 μm. The maximum ambient dose rate measurements at SCHOTT AG were conducted at a distance of 475 mm from the laser processing location. The maximum ambient dose rate at the GL.evo machine of approximately 2.1 mSv/h was measured at a distance of 160 mm during a laser turning process. The uncertainties listed for the dose rate values were calculated via error propagation of the calibration factor of the dosemeter.

USP-Laser Machine		Repetition Rate (kHz)	Pulse Energy (μJ)	Distance of Laser Spots (μm)	Peak Intensity (W/cm²)	Max. $dH^*(10)/dt$ (μSv/h)
Model	**Operator**					
3D-Micromac microSHAPE™	SCHOTT AG	400	550	0.25	9.1×10^{14}	128 ± 8
		200	550	0.5	9.1×10^{14}	24 ± 2
		50	1980	2	3.3×10^{15}	0.90 ± 0.06
Second setup, Coherent HR50 laser	SCHOTT AG	200	175	5	1.0×10^{13}	Natural background radiation
GFH GmbH GL.evo	PTB	50	371	10	3.0×10^{14}	Up to 2.1×10^{3}

In the first processes, to carefully test how the material behaves and not to damage the optics, only a fraction of the power was used. The processing areas produced by the microSHAPETM USP-laser machine were visibly deeper than those of the GL.evo USP-laser-machine (Figure 2a, top left) due to the high average laser output power of 220 W of microSHAPETM USP-laser machine, the large available peak intensities, and the large overlap between adjacent laser spots at a scan velocity of 100 mm/s. The GL.evo machine located at PTB uses a laser with a much smaller average power and peak intensity. For example, during one measurement with the microSHAPETM USP-laser machine, the overall removed material thickness of tungsten was approximately 500 μm (Figure 2a, bottom right) after four processing layer repetitions (z-direction in Figure 2c) in comparison to only a few μm after an equal number of repetitions at the GL.evo machine.

Due to the significantly higher removal rate in the reference scenario compared to the rates typically employed by SCHOTT AG for glass processes utilizing this machine, the exhaust system required temporary reinforcement during the measurements by repositioning the suction aperture nearer to the processing location. This was achieved through the incorporation of an additional vacuum pipe and the calibration of the machine's suction system to position it as near as feasible to the plasma zone. After each processing stage, which involved two to four repeats of the ablation layer, it was necessary to clean the lens of the laser focusing optics to eliminate smoke residues that could impair focusing quality and diminish the laser energy directed at the target. Figure 3 presents a detailed depiction of the experimental setup. In order to assess the direction and magnitude of the X-ray radiation field, a radiographic imaging plate from a computer radiography scanner was affixed to the interior of the USP-laser system in proximity to the processing site, as illustrated in Figure 3b.

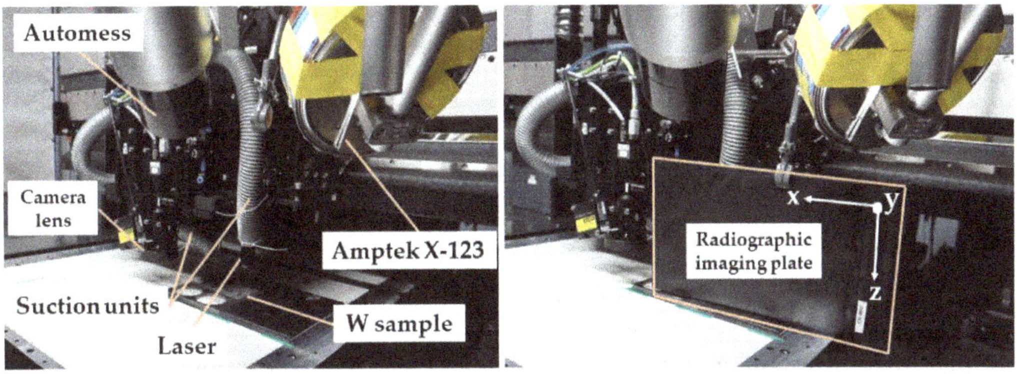

(a) (b)

Figure 3. (a) Experimental setup and positioning of the measuring devices in the interior of the USP-laser machine. The Amptek X-123 spectrometer was placed in a tinplate box with an opening in front of the window and additionally covered with seven layers of 13 µm thick aluminum foil for shielding to prevent electromagnetic compatibility influences. (b) Placement of the radiographic imaging plate to detect the direction and homogeneity of the X-rays emission during USP-laser processing.

2.3. Placement of the Measurement Equipment

As mentioned in Section 2.1, the existence of a coupled safety scanner monitoring system in the investigated microSHAPETM USP-laser machine made the placement of the measuring instruments challenging. The area monitored by the laser safety scanner encompassed a region with a height of about 10 cm to 30 cm above the granite table and reaching close to the laser head in y-direction of the machine. Thus, if any object is detected by the monitoring laser scanner in this region (from now on called the "safety monitored area"), either the USP-laser itself would not start, or it would immediately be switched off as an emergency case during operation. Due to these restrictions, the standard constructions of PTB, which normally include the usage of laboratory tripods or mobile lifting bases, could not be applied here. Instead, an aluminum beam was fixed to the top of the USP-laser machine creating a gantry to hold the X-ray spectrometer and the dosemeter in a way to avoid the safety monitored area of the Sick laser safety scanner. Furthermore, because of the safety monitored area and the shielding by optical components, both devices could only be set up in approximately 45° to 60° angles to the direction of the movement of the laser relative to the sample workpiece, as shown in Figure 3. The distance from the dosemeter to the focus point of the USP-laser was 475 mm, whereas the distance from the spectrometer to the focus point was 385 mm.

2.4. Instrumentation

Table 2 gives an overview of the instruments used in this study to quantify laser-induced X-ray emissions while operating the USP-laser machines.

Table 2. Overview and description of the measurement instruments. The optimal energy range of the Automess 6150 dosemeter lies above 20 keV. However, it is calibrated in an ISO radiation quality N-15 reference field at PTB.

Instrument	Dosemeter	Spectrometer	Radiography Scanner
Model	Automess 6150AD-b/E	Amptek X-123 CdTe	DÜRR CR 35 NDT
Optimal energy range	20 keV to 7 MeV	8 keV to 70 keV	Not specified
Measured property	Ambient dose rate dH*(10)/dt	X-ray pulse height spectrum	X-ray intensity

For measuring the ambient dose rate during USP-laser processing, the dose rate meter 6150AD-b/E from Automess—Automation and Messtechnik GmbH (Ladenburg, Germany) [22] was used. The dosemeters employed were calibrated in reference X-ray fields at PTB. As the lowest point for the energy calibration, a radiation quality of the narrow ISO X-ray series N-15 [23] with a mean energy of approximately 12.4 keV was used [24].

For measuring the laser-induced X-ray energy spectrum using the tungsten plate as a workpiece, an X-123 CdTe spectrometer from Amptek Inc. (Bedford, MA, USA) [25] was employed. It was placed in a tinplate box and additionally covered with seven layers of 13 μm thick aluminum foil to assure electromagnetic compatibility (EMC) and prevent interference through unwanted electrical or electromagnetic effects in the readout electronics. The transmission as a function of photon energy for 90 μm of aluminum is shown later. The Amptek X-123 detector is based on a CdTe crystal with a thickness of 1 mm used as an X-ray detector [26]. The interaction probability (intrinsic efficiency) as a function of the photon energy for different Amptek detector systems is depicted in **Figure 4**. The dashed violet curve corresponds to the device used for the measurements at SCHOTT AG. As can be seen, it encompasses the energy region of interest for radiation protection on USP-laser machines up to 70 keV. For low energies, the efficiency is limited by a 100 μm Beryllium window, which is not included in **Figure 4**, and the additional 90 μm of aluminum in front of the device. Due to absorption in the window material and the aluminum, a slightly decreased efficiency of the detector can be expected for energies lower than 8 keV.

For monitoring the homogeneity and the location of higher concentrations of emitted X-rays during USP-laser machining, a 2D Radiography Scanner CR 35 NDT Plus system from DÜRR (Stuttgart, Germany) [27] was used. As shown in **Figure 3b**, the radiographic plate could be placed adjacent to the tungsten sample, around 10 cm away from the laser-induced plasma and just outside the safety monitored area.

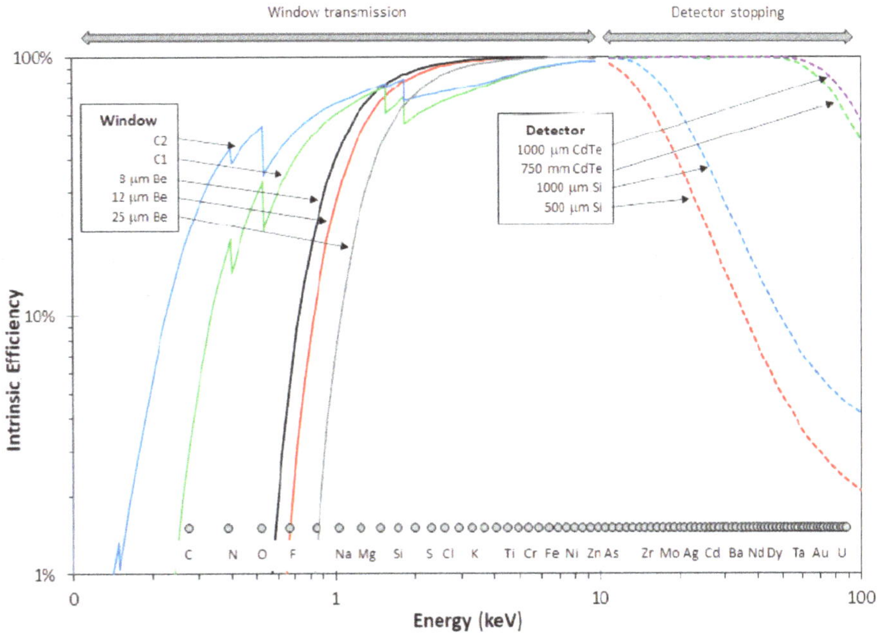

Figure 4. Comparison of the intrinsic efficiency of the X-123 detectors from Amptek. At low energies, the Be window limits sensitivity, and the 1 mm thick CdTe crystal (dashed violet curve) has a good efficiency up to 70 keV. © Amptek Inc 2021. Reprinted with permission.

3. RESULTS

3.1. Monitoring the Generation of X-ray Emissions

The recorded grayscale image of the radiography image plate is depicted in Figure 5a. The area of each pixel of the grayscale image is approximately 200 × 200 μm2. As the radiographic imaging plate was located very close to the source of the laser-induced X-ray emission (see Figure 3b), the grayscale image pixels had to be corrected for different distances from the plasma (see Figure 5b).

The shortest distance from the plasma was 10 cm, and the position of this point of shortest distance on the imaging plate is marked with a red dot on the lower left side of the grayscale image in **Figure 5b**. As the grayscale values are proportional to the local intensity of the X-ray emission and the intensity of a point source decreases with the square of the distance from the source, the distance-corrected grayscale values can be calculated as follows:

$$G(r_0) = G(r)\, r^2/r_0^2, \qquad (1)$$

where $G(r_0)$ is the corrected grayscale value at the reference distance r_0 (in this case, the shortest distance of 10 cm), and $G(r)$ is the measured grayscale value at a distance r. The distance r of each grayscale value can be calculated from the row and column position (u, v) on the imaging plate, r_0 and the row and column values (u_0, v_0) of the reference distance on the grayscale image (in this case $(u_0, v_0) = (371, 1021)$), and the pixel pitch p.

$$r^2 = r_0^{\,2} + p^2 (u - u_0)^2 + p^2 (v - v_0)^2. \tag{2}$$

Further corrections due to additional absorption in air for larger distances from the plasma and a possible angle dependency of the radiographic imaging plate have not been applied. The area with the highest X-ray emissions, characterized by bright gray pixels, is located close to the lower edge of the imaging plate. Accordingly, the X-ray radiation is emitted in a very flat angle to the horizontal plane. In addition, as expected, the emission also appears to be largely parallel to the line of movement of the USP-laser (y-direction).

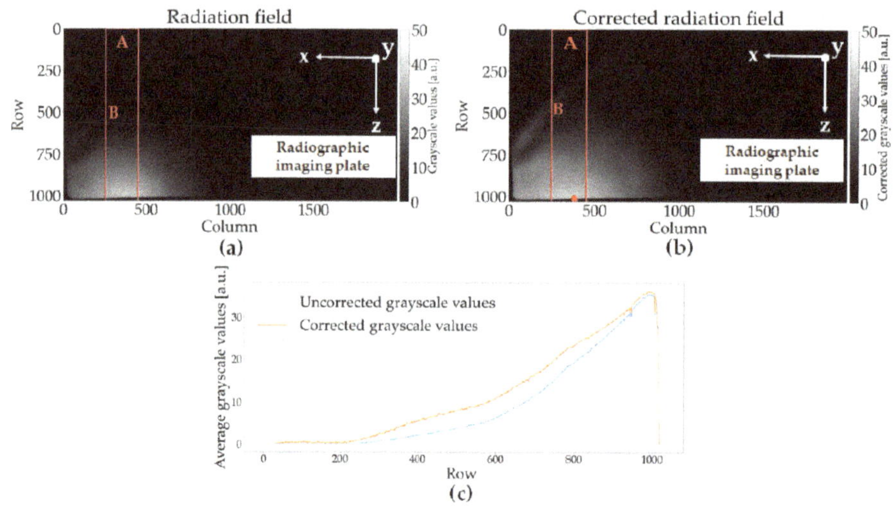

Figure 5. (a) Uncorrected and (b) distance-corrected grayscale image captured with the radiographic imaging plate. Both images show that a large part of the X-ray radiation is emitted very flat and almost exclusively parallel to the sample plate in the *y*-direction. The edge of the imaging plate can be seen very well as a sudden drop in the grayscale values in the lower and left part of the image. The red rectangles in (a,b) show the area of the pixels used for the profile curve plotted in (c). The red dot in (b) marks the position of the shortest distance to the plasma used for the calculation of the distance correction. (c) Cross-section of the grayscale values along the rows of the corrected (orange curve) and uncorrected (blue curve) grayscale images. The continuous decrease in grayscale value with a decreasing row number in the profiles indicates that a large part of the laser-induced X-rays is emitted in a very flat angle close to the horizontal plane.

The edge of the radiographic imaging plate is located in the lower and left part of the grayscale image. The different areas of the radiation field can also be seen in the comparison between the distance-corrected and uncorrected *z*-profile (the image pixels employed to generate the profile are indicated by the red rectangle in **Figure** 5a,b) of the grayscale values depicted in **Figure** 5c. From the measurements performed with the radiography imaging plate, one can conclude that the placement of the dosemeter and the spectrometer devices was not optimal regarding the maximum emission of the X-ray radiation, i.e., they were not directly installed in the maximum region but at the edge of the radiation field. However, as discussed in **Section 2.3**, this was the best arrangement possible for this USP-laser machine.

3.2. Obtained Energy Spectra and Dose Rates

Figure 6 depicts pulse height spectra from laser-induced X-ray emissions with and without a 50 μm thick copper filter in addition to the default aluminum filter with an overall thickness of approximately 90 μm. The spectra were measured with a laser pulse energy of 550 μJ, a pulse duration of 1 ps, and a repetition rate of 400 kHz. The mean X-ray photon energy values obtained during the USP-laser processing with and without a 50 μm thick copper foil were approximately 8.1 keV and 8.9 keV, respectively.

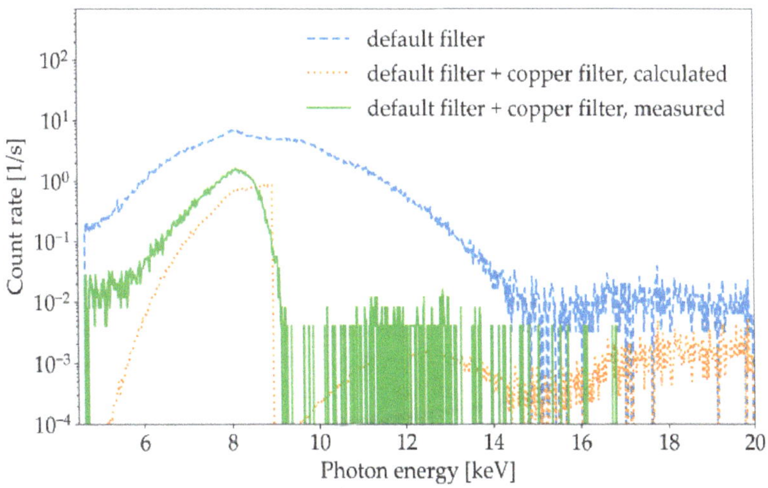

Figure 6. Pulse height spectra measured with the Amptek X-123 CdTe spectrometer during tungsten flat surface processing using a laser pulse energy of 550 μJ, a repetition rate of 400 kHz, a pulse duration of 1 ps, and a focus diameter of 10 μm. Shown are the measured spectra without any additional copper filter (blue curve) and with a copper filter (green curve), as well as the calculated spectrum based on the convolution of the spectrum without copper filter with the transmission function of 50 μm copper (orange curve).

Moreover, for comparison, the expected spectrum after transition of a 50 μm thick copper filter was calculated. This calculated spectrum (orange curve in Figure 6) is based on the measured spectrum without a copper filter (blue curve) multiplied by a transmission function T(E, d) in each energy channel. The function T(E, d) describes the exponential attenuation of photons traversing a material [28] and is given by

$$T(E, d) = e{-}(\mu(E)/\rho) \times \rho d, \quad (3)$$

where d and ρ are the thickness and density of the material (in this case 50 µm of copper foil), respectively, and µ(E)/ρ is the energy-dependent mass attenuation coefficient of the material. For the calculation, the list of mass attenuation coefficients from the National Institute of Standards and Technology was used [28]. These tabulated values were logarithmically interpolated to obtain a corresponding value of the mass attenuation coefficient for each energy channel of the Amptek X-123 spectrometer. The T(E, d) for 50 µm of copper and 90 µm of aluminum as a function of the photon energy, together with the curve for no attenuation, i.e., T(E, d) = 1, is depicted in Figure 7.

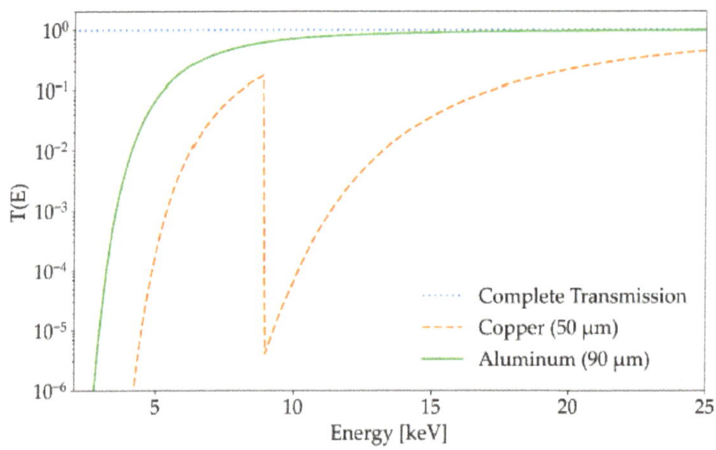

Figure 7. Transmission function T(E, d) (see Equation (1)) for 50 µm of copper (orange dotted curve) and 90 µm of aluminum (solid green curve) as a function of the energy. For the calculation, the list of mass attenuation coefficients from the National Institute of Standards and Technology was used. The values were logarithmically interpolated to get a value for each energy channel of the X-123 spectrometer. The rapid decrease at approximately 9 keV is caused by the sudden increase in the attenuation coefficient of copper due to the Cu-K edge.

As seen in **Figure 6**, the pulse height spectrum determined using the transmission function agrees very well with the pulse height spectrum measured with a 50 µm thick copper filter. The sharp drop in the spectrum at around 9 keV, which is caused by the absorption edge of the K-shell of copper, is clearly visible. The deviation for higher energies exceeding 15 keV is most likely caused by background noise and the pileup effect. This is also supported by the fact that the pulse height spectrum with an additional copper filter has no entries in this energy range. The pileup effect arises if multiple X-ray photons hit the single photon detector in the timeframe of the processing time and are registered as one single photon with correspondingly higher energy; this can be minimized by positioning the spectrometers at a large distance to the X-ray radiation source. As verified by the good agreement between the calculated and measured curves in **Figure 6**, further X-ray emission spectra were measured without the copper filter in front of the spectrometer window.

Figure 8 shows the comparison of the X-ray emission spectra for a fixed average laser power but two different combinations of pulse energy and repetition rate configured in the microSHAPETM USP-laser machine. The first measurement (orange curve) took place at a laser repetition rate of 50 kHz and a laser pulse energy of 1980 μJ, while the second measurement (blue curve) was carried out at a laser repetition rate of 200 kHz and a laser pulse energy of 550 μJ. The respective laser peak intensities of the laser pulses were 3.3 × 1015 W/cm2 and 0.9 × 1015 W/cm2. To be able to compare the measurements as plotted in Figure 8, the number of photons (counts) per channel with an energy bandwidth of 20 eV was divided by the accumulation time in seconds of each experiment.

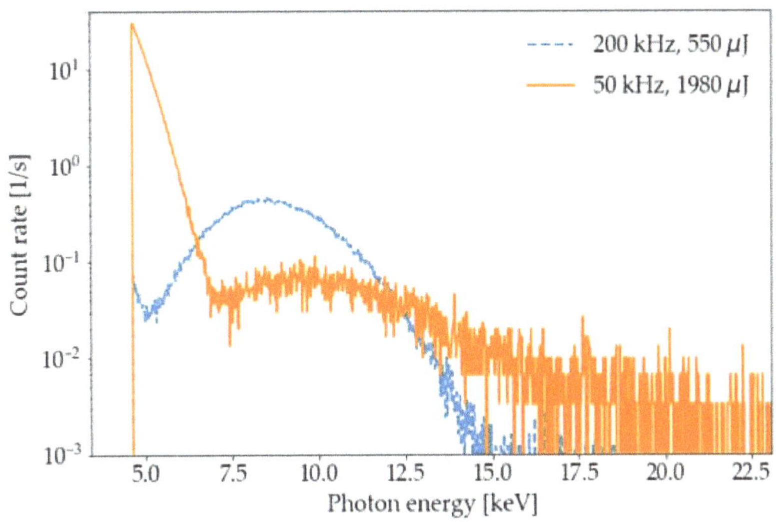

Figure 8. Pulse height spectra measured for the combination of two different USP-laser pulse energies and laser repetition rates: 550 μJ and 200 kHz (blue curve), 1980 μJ and 50 kHz (orange curve). The number of photons per channel with an energy bandwidth of 20 eV was divided by the total USP-laser processing time, to make the two spectra comparable. The increase in irradiance due to the higher pulse energy causes an energetic shift of the maxima of the spectra to higher photon energies. The increase of count rate in the low-energy region (between 5 keV and 7 keV) for the 50 kHz process is most likely caused by electronic noise due to an insufficient shielding of the CdTe detector. Overall, the recorded number of X-ray photons per second is very low, which was also confirmed in separate dose rate measurements with the Automess 6150AD-b/E dosemeter (see Table 1).

An increase in the laser pulse energy and the associated increase in the peak intensity of the laser pulses led to a shift of the spectrum to higher X-ray photon energies, as shown by the two combined spectral measurements in **Figure 8**. Accordingly, the average power of the laser, which was nearly identical for both configurations, cannot be employed as a simple scaling parameter for the emitted X-ray radiation as the photon energy spectrum is affected by the variation of the repetition rate and the pulse energy. The increase in counts in the low-energy region (between 5 keV and 7 keV) was most likely a consequence of an insufficient shielding of the CdTe detector and may have been caused by the pulsed intense electromagnetic laser plasma field. However, X-ray emissions at such low energies (here below 8 keV) would play only a minor role in relation to the hazard potential in enclosed USP-laser systems, because a large part of the photons would be absorbed in the ambient air and when passing through the protective housing and, thus, would not reach the operating personnel working on the machine.

Overall, the accumulated X-ray photon energy in the spectra was very low. This was also confirmed by the complementary dose rate measurements with the calibrated Automess 6150 dosemeter. The maximum measured ambient dose rates for the three different configurations at the microSHAPETM USP-laser system are listed in Table 1. During the surface process with a pulse energy of 1980 μJ, an ambient dose equivalent rate dH*(10)/dt of around 1 μSv/h was measured. These X-ray emission dose rates are much lower compared to the mSv/h reported recently in the literature for the same laser pulse duration and a lower laser pulse energy [11,12,18,19].

Further measurements were likewise carried out with the tungsten plate and a borosilicate glass sample, using a second experimental setup with a Coherent HR50 USP laser (Santa Clara, CA, USA). The Coherent HR50 laser has an average power of 35 W, a repetition rate of 200 kHz, and an emission wavelength of 1064 nm. For the radiation protection experiments, a pulse energy of 175 μJ and a laser pulse duration of 10 ps were applied; the focus diameter was approximately 20 μm, resulting in a peak intensity of approximately 1.0×10^{13} W/cm^2. Thus, the peak intensity of the laser is significantly lower than that of the microSHAPE™ USP-laser system with an output power of 220 W, reaching up to 10×3.3^{15} W/cm^2.

The second setup with the Coherent laser was selected for further measurements because of the distinctive features of the 2 mm steel protection housing. It has an opening station comprising two doors with windows made of PMMA polymer covered by a thin aluminum layer (see **Figure 9**). Additionally, in a small slit region where the two doors join when closed, rubber material without any additional steel or aluminum layers is located. In this region, transmission of X-ray radiation is much less suppressed when compared to the surrounding steel protection housing.

On one hand, they are vulnerabilities in the protective housing that must be taken into account concerning the transmission of X-rays. Conversely, during the experiments involving tungsten and glass targets, the Automess 6150 dosemeter detected no variation from the natural background radiation at a distance of 190 mm from the laser's focal point. Measurements utilizing the radiographic imaging plates and the X-123 CdTe spectrometer indicated an absence of signal from laser-induced X-ray emissions.

Figure 9. Photograph of the protection housing doors of the second experimental setup. The laser used in the system is a Coherent HR50 USP-laser with an average power of 35 W. The housing is made of 2 mm thick steel, and the windows are made of PMMA covered with a thin sheet of aluminum. Between the two doors there is a narrow area where, in the closed position, only rubber material lips join without any additional steel or aluminum layers.

4. DISCUSSION

In measurements conducted at the second experimental setup utilizing a Coherent HR 50 laser, with tungsten and borosilicate glass as target materials and a peak intensity of approximately 10^13 W/cm^2, no X-ray radiation emission was detected using the Automess 6150 dosimeter, the X-123 spectrometer, or radiographic imaging plates at a distance of approximately 19 cm from the plasma. This is not unexpected as the generation of X-rays with energies large enough to traverse short distances of air is expected to be very inefficient for laser peak intensities of 1013 W/cm^2 [11].

With the microSHAPETM USP-laser machine employing a tungsten workpiece, the measured dose rates are much smaller than expected for a laser-induced plasma generated at peak intensities of 3.3 × 1015 W/cm2 with tungsten as a target material. In the literature, typically, dose rates in the range of mSv/h are measured for such high peak intensities [18,19]. As can be seen from Table 1, the measured dose rate decreased with increasing pulse energy, which is unexpected. There are several possible explanations for these results.

The measurement apparatus was positioned suboptimally. The primary reason for this, as previously stated, was a safety system that monitored the interior of the USP-laser machine and inhibited the placement of objects in a substantial region right in front of the laser head and the focal point on the workpiece. The distance-corrected grayscale image (Figure 5b) illustrates that the radiation field of the laser-induced X-rays is situated near the horizontal plane and aligned with the y-axis parallel to the lines. Due to the space limitations, the dosemeter and spectrometer had to be installed at a large angle of approximately 45° from the horizontal plane and approximately 45° to 60° to the processing grooves in the *y*-direction. Consequently, the radiometric measurement equipment was located at the fringes of the radiation field, where the X-ray dose rate is expected to be much lower than in the maximum of the field.

Furthermore, the substantial rate of material removal generated significant quantities of airborne debris in proximity to the workpiece and focusing optics. The exhaust system lacked sufficient power to eliminate all traces of the laser-ablated material. Consequently, residual material accumulated on the focusing optics, impairing the quality of the laser spot and subsequently reducing the energy imparted to the workpiece. Supplementary heat effects that influence the focusing characteristics of the optical lenses may have also exerted an impact. The deteriorating impact is likely exacerbated by an increase in pulse energy, resulting in a greater volume of material eliminated each laser pulse. This may somewhat elucidate the low dose rate readings at a pulse energy of 1980 µJ in comparison to those at 550 µJ.

Moreover, the scanning speed of the laser spot on the workpiece was comparably low with 100 mm/s. Other experiments in the literature used a much higher scanning speed ranging from 400 mm/s [19] to 1000 mm/s [11] to prevent large overlap of adjacent laser spots and, thus, the formation of very deep ablated groove structures. For the measurements at SCHOTT AG, the combination of a large laser pulse energy of 550 µJ and 1980 µJ and the slow movement speed of the laser spot led to the formation of very deep ablated groove structures. During the laser material processing, the plasmas at the bottom of these structures were deep within the material; therefore, most of the emitted low-energy X-ray photons were absorbed by the surrounding material. This effect is especially important here because of the placement of the measurement equipment due to safety restrictions. As can be seen from **Figure 8**, the difference between the high energy tails of the spectrum with pulse energies of 550 µJ and 1980 µJ is in the range of only a few keV. Accordingly, on one hand, increasing the pulse energy will only have a minor effect on the transmission of

X-ray photons through the workpiece material, i.e., tungsten. On the other hand, increasing the pulse energy will lead to deeper groove structures and will, therefore, increase the amount of material that has to be traversed by the X-ray photons to reach the measurement equipment, resulting in an increased rate of absorption. This effect might explain the measurement of a lower dose rate at a higher pulse energy.

The acquired X-ray spectra from the highest investigated USP-laser pulse (1980 µJ) showed energies of up to 25 keV, which is consistent with other measurements at lower and similar peak intensities [**11,14**]. The measured spectra were tested by applying an additional copper filter with a thickness of 50 µm. In a comparison between the measured spectrum with the copper filter and the expected spectrum calculated from the spectrum and the transition function of 50 µm copper, a good agreement could be observed. Additionally, spectra of two different combinations of laser pulse energy and repetition rate were compared. As expected, a higher pulse energy and, consequently, a higher peak intensity led to a spectrum with higher maximum photon energies [**14,29,30**]. These measurements with nearly identical average laser powers but different pulse energies and repetition rates confirm that the average laser power cannot be employed as a simple scaling factor for the expected photon energy of laser-induced radiation.

The distribution of the distance-corrected radiation field illustrated in Figure 5b has a distinct maximum in the direction aligned with the laser movement along the processed lines in the material (y-direction). This outcome aligns well with prior measurements [11,13,14]. Nonetheless, as illustrated in Figure 5c, the distribution of the X-ray emission exhibits a peak in proximity to the horizontal plane. Previous results indicated that the peak of the radiation field for tungsten processing occurred at an angle of roughly 30° [11]. Two potential explanations for this divergence are provided herein. The anomalous angular distribution of the radiation field may be attributed to variations in laser peak intensity and, subsequently, the substantial quantity of residual material in the air surrounding the plasma. This may subsequently influence the angular properties of the plasma emission, owing to the absorption of low-energy X-rays in the material cloud. Another hypothesis is that, due to the profound groove structures and the extensive overlap between adjacent laser spots, the plasma resides on the side wall of the groove rather than at the bottom, parallel to the surface of the target material. This would result in a distinct alignment of the plasma density gradient, which considerably influences the emission characteristics of laser-induced X-rays. A rapid, dependable, and reproducible examination of the machine's protective housing or structural radiation shielding attributes may be impractical with laser-induced X-rays. Therefore, alternative inspection methodologies for forthcoming radiation protection evaluations of USP-laser equipment must be contemplated. An alternate notion may involve utilizing a steady and continuous X-ray source, such as an X-ray tube, as a replacement for laser-induced radiation. The prerequisites for this method include that the X-ray tube possesses an energy spectrum analogous to that of the laser-induced X-ray emission. By adjusting the high voltage and current of the X-ray tube, one may ascertain the ambient dose rate and energy spectrum of the radiation, ensuring that the transition radiation through the protective housing remains within legal dose rate limitations. The imitation of laser-induced X-ray generation by ordinary X-ray technology and current inspection protocols seems plausible and would greatly expedite the legal clearance process of USP-laser processing equipment by local authorities.

5. CONCISE SUMMARY

The findings and challenges encountered during dosimetric and spectrometric assessments of laser-induced X-ray emissions in a genuine industrial setting unequivocally illustrate that configuring a standard USP-laser apparatus for radiation protection evaluations under reference conditions with maximal X-ray emissions is exceedingly complex and labor-intensive. Enhancing the laser processing parameters, including pulse energy and focus diameter in relation to the generation of laser-induced X-ray radiation, may result in damage to the focusing optics or other essential components of the apparatus. Installed safety mechanisms may hinder the individual responsible for the radiation protection inspection from conducting measurements under optimal conditions. The redeposition of ablated material on the optics of the laser processing system may further influence measurements if the exhaust system lacks sufficient power to eliminate excessive amounts of ablated material generated during processing scenarios characterized by high pulse energies and repetition rates. The adaptations and challenges encountered during measurement indicate that, despite considerable experience with X-ray generation on USP-laser systems, identifying a machining process on a third-party machine suitable as a reference scenario with significant X-ray emissions (or even the worst-case scenario) is not straightforward. Given the extensive advancements of USP-laser machines in industrial applications in recent years and the prevalence of such machines in current industrial use, there is an urgent necessity for the establishment of a reliable and reproducible radiation protection inspection for USP-laser machines. An alternate inspection approach for USP-laser machines may utilize a steady and continuous X-ray source, such as an X-ray tube, as a replacement for the laser-induced radiation while examining the machine housing.

AT A GLANCE

Interactions between ultrashort laser pulses with intensities above 10^13 W/cm^2 and solids during material processing might result in the generation of X-rays with photon energies surpassing 5 keV, posing radiation risks to operators. A framework for assessing X-ray emission risks in laser material processing has not yet been established. A prerequisite for performing radiation protection inspections is the utilization of a reference scenario, namely laser settings and process parameters that result in a nearly constant and elevated level of X-ray emissions. To assess the practicality of establishing a reference scenario, ambient dose rates and photon energies were quantified using calibrated measurement instruments in an industrial environment at SCHOTT AG. Ultrashort pulsed (USP) lasers with a maximum average power of 220 W provided the opportunity to measure X-ray emissions at laser peak intensities of up to 3.3 × 1015 W/cm2 at pulse durations of ~1 ps. The findings demonstrate that augmenting the laser peak intensity alone is inadequate for producing elevated dose rates. The research were hindered by several restrictions that obstructed the measurement of elevated ambient dose rates. This chapter identifies a list of difficulties that may arise while doing measurements on USP-laser equipment in industrial environments.

REFERENCES

1. Bäuerle, D.W. *Laser Processing and Chemistry*, 4th ed.; Springer: Berlin/Heidelberg, Germany, 2011. **[Google Scholar]**
2. Saraceno, J.C.; Sutter, D.; Metzger, T.; Ahmed, M.A. The amazing progress of high-power ultrafast thin-disk lasers. *J. Eur. Opt. Soc.* **2019**, *15*, 15. **[Google Scholar] [CrossRef] [Green Version]**
3. Kühlke, D.; Herpers, U.; Von Der Linde, D. Soft X-ray emission from subpicosecond laser-produced plasmas. *Appl. Phys. Lett.* **1987**, *50*, 1785–1787. **[Google Scholar] [CrossRef]**
4. Stearns, D.G.; Landen, O.L.; Campbell, E.M.; Scofield, J.H. Generation of ultrashort X-ray pulses. *Phys. Rev. A* **1988**, *37*, 1684–1690. **[Google Scholar] [CrossRef] [PubMed]**
5. Harris, S.E.; Kmetec, J.D. Mixed-Species Targets for Femtosecond-Time-Scale X-ray Generation. *Phys. Rev. Lett.* **1988**, *61*, 62–65. **[Google Scholar] [CrossRef] [PubMed]**
6. Zigler, A.; Burkhalter, P.G.; Nagel, D.J.; Boyer, K.; Luk, T.S.; McPherson, A.; Solem, J.C.; Rhodes, C.K. High intensity genera-tion of 9–13 Å X-rays from BaF_2 targets. *Appl. Phys. Lett.* **1991**, *59*, 777–778. **[Google Scholar] [CrossRef]**
7. Murnane, M.M.; Kapteyn, H.C.; Rosen, M.D.; Falcone, R.W. Ultrafast X-ray Pulses from Laser-Produced Plasmas. *Science* **1991**, *251*, 531–536. **[Google Scholar] [CrossRef]**
8. Giulietti, G.; Gizzi, L.A. X-ray emission from laser-produced plasmas. *Riv. Nuovo Cim.* **1998**, *21*, 1–93. **[Google Scholar] [CrossRef]**
9. Thogersen, J.; Borowiec, A.; Haugen, H.K.; McNeill, F.E.; Stronach, I.M. X-ray emission from femtosecond laser micromachining. *Appl. Phys. A* **2001**, *73*, 361–363. **[Google Scholar] [CrossRef]**
10. Bunte, J.; Barcikowski, S.; Puester, T.; Burmester, T.; Brose, M.; Ludwig, T. Secondary Hazards: Particle and X-ray Emission. *Topics. Appl. Phys.* **2004**, *96*, 309–321. **[Google Scholar] [CrossRef]**
11. Legall, H.; Schwanke, C.; Pentzien, S.; Dittmar, G.; Bonse, J.; Krüger, J. X-ray emission as a potential hazard during ultrashort pulse laser material processing. *Appl. Phys. A* **2018**, *124*, 407. **[Google Scholar] [CrossRef] [Green Version]**
12. Behrens, R.; Pullner, B.; Reginatto, M. X-ray emission from materials processing lasers. *Radiat. Prot. Dosim.* **2019**, *183*, 361–374. **[Google Scholar] [CrossRef]**
13. Legall, H.; Schwanke, C.; Bonse, J.; Krüger, J. The influencc of processing parameters on X-ray emission during ultra-short pulse laser machining. *Appl. Phys. A* **2019**, *125*, 570. **[Google Scholar] [CrossRef] [Green Version]**

14. Weber, R.; Giedl-Wagner, R.; Förster, D.J.; Pauli, A.; Graf, T.; Balmer, J.E. Expected X-ray dose rates resulting from industrial ultrafast laser applications. *Appl. Phys. A* **2019**, *125*, 635. **[Google Scholar] [CrossRef] [Green Version]**

15. Legall, H.; Schwanke, C.; Bonse, J.; Krüger, J. X-ray radiation protection aspects during ultrashort laser processing. *J. Laser Appl.* **2020**, *32*, 022004. **[Google Scholar] [CrossRef]**

16. Freitag, C.; Giedl-Wagner, R. X-ray Protection in an Industrial Production Environment. *PhotonicsViews* **2020**, *17*, 37–41. **[Google Scholar] [CrossRef]**

17. Legall, H.; Bonse, J.; Krüger, J. Review of X-ray exposure and safety issues arising from ultra-short pulse laser material pro-cessing. *J. Radiol. Prot.* **2021**, *41*, R28–R42. **[Google Scholar] [CrossRef]**

18. Mosel, P.; Sankar, P.; Düsing, J.F.; Dittmar, G.; Püster, T.; Jäschke, P.; Vahlbruch, J.-W.; Morgner, U.; Kovacev, M. X-ray Dose Rate and Spectral Measurements during Ultrafast Laser Machining Using a Calibrated (High-Sensitivity) Novel X-ray Detec-tor. *Materials* **2021**, *14*, 4397. **[Google Scholar] [CrossRef] [PubMed]**

19. Schille, J.; Kraft, S.; Pflug, T.; Scholz, C.; Clair, M.; Horn, A.; Loeschner, U. Study on X-ray Emission Using Ultrashort Pulsed Lasers in Materials Processing. *Materials* **2021**, *14*, 4537. **[Google Scholar] [CrossRef] [PubMed]**

20. Metzner, D.; Olbrich, M.; Lickschat, P.; Horn, A.; Weißmantel, S. X-ray generation by laser ablation using MHz to GHz pulse bursts. *J. Laser Appl.* **2021**, 33, 032014. **[Google Scholar] [CrossRef]**

21. Sick Sensor Intelligence. Available online: **https://www.sick.com/de/de/optoelektronische-schutzeinrichtungen/sicherheits-laserscanner/s300-mini-standard/c/g214851** (accessed on 10 September 2021).

22. Automess 6150AD. Available online: **https://www.automess.de/produkte/produktfamilie-6150ad/szintillatorsonde** (accessed on 15 August 2021).

23. International Organization for Standardization. X and gamma reference radiation for calibrating dosemeters and doserate meters and for determining their response as a function of photon energy—Part 1: Radiation characteristics and production methods. In *International ISO 4037-1*; International Organization for Standardization: Geneva, Switzerland, 2019. **[Google Scholar]**

24. Ankerhold, U.; Berens, R.; Ambrosi, P. X ray Spectrometry of Low Energy Photons for Determining Conversion Coefficients from Air Kerma, K_a, to Personal Dose Equivalent, $H_p(10)$, for Radiation Qualities of the ISO Narrow Spectrum Series. *Radiat. Prot. Dosim.* **1999**, *81*, 247–258. **[Google Scholar] [CrossRef]**

25. X-123 CdTe Complete X-ray & Gamma Ray Spectrometer. Available online: **https://www.amptek.com/internal-products/x-123-cdte-complete-X-ray-gamma-ray-spectrometer-with-cdte-detector** (accessed on 27 October 2021).

26. Redus, R.; Huber, A.; Pantazis, J.; Pantazis, T.; Sperry, D. Design and performance of the X-123 compact X-ray and gamma-ray spectroscopy system. In Proceedings of the IEEE Nuclear Science Symposium, IEEE Xplore, Piscataway Township, NJ, USA, 29 October–1 November 2006; Volume 6, pp. 1–40. **[Google Scholar]**

27. DÜRR NDT 2D-Scanner CR 35 NDT. Available online: **https://www.duerr-ndt.de/produkte/computer-radiographie/cr-35-ndt.html** (accessed on 28 October 2021).

28. Hubbell, J.H.; Seltzer, S.M. Tables of X-ray Mass Attenuation Coefficients and Mass Energy-Absorption Coefficients from 1 keV to 20 MeV for Elements Z = 1 to 92 and 48 Additional Substances of Dosimetric Interest (version 1.4, 2004). Available online: **http://physics.nist.gov/xaamdi** (accessed on 29 September 2021).

29. Eliezer, S. *The Interaction of High-Power Lasers with Plasmas*, 2nd ed.; Institute of Physics: Bristol, UK, 2002. **[Google Scholar]**

30. Salzman, D. Theory of energy deposition by suprathermal electrons in laser-irradiated targets. *Phys. Rev. E* **2002**, *65*, 056409. **[Google Scholar] [CrossRef] [PubMed]**

CHAPTER

INVESTIGATION OF XRAY EMISSION USING ULTRASHORT PULSED LASERS IN MATERIAL PROCESSING

7

1. INTRODUCTION: AN OVERVIEW

Ultrashort pulse lasers feature excellent performances for high-efficient and precision machining that also has attracted increasing attention for innovations in micro fabrication and surface engineering. Moreover, the recent power scaling of femtosecond (fs) and picosecond (ps) lasers reaching multi-kW class level [1,2] will pave the way of the ultrashort pulse laser technology from the laboratory into industrial production. Actually, in typical material processing scenarios, ultrashort laser pulses of micro to milli joules optical energy and hundreds of femtoseconds to picoseconds pulse durations are employed, for example, in laser micro drilling [3,4], laser engraving [5,6], or laser surface texturing and modifications [7,8]. By tightly focusing the laser radiation, high peak intensities in the range from 1013 to 1016 W·cm-2 can be reached on the material surface. However, when using laser pulses of such high intensity in material ablation, high-intense electron plasmas will be produced, itself emitting X-ray photons at a several keV level. The X-ray photons result from Bremsstrahlung and characteristic line emissions when accelerated electrons lose their kinetic energy by scattering, acceleration and recombination with other charged particles, in free–free, free–bound, and bound–bound processes.

However, recent works have shown that the X-ray photon emission increases proportionally with higher average laser power, and actual X-ray emission dose rates can cause serious health risks for the laser operators [**9,10,11,12,13,14**]. In particular, this is the case when high-pulse repetition frequency lasers will be used in materials ablation as the X-ray emission per pulse can accumulate to harmful X-ray photon dose over time. On the one hand, it is verified that X-ray photon emission increases with peak intensity, pulse repetition rate, and average laser power. On the other hand, there is clear evidence that little variations of the operational conditions can have a considerable effect on the X-ray emission, causing unexpected high dose levels. This is much more relevant, as a great number of parameters (more than 20) have been experimentally validated as influencing factors on X-ray emission dose level, including laser beam characteristics, processing conditions, and the irradiated material, further mutually affecting each other. Moreover, the specific effect of each parameter and their variations is not sufficiently investigated and understood so far, making it difficult for the reliable assessment of the X-ray emission and safety precautions in the operational area.

In the following, the significant impact of the pulse repetition frequency and the intra-line pulse distance during laser beam scanning on the spectral X-ray photon flux and the X-ray emission dose is discussed by comparing low-intensity and high-intensity ultrashort pulses irradiating AISI 304 stainless steel. It will be shown that an unexpected high level of X-ray emission dose rate of tens of mSv·h-1 could be produced when low-intensity pulses irradiate at 1.6 MHz pulse repetition frequency (PRF). This resulted potentially from plasma resonance absorption as an efficient mechanism for optical energy transfer, also yielding high X-ray emissions in the burst- and bi-burst pulse regime.

2. MATERIALS AND METHODOLOGY

All experiments were performed in a laser safety enclosure to protect the experimental operators from potential risks and hazards arising from the powerful laser beams and X-ray emissions. A schematic view of the laser machining setup used in this study is presented in Figure 1.

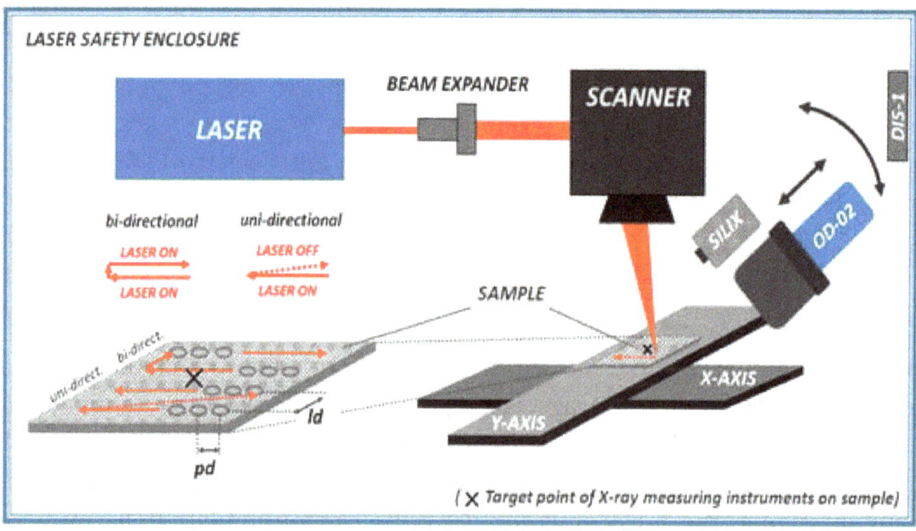

Figure 1. Schematic view of the laser safety enclosure including laser machining setup.

First of all, the efficacy of the enclosure material was verified by initial testing in order to ensure reasonable assurance of the enclosure against the laser-induced

X-ray emissions. In addition, three electronic Direct Ion Storage memory cells (DIS-1, Mirion Technologies (Canberra) GmbH, Ruesselsheim, Germany) were applied as safety precautions. Two of the X-ray badge detectors were placed at the inner and outer sides of the laser safety enclosure. A third one was worn by the operator throughout the experiments. The actual received X-ray dose values were read out at the beginning and the end of each working day from the badges. It should be mentioned here that no increase of the X-ray dose could be detected, whether from the badges placed at the enclosure outside nor the one worn by the operator, confirming the safe working conditions at all time.

In total, four different laser systems were employed in this study for providing a wide range of complementary laser beam characteristics; see Table 1. The studied parameter settings include up to 5.2 × 1016 W·cm−2 peak intensity, pulse repetition frequencies in the range between 1 and 2 MHz, maximum 72.2 W average laser power, as well as laser burst and bi-burst processing mode. The energy of the irradiating pulses QP was defined by the ratio of the average laser power Pav and the pulse repetition frequency fP described by:

$$Q_P = \frac{P_{av}}{f_P}. \quad (1)$$

Table 1. Summary of the laser systems and studied laser parameter settings.

Parameter / Laser System	Symbol Unit	Astrella Coherent Inc.	FX 200 Edgewave GmbH	UFFL 100 AFS GmbH	Carbide Light Conversion
wavelength	λ in nm	800	1030	1030	1030
pulse duration	τ_H in fs	40	600	340	200
focus spot diameter	$d_{0,86}$ in µm	13.5 **	30.0 ***	30.0 ***	49.7 **
pulse rep. frequency intra-burst frequency bi-burst frequency	f_P in MHz f_{IB} in MHz f_{BB} in GHz	0.001	0.2–2.0	0.51	0.4, 1.0 66.7 2.3
max. average power *	P_{av} in W	1.6	72.2	18.7	33
max. pulse energy *	Q_P in µJ	1600	36.1	37	82.5
max. peak power *	P_0 in W	4×10^{10}	5.6×10^{7}	9.6×10^{8}	3.6×10^{8}
max. intensity *	I_0 in W cm^{-2}	5.2×10^{16}	1.6×10^{13}	2.7×10^{13}	3.8×10^{13}

** on target, ** LIU-plot method, *** MSM measurement.*

The focus spot diameters d86 (86.5% power enclosure method) presented in Table 1 have been determined by analyzing the laser beam caustic (MSM Micro Spot Monitor, Primes GmbH, Pfungstadt, Germany) or rather from the established semi-log plot method of squared ablation crater diameter versus peak fluence [15].

The X-ray emission induced by the ultrashort pulsed laser irradiations was studied on technical-grade AISI 304 stainless steel metal sheets. The chemical composition of the alloying elements is presented in Table 2. The dimensions of the laser processed substrates were 50 × 50 × 2 mm³ in width, length, and sample thickness. In each test, the substrates were placed on a sample holder in the focal plane with the help of an XYZ stage assembly. Two different machining strategies were applied depending on the prevailing machinery conditions and setups of the individual laser systems: (i) relative movements between the laser beam and substrate by accelerating the sample in XY directions; and (ii) scanning the laser radiation across the substrate surface via a galvanometer scan system (intelliScan 30, Scanlab GmbH, Puchheim, Germany). As illustrated in Figure 1, the geometrical distance of the irradiating pulses along and between the laser processed lines is defined by the intra-line pulse distance pd (as a function of laser beam moving speed and pulse repetition frequency) and the hatch distance ld, respectively. The laser pulse peak power P0 provided in Table 1 for the investigated laser beams was calculated according to

$$P_0 = \frac{Q_P}{c \cdot \tau_H} \quad (2)$$

where the pulse energy QP is divided by the measured pulse duration τH and by taking into account the correction factor c considering the temporal pulse shape of the respective irradiating pulses, either being Gaussian or sech2. Following

$$I_0 = \frac{2 \cdot P_0}{\pi \cdot w_{0.86}{}^2}, \quad (3)$$

the peak intensity I0 was calculated from the peak power and the focus spot radius w0.86.

Table 2. Chemical composition of the AISI 304 stainless steel alloying elements.

Element	C	Si	Mn	P	S	Cr	Ni	N
Element Mass Fraction in %	0.07	1.0	2.0	0.045	0.03	17.5–19.5	8.0–10.5	0.1

Two different detector systems were used for the X-ray photon emission measurements: a survey meter OD-02 (STEP Sensortechnik und Elektronik Pockau GmbH, Pockau-Lengefeld, Germany) measuring the directional X-ray emission dose H'(0.07) and the X-ray emission dose rate equivalent *H*''(0.07) = dH'(0.07)/dt in the soft-radiation region above 6 keV. Additionally, a SILIX lambda detector (Ingenieurbüro Prof. Dr.-Ing. Guenter Dittmar, Aalen, Germany) monitored the X-ray emission spectra and X-ray emission dose rate in the photon energy range between 2 and 20 keV. The X-ray detectors were aligned to the center of the samples, which is indicated in Figure 1 by the target point (X). The detector's view was along the X-axis equivalent to the laser processing direction. The nominal detector distance D and the detection angle α were varied with reference to the target point and the horizontal plane. Thereby, the nominal detector distance represented a mean measure because the detector target points were set in the center of the laser processed area. Hence, in the detector's view direction (X-axis), the closest point and most distant point from the sensitive detector surface were at the start and end of a laser scanned line, and likewise the greatest distance in the Y-direction was at the edges of the laser processed fields.

Laser beam scanning was applied either in uni-directional or bi-directional mode depending on the specifics of the studied laser setups in terms of laser beam control and synchronization. In uni-directional processing, the laser beam departs from the X-ray detector, thus receiving maximum X-ray dose levels [**13**]. The "laser-on" duty cycle was 46% in uni-directional processing controlled by oscilloscope measurements throughout the study. As a consequence, the monitored X-ray emissions represented less than half of the process-relevant emission dose and were corrected (by a factor of 2.17) in order to compare and discuss the process-typical dose rates in the following. In bi-directional mode, by contrast, the laser beam was switched on during all the processing. Hence, the monitored X-ray emissions represent average readings from the bi-directional laser material processing. This is due to the fact that higher X-ray emissions have been reported for a laser beam running away instead of moving toward the sensitive detector surface [**13**].

3. RESULTS AND DISCUSSIONS

3.1. Effect of the Detection Angle and Detector Distance on X-ray Dose

At first, the dependencies of detector's angle and distance on the monitored X-ray emission dose rate were studied. This was in order to identify the optimum measuring conditions for the assessment of X-ray emissions in the accomplished study. On the one hand, this was encouraged, as the authors′ literature review revealed a dependence between the X-ray emission dose and the detection angle [**9**]. So far, a maximum X-ray dose rate level was reported at a 30° detection angle, as investigated in the range between 10° and 40°. On the other hand, the correct detector–substrate distance is of relevance because, firstly, the X-ray photons are absorbed in air, and secondly, to ensure that the active detector area is homogeneously exposed by the laser-induced X-ray emissions. Both have a great impact on the detector readings.

The X-ray emissions recorded by the survey meter OD-02 and the SILIX X-ray monitor at a 35° detector angle and different distances from the substrate can be seen in Figure 2a at similar dose values. A fitting curve (dashed line) is included in the plot, showing the recorded X-ray emission dose rate inversely proportional to the square of the distance D of the detector from the substrate target point. However, at the larger distances, the experimental data differ to a greater extent from the inverse square-root relationship resulting from X-ray beam absorption in air. This is pointed out in Figure 2a by the modified fitting curve (solid line) emphasizing good agreement for experimental and computed data when the attenuation coefficient μ is considered in the photon flux Φphot, calculations

$$\Phi_{\text{phot}}(D,\lambda) \propto D^{-2} e^{-\mu(\lambda)D}. \quad (4)$$

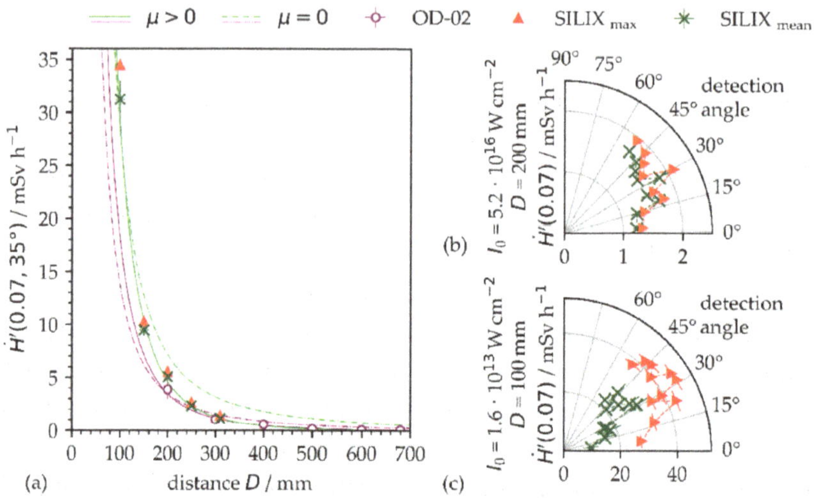

Figure 2. (a) OD-02 and SILIX readings of X-ray emission dose rate on AISI 304 as a function of detector distance D; the fitting curves are representative for vacuum (dashed line) and air (solid line) ambient conditions at I0 = 1.6 × 1013 W·cm-2 and 35° detection angle. (b) X-ray emission dose rate as a function of detection angle for pulses of I0 = 5.2 × 1016 W·cm-2 at 200 mm SILIX detector distance. (c) X-ray emission dose rate as a function of detector angle for pulses of I0 = 1.6 × 1013 W·cm-2 at 100 mm SILIX detector distance.

Thereby, the attenuation coefficient depends strongly on the photon energy and varies in air between 0.94 cm−1 ≤ μ ≤ 6.36 × 102 cm−1 for the X-ray photons between 2 keV ≤ Ephot ≤ 20 keV [10]. The attenuation coefficients derived from the fitting curves presented in Figure 2a amount to μ = 0.05 ± 0.02 cm−1 (SILIX) or μ = 0.03 ± 0.02 cm−1 (OD-02), which is about two orders of magnitude lower than the literature data. However, this discrepancy is not clarified yet; the steadily changing distance of the actual laser processing zone to the detector resulting from laser beam scanning, the X-ray dose rate as an integral measure from the X-ray photon spectrum, and measurement uncertainties for X-ray emissions all had a great impact on the experimental data.

In addition, a significant effect of the detector angle on the X-ray emission dose rate has been found by varying the angular orientation of the SILIX detector in the range between 7.0° < α < 51.0° with respect to the specified target point. Most likely independent from the peak intensity, whether I0 = 5.2 × 1016 W·cm−2 (Figure 2b) or I0 = 1.6 × 1013 W·cm−2 (Figure 2c), the maximum X-ray emission dose rate was detected at about 30°. It is noteworthy that a considerably higher X-ray emission dose rate was measured for the pulses of lower intensity, 43 mSv·h−1 vs. 2.1 mSv·h−1, potentially resulting from the higher average laser power, 72.2 W vs. 1.6 W, and shorter distance of the detector from the target point, 100 mm vs. 200 mm.

3.2. Spectral X-ray Emission

The spectral X-ray photon flux was analyzed by placing the SILIX X-ray monitor at a fixed 35° detection angle and 100 mm distance from the target point at the substrates. Two representative X-ray photon emission spectra in **Figure 3** verify a broad continuous Bremsstrahlung emission in the photon energy range below 10 keV. The detected characteristic line emissions can be related to the main AISI 304 alloying elements. Therefore, as a reference, the principal $K_{\alpha 3\text{-}1}$ and $K_{\beta 5\text{-}1}$ shell interband emissions are included in **Figure 3**, such as for iron (Fe), chromium (Cr), and nickel (Ni), as provided by the NIST database [**16**]. By comparing the individual characteristic lines, the photon energy peaks of the alloying elements are slightly offset from the principal X-ray transition energies ($\Delta E_{ph} < 0.09$ keV) to higher photon energy levels being within the measurement accuracy of the SILIX X-ray monitor.

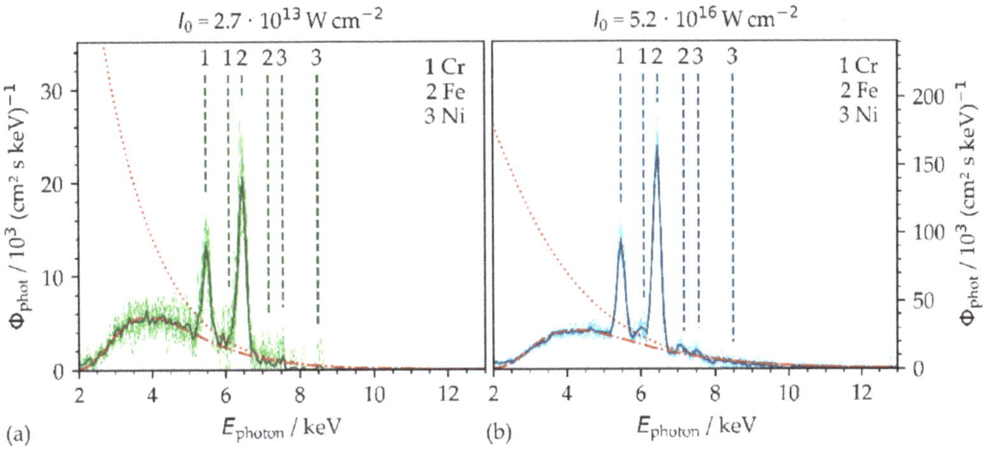

Figure 3. X-ray photon flux Φphot at a °35 detection angle and 100 mm distance for low peak intensity (a) I_0 = 10 × 2.7^{13} W·cm^{-2} and high peak intensity (b) I_0 = 10 × 5.2^{16} W·cm^{-2}. The characteristic X-ray emission lines of the correlating AISI 304 alloying elements are included. The recorded spectral X-ray photon distribution follows a Maxwell–Boltzmann distribution computed for vacuum (dotted line, non-attenuated) and ambient air (dashed line including 3 σ confidence interval, 100 mm distance) conditions.

The monitored Bremsstrahlung spectra represent qualitatively the energy distribution of the X-ray photon flux at 100 mm distance and can be approximated by a Maxwell–Boltzmann distribution [**9,12**] according to

$$f_{Maxwell}(E)dE = \sqrt{\frac{4E}{\pi (k_B T_e)^3}}\, e^{-\frac{E}{k_B T_e}}\, dE \tag{5}$$

where k_B represents the Boltzmann constant, T_e is the electron temperature, and E is the energy of the emitted X-ray photons.

The recorded Bremsstrahlung continuum can be seen in good correlation with the computed X-ray emissions when considering the transmission of the X-rays through air at ambient pressure and 100 mm distance in the Maxwell–Boltzmann modeling procedure (**Figure 3**, dashed line). In addition, the non-attenuated spectrum computed for vacuum surrounding conditions is presented (dotted line). The best fit between the empirically observed and theoretically modeled X-ray photon distribution (in air) was obtained by the electron temperature of either 1.18 ± 0.1 keV (**Figure 3a**) for the low-intensity or 1.56 ± 0.1 keV (**Figure 3b**) for the high-intensity pulses. These electron temperatures are in good agreement with the temperatures reported elsewhere for hot electrons and respective laser peak intensities ranging between 0.82 and 1.76 keV [**9,15,17**].

The averaged X-ray emission dose rate of the above given spectra was measured to $H''(0.07))$ = 0.2 ± 1.1 mSv·h^{-1} with the low peak intensity pulses (I_0 = 10 × 2.7^{13} W·cm^{-2}), as shown in **Figure 3a**, and $H''(0.07)$ = 11.4 ± 2.3 mSv·h^{-1} with the high peak intensity pulses (I_0 = 10 × 5.2^{16} W·cm^{-2}); see **Figure 3b**. However, for the high peak intensity pulses, the X-ray emission rate was about one order of magnitude higher in spite of the considerably lower applied average laser power (22.4 W vs. 1.6 W) and PRFs (506 kHz vs. 1 kHz). This finding proves the great impact of the laser peak power and intensity on the X-ray emission and will be discussed more in detail in the following section.

3.3. X-ray Emission in the Low-Intensity and High-Intensity Pulse Regime

The effect of the peak intensity on the X-ray emission was investigated by recording the X-ray photon spectra at 100 mm distance and 35° detection angle of the SILIX monitor from the reference target point at the substrates. The peak intensity of the impinging laser pulses was varied in the range between 8 × 10^{12} W·cm^{-2} < I_0 < 10 × 1.6^{13} W·cm^{-2} and 10 × 0.3^{16} W·cm^{-2} < I_0 < 10 × 5.2^{16} W·cm^{-2} that also refers to the low-intensity and high-intensity pulse regime in the following.

A closer view to the X-ray emission spectra in **Figure 4** reveals a notably higher extent of Bremsstrahlung emissions for pulses of lower peak intensity, while the greater proportional amount of characteristic X-ray emissions can be seen for pulses of higher intensity. The corresponding X-ray emission dose rates were measured of $H'(0.07)$ = 4.6 ± 38.9 mSv·h^{-1} for pulses of 10 × 1.6^{13} W·cm^{-2} and $H'''(0.07)$ = 2.3 ± 11.4 mSv·h^{-1} for 10 × 5.2^{16} W·cm^{-2} peak intensity. This was achieved by irradiating a laser beam of 72.2 W and 1.6 MHz (low-intensity regime) or rather 1.6 W and 1 kHz average laser power and pulse repetition frequency (high-intensity regime), respectively.

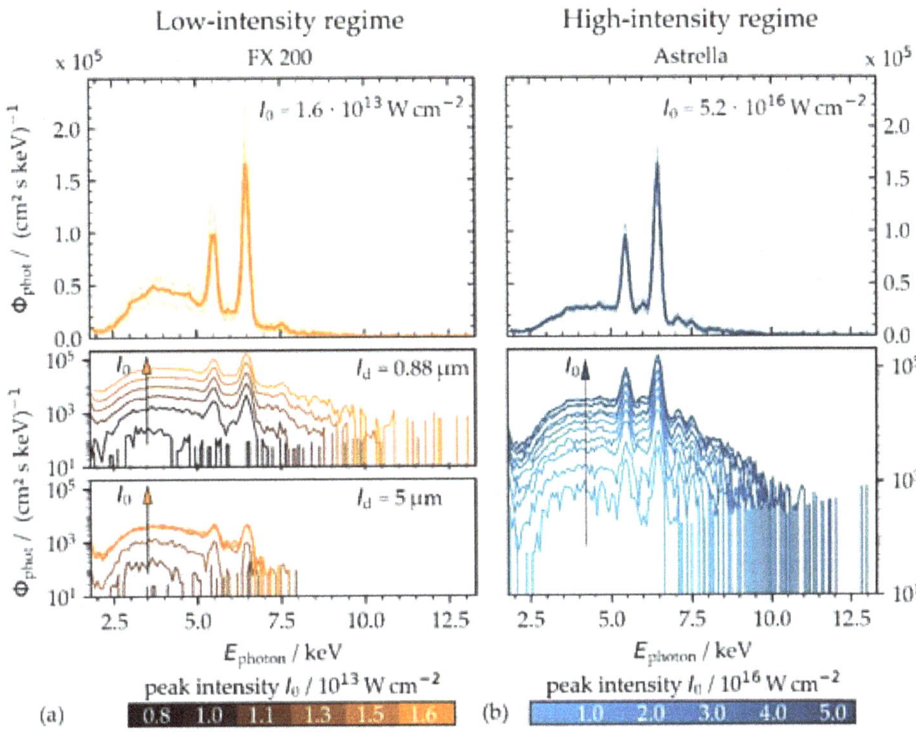

Figure 4. X-ray photon flux Φphot detected on AISI 304 in the low-intensity (a) and high-intensity pulse regime (b). The X-ray emission spectra were recorded by using the SILIX X-ray monitor at 35° detection angle and 100 mm distance. In the low-intensity regime, the intra-line pulse distance was varied from 0.88 µm (left, center) to 5 µm (left, bottom). Increasing peak intensity is indicated by the arrow.

The differences of the X-ray photon spectra can be attributed to the different mechanisms of laser plasma interaction and electron plasma heating that largely depend on the irradiating conditions. So, pulses of lower intensity ($I_0 < 10^{14}$ W·cm^{-2}) will be primarily absorbed in the underdense corona region of the plasma by inverse Bremsstrahlung absorption processes [18]. The excited electrons will be accelerated in the Coulomb field that resulted in the emission of the Bremsstrahlung continuum. This free–free generation of Bremsstrahlung X-ray photons is inefficent and yields only a low proportion of X-ray photons per laser pulse. Certainly, a great amount of low-intensity pulses can induce X-rays at a high dose level, as can be seen in Figure 4a (top) by the example of low-intensity pulses ($I_0 = 10 \times 1.6^{13}$ W·cm^{-2}) irradiated at 1.6 MHz pulse repetition frequency.

By contrast, resonance absorption is suggested as the dominant mechanism for pulses of high peak intensity, $I_0 > 10^{15}$ W·cm^{-2}. This collision-less process is most efficient for parallel polarized laser beams irradiating at a large incident angle. The laser energy will be transferred to the electron plasma in the region of the critical plasma density, where the electron plasma frequency is equal to the laser frequency [19]. The highly excited plasma electrons interact with the bound electrons of the target atoms by collisional impact ionization, leaving vacant energy levels. Nevertheless, resonance absorption might also be effective when pulses of lower peak intensity irradiate at megahertz PRF and small spatial pulse distance. This can be supposed from the X-ray emission spectra presented in **Figure** 4a (center and bottom) recorded for pulses of the same intensity (1.6×10^{13} W·cm^{-2}) and PRF (1.6 MHz) but different intra-line pulse distances varied by the laser beam moving speed. As characteristic X-ray emission lines are clearly distinguishable in the spectra recorded at an 0.88 μm intra-line pulse distance, the Bremsstrahlung continuum is dominant at the larger 5 μm intra-line pulse distance.

In addition, there is a trend toward higher X-ray photon flux when pulses of higher peak intensity irradiate, as clearly observable in **Figure 4** for the low-intensity (left) and the high-intensity (right) pulse regime. On the one hand, the displayed spectra indicate no significant dependence of the characteristic X-ray emissions from the iron and chromium alloy elements from the peak intensity. On the other hand, the spectral width of the Bremsstrahlung continuum enlarged to higher X-ray photon energy with pulses of higher peak intensity, such as 10.0 keV or rather 12.0 keV in maximum for the low-intensity ($I_0 = 10 \times 1.6^{13}$ W·cm^{-2}) or high-intensity ($I_0 = 10 \times 5.2^{16}$ W·cm^{-2}) pulse regime. Here, too, the effect of the intra-line pulse distance on the X-ray photon emission spectrum can be recognized in **Figure** 4a, showing 11.0 keV maximum X-ray photon energy at 0.88 μm, while the maximum of about 7.5 keV can be seen for the 5.0 μm intra-line pulse distance.

The emitted X-ray dose per pulse (**Figure** 5, green data points) was derived by dividing the monitored X-ray emission dose rate by the number of impinging pulses. In addition, as another evaluation measure, the X-ray emission efficiency represents the X-ray dose per pulse in relation to the applied pulse energy; see **Figure** 5 (blue data points). In the low-intensity pulse regime, the X-ray dose per pulse increases exponentially with higher peak intensity, and in the studied range, a maximal value of 6.7 pSv/pulse was measured at $I_0 = 10 \times 1.6^{13}$ W·cm^{-2} and $f_P = 1.6$ MHz; see **Figure** 5a. This is in contrast to the high-intensity regime, as shown in **Figure** 5b, where the X-ray dose per pulse increases linearly with increasing peak intensity. Here, the determined maximum value of the X-ray dose per pulse of 3.2 nSv/pulse is almost three orders of magnitude larger than for low-intensity pulses. In terms of X-ray emission efficiency, **Figure** 5a indicates an exponential dependence for the low-intensity regime to 150 fSv·μJ^{-1} maximum, while saturation is given in **Figure** 5b of about 2.2 pSv·μJ^{-1}.

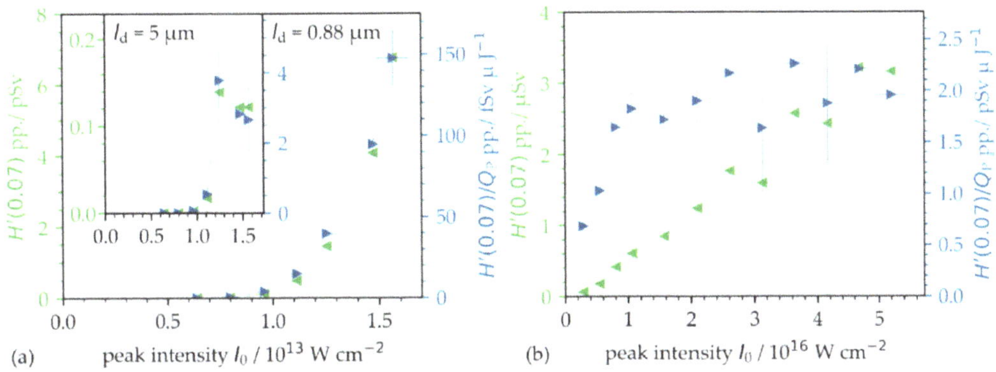

Figure 5. X-ray dose per pulse (green data points) and the X-ray emission efficiency (blue data points) on AISI 304 in the (a) low-intensity and (b) high-intensity regime. The given data were derived from the X-ray emission dose monitored by the SILIX detector at a 35° detection angle and 100 mm distance. The dotted lines guide the eye.

For pulses of low peak intensity irradiated at megahertz PRF, a remarkable dependence of the X-ray emission efficiency on the spatial pulse distance could be identified. The X-ray dose per pulse is at a considerably lower level with a 5.0 µm intra-line pulse distance under otherwise similar processing conditions; see **Figure 5a** (imprint). This might be due to the fact that the interaction of the following pulse(s) with the still prevailing laser ablation particle/plasma plume changed dramatically depending on the lateral pulse distance. The underlying physical effects are not clear yet; however, plasma resonance absorption was introduced above as a potential mechanism for the transfer of optical energy to the plasma even when pulses of low peak intensity irradiate at high PRF. This is supported by the fact that the angle of incidence between the incoming laser beam and the critical plasma density layer is continually changing with time. Hence, the actual plasma density profile (affected by the time delay between the pulses determined by the pulse repetition frequency) and the lateral offset of the next impinging pulse determine the optimum conditions for laser beam coupling and energy transfer to the plasma with a consequence of maximum X-ray photon emission. This hypothesis will be tested in the ongoing study.

For the presented data, it should be mentioned that scaling of the X-ray dose per pulse or X-ray emission efficiency with the irradiated pulse number and pulse intensity can underestimate the actual emitted X-ray dose. This is due to the fact that laser plasma and X-ray emission influencing effects are not adequately considered in the proposed assessments [**12**]. For example, this includes the temporal pulse distance, geometrical pulse distance, angle of incidence as well as surface inhomogeneity, and roughness effects that steadily change with ongoing laser processing and scan numbers, and all of them mutually reinforce each other. To underline this, a more detailed view on the particular effect of the geometrical and temporal distance between the irradiating pulses on X-ray emission is provided in the following sections.

3.4. Effect of Intra-Line Pulse Distance on X-ray Emission in the Low-Intensity Pulse Regime

A remarkable influence of the intra-line distance between the impinging ultrashort laser pulses on X-ray emission has already been reported for tungsten studied with PRFs in the range between 50 kHz < PRF < 400 kHz [10]. Pronounced X-ray emission maxima and minima have been detected where the amplitude of the X-ray doses varied by a factor up to two depending on the spatial distance of the next impinging pulse. These previous findings are supported by the results of our study as a greater effect of the intra-line pulse distance on X-ray emission was measured for AISI 304 irradiated with low-intensity pulses of $I_0 = 10 \times 1.6^{13}$ W·cm^{-2} at 1.6 MHz PRF (Figure 6a). The intra-line pulse distance was varied between 0.5 µm ≤ *pd* ≤ 6.5 µm by adjusting the laser beam moving speed from 0.8 to 10.4 m·s^{-1}.

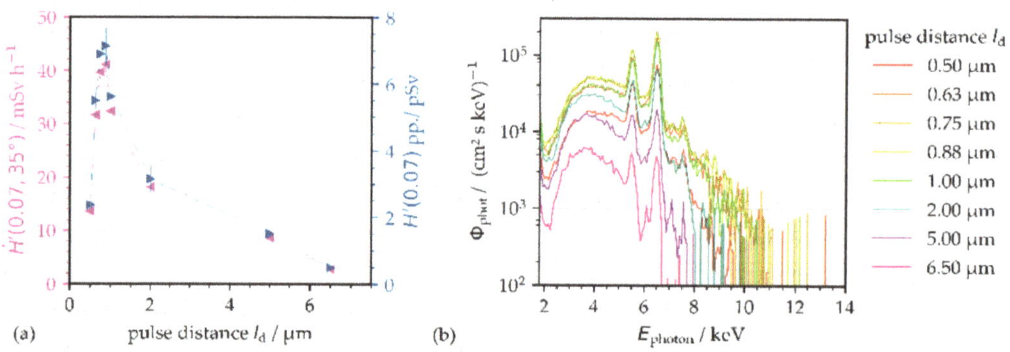

Figure 6. X-ray emissions on AISI 304 at 100 mm distance and °35 detection angle as induced by low-intensity pulse irradiations at $I_0 = 10 \times 1.6^{13}$ W·cm^{-2}, $f_p = 1.6$ MHz PRF, and $P_{av} = 72.2$ W average laser power. (a) X-ray emission dose rate and (b) X-ray photon spectra as a function of intra-line pulse distance. The dotted lines in (a) guide the eye.

The highest X-ray emission dose rate was measured of $H''(0.07)=44.9$ mSv·h−1 (0.07)=44.9 mSv·h−1 at 0.88 µm intra-line pulse distance. At smaller or larger intra-line pulse distances, the maximum X-ray emission dose reduced significantly to $H''(0.07)=18.3$ mSv·h−1 (0.07)=18.3 mSv·h−1 at 0.5 µm or rather $H''(0.07)=3.8$ mSv·h−1 (0.07)=3.8 mSv·h−1 at 6.5 µm, respectively.

For tungsten, it is reported in the literature that the X-ray emission maxima and minima resulted from multi-beam reflections at the sidewalls of the ablated grooves [10], which (influenced by the intra-line pulse distance) originated with different micro groove wall angles. Here, in our study on AISI 304, a substantial effect of the surface topography on the X-ray emission dose can largely be ruled out. This is because the pronounced X-ray emission maximum at 0.88 µm intra-line pulse distance could also be confirmed for different surface topographies (not presented here), such as those produced with the intra-line pulse distances specifically addressed in **Figure 6**a. A more likely explanation for the observed great X-ray emission at small intra-line pulse distance is a strong interaction of the next irradiating pulses with the still existent laser ablated plasma/nanoparticle plume. As a matter of fact, an evident laser plasma/nanoparticle ablation plume has been verified in pump-probe shadowgraph analyses for the times 0.5–1.0 µs after pulse irradiation [**20**]. These time intervals correlate very well to the time domain of 1.6 MHz repetitive pulses presented in **Figure 6**.

This can be linked to the effect of the intra-line pulse distance on X-ray emission, as the geometrical spacing defines the angle of incidence for the next impinging laser pulse in relation to the flank of the apparent laser ablation plume; see **Figure 7**.

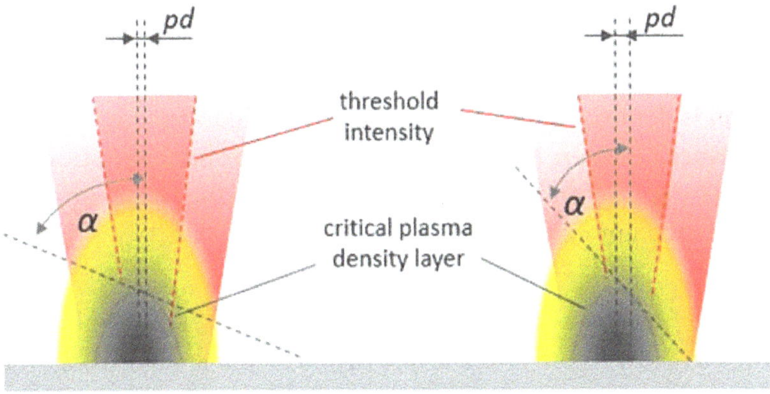

Figure 7. Schematic view on the effect of the intra-line pulse distance *pd* on the angle of incident of the laser beam α with respect to the flank of the critical plasma density layer.

Nevertheless, the laser ablation plume is on a much larger dimension than the very little intra-line pulse distance of only a few micro meters, showing the greatest impact on X-ray emission in **Figure 6**. Therefore, on the one hand, a reasonable explanation might be that an efficient laser pulse with plasma interaction takes place close to the solid density state [**21**]. On the other hand, up to 80% absorption levels were reported for p-polarized laser beams under striking incidence at steep plasma density profiles by collision-less resonance absorption mechanisms [**21**]. Highly excited plasma states and steep plasma density gradients emerge in regions of intense laser excitation at the highest laser peak intensity being in the narrow center of the Gaussian laser beam. So, it is obvious that the effective cross-section area of the high-intensity center part of the laser beam with the critical plasma layer is considerably smaller than the interaction area of the laser beam with the induced plasma/nanoparticle plume. Hence, a variation of a few micrometers of the intra-line pulse distance changes the position of the next irradiating pulse on the flank of the critical plasma layer. In fact, it has been demonstrated so far that plasma resonance absorption is very sensitive to the specific conditions of the resonant field near the critical density surface [22]. From this, it becomes clear that the distance of the next following pulse from the critical plasma density layer could significantly affect the incident angle of the laser beam relative to the critical plasma layer flank, in turn, with a high impact on the laser plasma resonance absorption and thus optical energy transfer.

However, additional collision-less absorption processes have been identified so far for intense laser pulses interacting with overdense plasma; for example, sheath inverse Bremsstrahlung [23], non-linear mechanisms [24] or anharmonic resonance effects [25]. These absorbing processes depend strongly on the angle of incident and the polarization of the incident laser beam and cannot be ruled out to have an effect on the X-ray photon emissions detected in this study.

The higher level of X-ray emission at the smaller intra-line pulse distances is also confirmed by the X-ray emission spectra presented in **Figure 6b**. It can be seen that X-ray emission is dominated by the Bremsstrahlung continuum at the larger intra-line pulse distance between 2 µm $< pd <$ 6.5 µm, while characteristic X-ray emission and increased X-ray photon flux is obvious for the smaller intra-line pulse distance ranging between 0.5 µm $< pd <$ 1.0 µm.

The highest X-ray photon yield and most prominent characteristic line emission were measured for pulses of 0.75 µm and 0.88 µm intra-line pulse distance corresponding to the position of highest X-ray dose rate in Figure 6b. In addition, the spectral width of the monitored X-ray photon fields enlarged to higher energy with pulses of smaller intra-line pulse distance. In fact, 8.0 keV maximum photon energy was detected at a 5 µm intra-line pulse distance compared to 11.0 keV at 0.88 µm, respectively. These enlarged X-ray spectra indicate even for the low-intensity pulses irradiated at small intra-line pulse distance a higher degree of plasma ionization and, therefore, a more-efficient coupling of the irradiating low-intensity ultrashort pulses. This is supported by the fact that the enlarged X-ray photon spectra of Figure 6b could be observed similar to the ones shown above in Figure 4 for higher electron temperatures induced by high-intensity pulse irradiations. As a consequence, regarding dependency on the applied intra-line pulse distance, a change of the underlying coupling mechanisms can be supposed for the low-intense pulse regime, thus from inverse Bremsstrahlung absorption at a larger intra-line pulse distance to high-efficient optical energy transfer by plasma resonance absorption for the smaller ones. However, the underlying physical principles for the strong impact of the intra-line pulse distance causing up to 12-fold enhanced X-ray emissions even in case of low-intensity ultrashort pulses could not be clarified so far. Even though the interaction of the next following laser pulse(s) with the still apparent laser ablated nanoparticle plume at 1.6 MHz time scale could be a potential effect for the observed higher X-ray emissions. This was not considered yet, but evidence therefore can be found in the literature reporting on enhanced X-ray emissions from nanoparticles irradiated by ultrashort laser pulses [26,27].

3.5. Effect of the Temporal Pulse Distance on X-ray Emission in the Low-Intensity Pulse Regime

It becomes obvious that the temporal delay between the pulses will also affect X-ray emission when considering plasma resonance absorption as the dominant optical transfer mechanism for low-intensity pulses [**28**]. As already stated above, this is due to the fact that the actual shape of both the laser ablation plume and the critical plasma density layer is continuously changing with time. From this, it can be supposed that the impinging ultrashort pulses will meet different conditions in the interaction cross-section, depending on the inter-pulse delay defined by the PRF. Hence, the operating plasma density and actual flank angle are directly affected by the PRF, in turn, having an immediate impact on the optical energy transfer and further on X-ray photon yield.

With increasing PRF, which corresponds to shorter temporal inter-pulse delays, the X-ray photon emission increases to maximum level, as can be seen in **Figure 8** for pulses of $I_0 = 10 \times 1.6^{13}$ W·cm^{-2} irradiated at PRFs in the range of 1.6 MHz. A further increase of the applied PRF to 2 MHz caused a prompt drop of the X-ray photon yield. Here, too, a considerable impact of the intra-line pulse distance is recognizable when comparing the maximum X-ray dose rates presented in **Figure 8**a,c. At a 5.0 µm intra-line pulse distance, the maximum X-ray dose rate was measured of $H'(0.07)$=4.6 mSv·h−1˙, while 8-fold higher X-ray emissions of $H'(0.07)$=37.7 mSv·h−1˙were detected at 0.88 µm, respectively. In addition, a significant difference of the X-ray emission characteristic can be seen in the corresponding X-ray photon spectra pointed out in **Figure 8**b,d. The X-ray photon spectra recorded at 5.0 µm intra-line pulse distance are characterized by broad continuous Bremsstrahlung emission in the photon energy range below 8 keV; see **Figure 8**b. At 0.88 µm in **Figure 8**d, by contrast, the X-ray photon spectra enlarged to 12 keV photon energy featured by dominant peaks of the characteristic X-ray emission lines. These results confirm the aforementioned change of the energy transfer mechanisms in the low-intense pulse regime from inverse Bremsstrahlung absorption to plasma resonance absorption. Thus, a strong influence of the actual conditions of the laser ablated plasma/nanoparticle plume on the effective laser beam absorption mechanism is suggested, which depends on both the temporal and spatial inter-pulse distance.

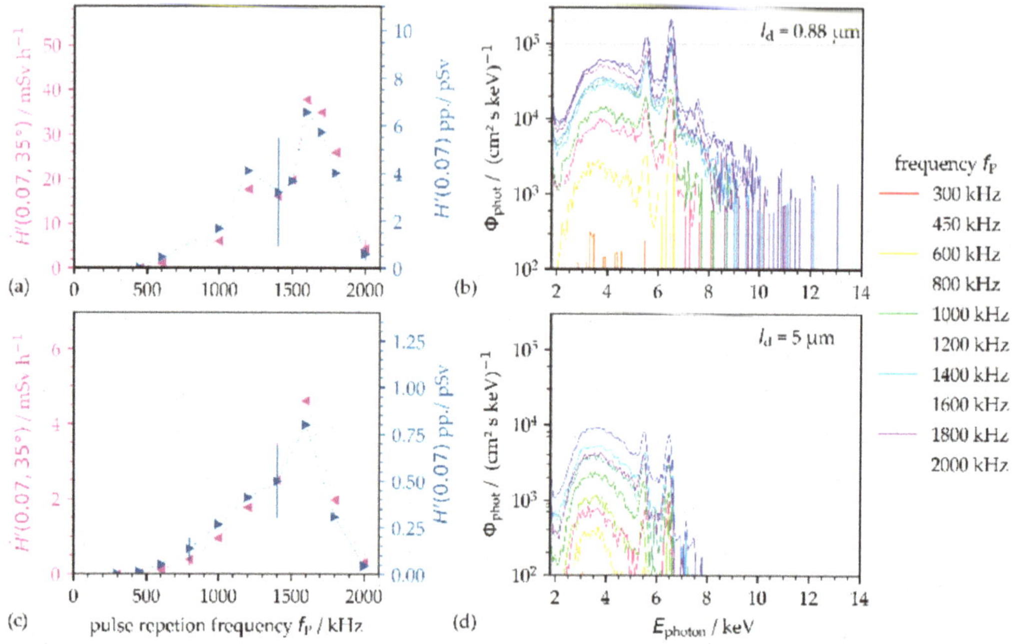

Figure 8. Effect of the pulse repetition frequency on X-ray emissions in the low-intensity pulse regime. Two intra-line pulse distances, (a,b) 0.88 µm and (c,d) 5.0 µm, were studied at 100 mm distance and 35° detection angle between the AISI 304 substrate and SILIX monitor. The dotted lines in (a,c) guide the eye.

3.6. X-ray Emissions in the Burst and Burst-in-Burst Laser Pulse Regime

Laser pulse train processing in the burst and burst-in-burst (bi-burst) regime offers more flexibility to adjust and tailor optical energy deposition into the material. This is mainly due to the fact that the optical energy of a single irradiation event (burst) is divided by the number of individual pulses within the laser burst, while in the bi-burst regime, the individual pulses were further separated in sub-pulses. In the burst regime applied here, the time delay between the individual pulses within the burst was 15 ns and between the burst sub-pulses in the bi-burst regime was 440 ps, respectively. Two burst repetition frequencies were studied: PFRB = 400 kHz and 1 MHz in bi-directional processing.

Initial studies on the burst mode in laser material processing showed a considerable drop of the material removal rate for each second irradiating pulse, especially when the next-following pulse was delayed to the previous one in the ten-nanosecond time domain [29]. Therefore, a reasonable explanation is the interaction of the second, fourth, sixth (and so on) intra-burst pulse with the ablation plasma/nanoparticle plume induced by the preceding pulse(s). Corresponding to the differences in material removal, a variance can also be observed for the X-ray emissions in the burst-mode regime monitored at a 35° detection angle and 100 mm distance between the substrate and SILIX detector; see **Figure 9**. For reference, the spectral X-ray photon flux and X-ray emission dose rate for 200 fs pulses of I_0 = 10 × 3.8^{13} W·cm^{-2} applied in the single pulse (-1pulse burst) regime is presented in **Figure 9**a. At 400 kHz PRF, the full available power P_{av} = 33.0 W of the burst-mode laser was irradiated by repeated crossing (bi-directional, up to 5 scans) the substrate at 0.4 m·s^{-1} laser beam moving speed and a corresponding 1 μm intra-line pulse distance. The given X-ray emission spectra with a maximum of 12.5 keV photon energy are featured by distinct characteristic X-ray emission lines, and the Bremsstrahlung photon yield is relatively low. For the first scan crossing, the X-ray emission dose rate was measured to *H*''(0.07)=0.6±0.2 mSv·h-1. This yields about 15 μSv·h^{-1}·W^{-1} and is in the range of the dose rate levels presented above in **Figure 8**c for pulses at similar PRF (450 kHz) but longer pulse duration (600 fs instead of 200 fs). With ongoing processing, initially, the X-ray emission dose rate increased significantly in the second scan to *H*''(0.07)=10.8±2.0 mSv·h−1'(0.07)=10.82.0± mSv·h−1 maximum. While in the further course of scanning, the X-ray emission reduced steadily to lower levels measured in the fifth scan of *H*''(0.07)=1.4±0.3 mSv·h−1.

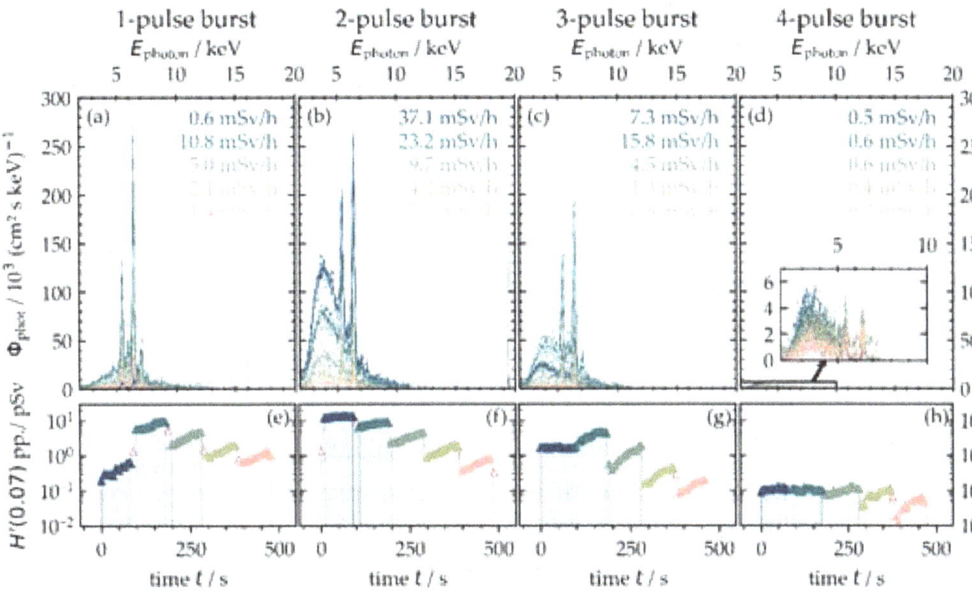

Figure 9. X-ray emission in the laser burst processing regime at a 400 kHz burst repetition frequency monitored at 100 mm distance, 35° detection angle, and five individual scans across the substrate. The intra-burst pulse number was varied between 1 (a) and 4 (d). The spectral X-ray photon flux and X-ray emission dose rate (a–d) as well as the X-ray dose per pulse (e–h) are presented for increasing number of scan crossings.

For two individual pulses in the laser burst (two-pulse burst in **Figure 9**b), a considerably higher amount of Bremsstrahlung emission was detected. This is potentially arising from the highly excited electron field induced by strong laser pulse with plasma interaction in the time domain of 15 ns, resulting in almost four-fold higher X-ray emissions than in the single pulse regime. It is noteworthy that such a high level of X-ray emission occurred with intra-burst pulses of half of the intensity of the single pulses (one-pulse burst, **Figure 9**a), because the impinging optical energy of the irradiation event is divided by the number of individual pulses in the two-pulse burst. By comparing these X-ray emissions with dose rates from pulses of similar intensity and PRF in **Figure 8**c, a major growth from 12.5 $\mu Sv \cdot h^{-1} \cdot W^{-1}$ for single pulses to 1120 $\mu Sv \cdot h^{-1} \cdot W^{-1}$ in the two-pulse burst regime can be recognized. These results confirm the strong plasma plume interaction in the laser burst regime with further evidence on plasma resonance absorption as the dominant energy transfer mechanism for pulses irradiating with very short time delay.

For a three-pulse burst, as shown in **Figure 9**c, the X-ray emission dose and Bremsstrahlung continuum originated somewhat in between single pulses and two-pulse burst irradiations. In this irradiation regime, the intensity of the individual intra-burst pulses reduced to one-third from the single pulses. With a higher number of pulses in a burst, the X-ray dose rates were found in the range of single pulses or below. This is exemplified in **Figure 9**d for laser bursts consisting of four intra-pulses. Accordingly, the intensity of the individual pulses reduced by a factor of four, which might be the reason for the Bremsstrahlung dominated X-ray emission spectra of low photon energy ranging between 2 and 7.5 keV.

For single-pulse processing at PRF = 1.0 MHz and P_{av} = 33.0 W average laser power, no X-ray emission could be detected, which might be due to the fact that the respective peak intensity $I_0 = 10 \times 1.6^{13}\ W \cdot cm^{-2}$ of the impinging pulses was too close to the threshold for X-ray photons generation on AISI 304. The observation is a bit in contrast to **Figure 5**, where the emission of X-ray photons can be seen for intensities even below of $I_0 = 10 \times 1.0^{13}\ W \cdot cm^{-2}$. However, this was achieved with pulses of 600 fs duration, while the pulse duration of the burst-mode laser presented here was 200 fs, which will also affect X-ray photon emission and equivalent dose rates.

However, a significant enhancement of the X-ray emission dose rate accompanied by pronounced characteristic X-ray emissions were detected for laser bursts irradiating at a 1.0 MHz burst repetition frequency and corresponding lower peak intensity of the intra-burst pulses. The maximum X-ray emissions were measured to $H''(0.07)$=32.8±3.6 $mSv \cdot h^{-1}$ (3rd scan) with the two-pulse burst and $H''(0.07)$=14.1±1.6 $mSv \cdot h^{-1}$ (2nd scan) with the three-pulse burst; see **Figure 10**a,b. AISI 304 processing with a higher number of intra-burst pulses yields much lower X-ray emissions, and the dominant Bremsstrahlung continuum shifted to X-ray photon energies below 7.5 keV, as can be seen in **Figure 10**d for the six-pulse bursts.

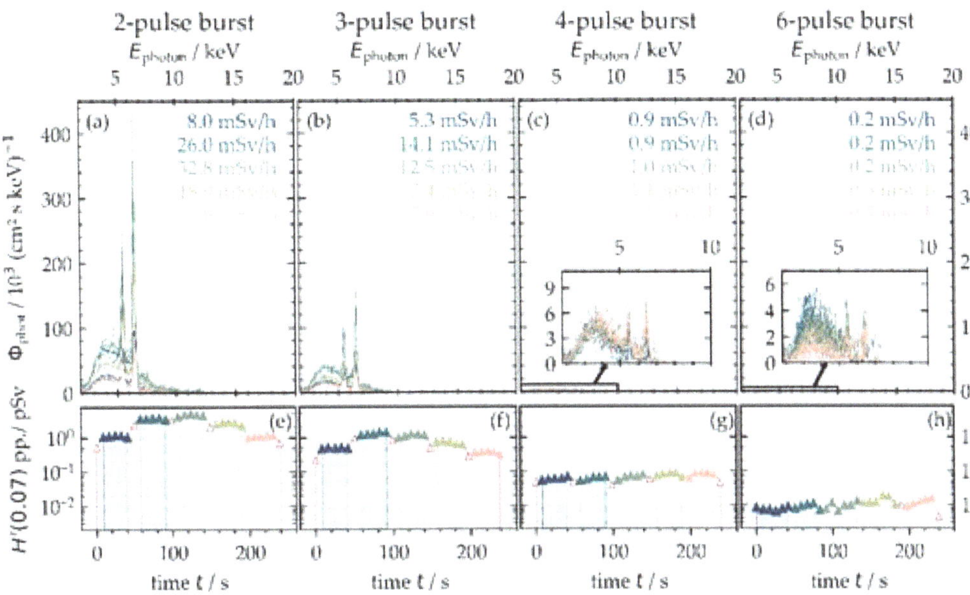

Figure 10. X-ray emission in the laser burst processing regime at 1.0 MHz burst repetition frequency monitored at 100 mm distance, 35° detection angle, and five individual scans across the substrate. The intra-burst pulse number was varied between 2 (a) and 6 (d). The spectral X-ray photon flux and X-ray emission dose rate (a–d) as well as the X-ray dose per pulse (e–h) are presented for an increasing number of scan crossings.

Another separation of the pulses in the bi-burst regime showed no further X-ray emission enhancement in the studied parameter range. For example, the X-ray emission dose rate was in the range between 0.6 mSv·h^{-1} < $H''(0.07)$ < 0.8 mSv·h^{-1} for a two-sub-pulse bi-burst of 440 ps inter-pulse delay at 400 kHz PRF_B and 33.0 W average laser power, as shown in **Figure 11**a. By irradiating two sub-pulses in a two-pulse burst processing scenario, such as shown in **Figure 11**b, the X-ray emission dose rate increased significantly to $H''(0.07)$=30.0±1.2 mSv·h-1' maximum level, even when the peak intensity of the operating sub-pulses reduced to a quarter of the original single pulses below $I_0 = 10 \times 1.0^{13}$ W·cm^{-2}. However, with a higher number of pulses within the burst and bi-burst pulse train, the X-ray emission reduced to a lower value, which is mainly attributed to the corresponding lower peak intensity of the sub-pulses limited by the maximum energy of a laser burst; see **Figure 11**c,d.

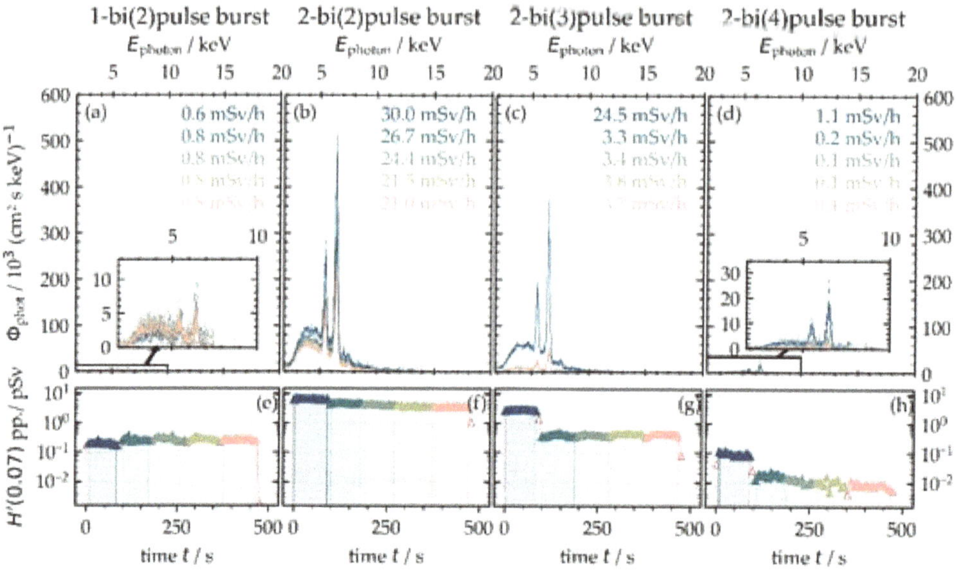

Figure 11. X-ray emissions in the laser bi-burst processing regime with notable X-ray dose rates at 400 kHz burst repetition frequency at 100 mm distance, °35 detection angle, and five individual scans across the substrate. The graphs display the high influence of the second MHz-burst pulse (two-pulse burst) with GHz bi-bursts (4–2 (b–d)) in contrast to the GHz bi-burst (a). The peak intensity of the sub-pulses was I_0 = 10 × 0.48/0.63/0.95/1.9[13] W·cm^{-2} The spectral X-ray photon flux and X-ray emission dose rate (a–d) as well as the X-ray dose per pulse (e–h) are presented for increasing number of scan crossings.

Nevertheless, the preliminary findings from the burst and bi-burst processing regime underscore the significant impact of laser pulse and plasma interaction on X-ray photon production. The second intra-burst pulse is particularly significant, resulting in a substantial increase in the X-ray emission dosage rate. This will be demonstrated in the current investigation, even for ultrashort pulses in GHz pulse trains at average laser powers of hundreds of Watts.

4. CONCISE SUMMARY

The presented study on laser-induced X-ray emissions on technical-grade AISI 304 stainless steel provides an overview on the laser processing under different conditions and parameter variations affecting the spectral X-ray photon flux and X-ray emission dose rate. In total, four different ultrashort pulse laser systems were applied providing complementary laser beam characteristics, including pulse peak intensities between 8 × 10^{12} W·cm^{-2} < I_0 < 10 × 5.2^{16} W·cm^{-2}, 2.0 MHz maximum PRF, up to 72.2 W average laser power as well as burst and bi-burst processing mode. The findings of this study confirm the general trend toward higher X-ray photon emission for pulses of higher peak intensity. Thereby, the monitored X-ray photon emissions are in good agreement with data reported previously by other groups; see

Figure 12. The presented efficient X-ray emission dose rates were calculated from the X-ray emission dose rates monitored at 100 mm distance and 35° detection angle divided by the applied average laser power.

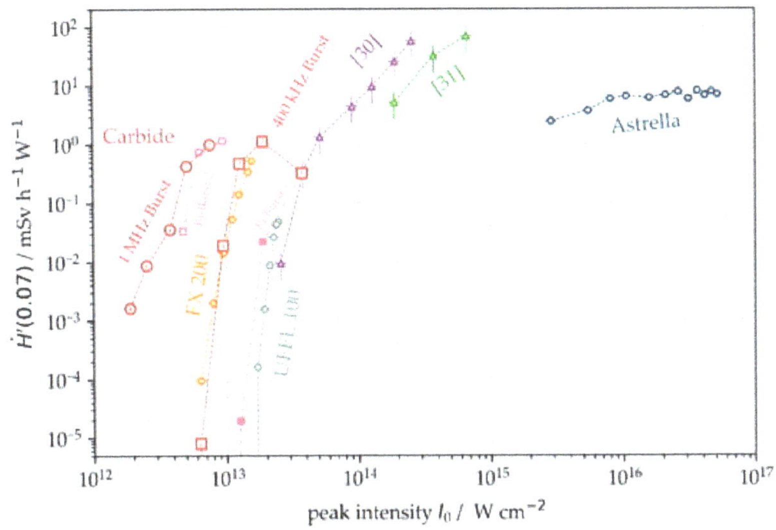

Figure 12. Summary of efficient X-ray emission dose rates as a function of peak intensity for different process regimes and laser systems. FX 200 (f_P = 1.6 MHz, ld = 0.88 µm); UFFL 100 (f_P = 506 kHz, ld = 0.88 µm); Astrella (f_P = 1 kHz, ld = 20 µm); [30] provided data recalculated for 100 mm detector distance, [31] provided data recalculated for 100 mm detector distance.

However, an unexpected high level of X-ray photon emission dominated by broad Bremsstrahlung emission could be detected for low-intensity pulses irradiation under specific processing conditions. For example, a maximum X-ray emission dose rate $H^{\cdot}(0.07)$ = 2.8 ± 41.3 mSv·h^{-1} and 0.4 mSv·h^{-1}·W^{-1} corresponding to an effective X-ray emission dose were detected for pulses of I_0 = 10 × 1.6^{13} W·cm^{-2} irradiated at 1.6 MHz PRF, 0.88 µm intra-line pulse distance, and 65 W average laser power. Therefore, a strong interaction of the next laser pulse(s) with the still apparent nanoparticle/plasma plume induced by preceding laser pulse ablations is suggested. As a consequence, plasma resonance absorption and effective electron plasma heating dominates the optical energy transfer, yielding broad Bremsstrahlung emission from the highly excited electron field. As a result, a considerable amount of X-ray photon emission could be detected even for low-intensity pulses at peak intensities considerably below the threshold value known so far. In summary, the following main conclusions can be stated on the basis of the presented results:

X-ray photon emission per pulse increases exponentially with peak intensity in the low-intense pulse regime (I_0 < 10 × 1.6^{13} W·cm^{-2}) and is on linear growth with high-intensity pulses (I_0 > 10^{16} W·cm^{-2});

Unexpected high X-ray emission dose rates ($H^{\cdot\prime}(0.07)$) > 20 mSv·h^{-1}) are achieved when low-intensity pulses irradiate at a small intra-line pulse distance and megahertz pulse repetition frequency (1.6 × 10^{13} W·cm^{-2}, 1.6 MHz, 0.88 µm intra-line pulse distance);

Resonance plasma absorption and strong electron plasma heating are suggested as underlying process for highly excited electron fields yielding broad Bremsstrahlung spectra up to 12.5 keV X-ray photon energies and pronounced characteristic X-ray emission lines;

Within burst and bi-burst pulse trains, in particular, the second intra-burst pulses enhance significantly the X-ray emission dose rate that is potentially induced by strong laser pulse with plasma interaction.

Concerning kW-class ultrashort pulse lasers in materials processing, this undesirable X ray emission might build to detrimental X-ray dose levels. To advance understanding in this domain, our current research concentrates on high-average power lasers and megahertz pulse repetition frequency and burst-mode laser beams to elucidate the relationships between the inherent properties of laser beams and X-ray emission. A future task is to devise and implement appropriate X ray protection solutions, which are essential for facilitating the use of ultrashort pulse lasers as a potent instrument in contemporary microfabrication.

REFERENCES

1. Roecker, C.; Loescher, A.; Delaigue, M.; Hoenninger, C.; Mottay, E.; Graf, T.; Abdou Ahmed, M. Flexible Sub-1 ps Ultrafast Laser Exceeding 1 kW of Output Power for High-Throughput Surface Structuring. In *Advanced Solid State Lasers*; Optical Society of America: San Diego, CA, USA, 2019; p. AM4A.2. **[Google Scholar]**

2. Mueller, M.; Aleshire, C.; Klenke, A.; Haddad, E.; Légaré, F.; Tuennermann, A.; Limpert, J. 10.4 kW coherently combined ultrafast fiber laser. *Opt. Lett.* **2020**, *45*, 3083–3086. **[Google Scholar] [CrossRef] [PubMed]**

3. Kamlage, G.; Bauer, T.; Ostendorf, A.; Chichkov, B.N. Deep drilling of metals by femtosecond laser pulses. *Appl. Phys. A* **2003**, *77*, 307–310. **[Google Scholar] [CrossRef]**

4. Gruner, A.; Schille, J.; Loeschner, U. Experimental Study on Micro Hole Drilling Using Ultrashort Pulse Laser Radiation. *Phys. Procedia* **2016**, *83*, 157–166. **[Google Scholar] [CrossRef] [Green Version]**

5. Mauersberger, S.; Schille, J.; Kujawa, K.; Schneider, L.; Million, C.; Hartung, K.; Oehlert, K.; Loeschner, U. High-precision surface profiling using multi-hundred watts ultrashort pulse lasers and ultrafast polygon-mirror based scanner. *J. Laser Micro Nanoeng.* **2020**, *15*, 1–9. **[Google Scholar]**

6. Gafner, M.; Kramer, T.; Remund, S.M.; Holtz, R.; Neuenschwander, B. Ultrafast pulsed laser high precision micromachining of rotational symmetric parts. *J. Laser Appl.* **2021**, *33*, 012053. **[Google Scholar] [CrossRef]**

7. Mueller, F.A.; Kunz, C.; Graef, S. Bio-Inspired Functional Surfaces Based on Laser-Induced Periodic Surface Structures. *Materials* **2016**, *9*, 476. **[Google Scholar] [CrossRef] [PubMed]**

8. Schille, J.; Schneider, L.; Mauersberger, S.; Szokup, S.; Hoehn, S.; Poetschke, J.; Reiss, F.; Leidich, E.; Loeschner, U. High-Rate Laser Surface Texturing for Advanced Tribological Functionality. *Lubricants* **2020**, *8*, 33. **[Google Scholar] [CrossRef] [Green Version]**

9. Legall, H.; Schwanke, C.; Pentzien, S.; Dittmar, G.; Bonse, J.; Krueger, J. X-ray emission as a potential hazard during ultrashort pulse laser material processing. *Appl. Phys. A* **2018**, *124*, 407. **[Google Scholar] [CrossRef] [Green Version]**

10. Legall, H.; Schwanke, C.; Bonse, J.; Krueger, J. The influence of processing parameters on X-ray emission during ultra-short pulse laser machining. *Appl. Phys. A* **2019**, *125*, 570. **[Google Scholar] [CrossRef] [Green Version]**

11. Behrens, R.; Pullner, B.; Reginatto, M. X-ray emissions from materials processing lasers. *Radiat. Prot. Dosim.* **2019**, *183*, 361–374. **[Google Scholar] [CrossRef]**

12. Weber, R.; Giedl-Wagner, R.; Foerster, D.J.; Pauli, A.; Graf, T.; Balmer, J.E. Expected X-ray dose rates resulting from industrial ultrafast laser applications. *Appl. Phys. A* **2019**, *125*, 635. **[Google Scholar] [CrossRef] [Green Version]**

13. Freitag, C.; Giedl-Wagner, R. X-ray Protection in an Industrial Production Environment. *Photonics Views* **2020**, *17*, 37–41. **[Google Scholar] [CrossRef]**

14. Legall, H.; Bonse, J.; Krueger, J. Review of X-ray exposure and safety issues arising from ultra-short pulse laser material processing. *J. Radiol. Prot.* **2021**, *41*, R28. **[Google Scholar] [CrossRef]**

15. Liu, J.M. Simple technique for measurements of pulsed Gaussian-beam spot sizes. *Opt. Lett.* **1982**, *7*, 196. **[Google Scholar] [CrossRef]**

16. X-ray Mass Attenuation Coefficients, NIST STANDARD Reference Database 126. Available online: **https://dx.doi.org/10.18434/T4D01F** (accessed on ٥ March ٢٠٢١).

17. Gibbon, P.; Foerster, E. Short-pulse laser—Plasma interactions. *Plasma Phys. Control. Fusion* **1996**, *38*, 769–793. **[Google Scholar] [CrossRef]**

18. Schumacher, D. Investigation of Laser Driven Hohlraum Radiation and Energy Loss of Heavy Ions in Indirectly Heated Plasma. Ph.D. Thesis, Technische Universitaet Darmstadt, Darmstadt, Germany, 6 February 2012. **[Google Scholar]**

19. Sheng, Z.M.; Weng, S.M.; Yu, L.L.; Wang, W.M.; Cui, Y.Q.; Chen, M.; Zhang, J. Absorption of ultrashort intense lasers in laser-solid interactions. *Chin. Phys. B* **2015**, *24*, 015201. **[Google Scholar] [CrossRef]**

20. Kraft, S.; Schille, J.; Mauersberger, S.; Schneider, L.; Loeschner, U. Pump-probe imaging for process control and optimization in high-speed laser micro machining. In *Laser-Based Micro-and Nanoprocessing XIV*; International Society for Optics and Photonics: Washington, WA, USA, 2020; Volume 11268, p. 112681H. **[Google Scholar]**

21. Cerchez, M.; Jung, R.; Osterholz, J.; Toncian, T.; Mulser, W.P.; Ruhl, H. Absorption of Ultrashort Laser Pulses in Strongly Overdense Targets. *Phys. Rev. Lett.* **2008**, *100*, 24500. **[Google Scholar] [CrossRef] [PubMed] [Green Version]**

22. Dragila, R. Resonance absorption in inhomogeneous plasma with randomly rippled critical surface. *Phys. Fluids* **1983**, *26*, 1682–1687. **[Google Scholar] [CrossRef]**

23. Catto, P.J.; Moore, R.M. Sheath inverse bremsstrahlung in laser produced plasmas. *Phys. Fluids* **1977**, *20*, 704. **[Google Scholar] [CrossRef]**

24. Chen, L.M.; Forget, P.; Fourmaux, S.; Kieffer, J.C.; Krol, A.; Chamberlain, C.C.; Hou, B.X.; Nees, J.; Mourou, G. Study of hard X-ray emission from intense femtosecond:sapphire laser–solid target interactions. *Phys. Plasma* **2004**, *11*, 4439–4445. **[Google Scholar] [CrossRef] [Green Version]**

25. Mulser, P.; Bauer, D. Anharmonic resonance in intense laser-matter interaction: Key to collisionless absorption. *AIP Conf. Proc.* **2010**, *1308*, 63–65. **[Google Scholar]**

26. Mondal, S.; Chakraborty, I.; Ahmad, S.; Carvalho, D.; Singh, P.; Lad, A.D.; Narayanan, V.; Ayyub, P.; Ravindra, G.K.; Zheng, J.; et al. Highly enhanced hard X-ray emission from oriented metal nanorod arrays excited by intense femtosecond laser pulses. *Phys. Rev. B* **2011**, *83*, 035408. **[Google Scholar] [CrossRef]**

27. Sankar, P.; Thomas, J.; Shashikala, H.D.; Philip, R. Enhanced bremsstrahlung X-ray emission from Ag nanoparticles irradiated by ultrashort laser pulses. *Opt. Mater.* **2019**, *92*, 30–35. **[Google Scholar] [CrossRef]**

28. Horn, A.; Kaiser, C.; Ritschel, R.; Mans, T.; Russbüldt, P.; Hoffmann, H.D.; Poprawe, R. Si-Kα radiation generated by the interaction of femtosecond laser radiation with silicon. *J. Phys. Conf. Ser.* **2007**, *59*, 159–163. **[Google Scholar] [CrossRef]**

29. Jaeggi, B.; Cangueiro, L.; Bruneel, D.; de Ramos Campos, J.A.; Hairaye, C.; Neuenschwander, B. Micromachining using pulse bursts: Influence of the pulse duration and the number of pulses in the burst on the specific removal rate. In *Laser Applications in Microelectronic and Optoelectronic Manufacturing (LAMOM) XXIII*; International Society for Optics and Photonics: Washington, WA, USA, 2018; Volume 10519, p. 1051905. **[Google Scholar]**

30. Legall, H.; Bundesanstalt für Materialforschung und -prüfung (BAM), Berlin, Germany. Personal communication, 8 December 2020.

31. Giedl-Wagner, R.; GFH GmbH, Deggendorf, Germany. Personal communication, 17 June 2021.

CHAPTER

8 USING A CALIBRATED (HIGH-SENSITIVITY) NOVEL X-RAY DETECTOR TO MEASURE THE X-RAY DOSE RATE AND SPECTRUM DURING ULTRAFAST LASER MACHINING

1. INTRODUCTION: AN OVERVIEW

Since the late 1960s, the laser has transitioned from a laboratory novelty to an advanced industrial instrument. Currently, the peak intensity of high-power laser systems can be utilized in a regulated manner as a machining tool for many metallurgical applications, such as welding, cutting, drilling, and surface hardening [1]. The primary benefit of laser machining is its ability to produce a superior quality product with less distortion or damage at an accelerated pace, which is essential for industrial requirements [2]. Recent evidence indicates that ultrashort laser machining can generate X-ray radiation at elevated repetition rates, provided the peak intensity surpasses 10^13 W/cm^2 [3,4,5,6]. Industrial laser sources employed for materials processing generally possess pulse lengths ranging from picoseconds to femtoseconds. Utilizing these pulse durations, it is straightforward to produce the aforementioned intensity. The elevated laser intensity causes electron temperatures in the generated plasma to reach several keV, leading to bremsstrahlung emissions in the keV X-ray range [7]. The transfer of laser energy to a plasma electron is contingent upon the prevailing mechanism of the laser-plasma interaction. When the kinetic energy of plasma electrons reaches a sufficient threshold, distinctive X-ray radiation may be produced.In recent publications, it has been shown that during ultrafast laser machining at high repetition rates of several 100 kHz, the amount of emitted X-ray radiation can exceed the regulatory exposure limits for members of the public [8]. Low X-ray doses per pulse can accumulate through high repetition rates and exceed radiation protection limits. Therefore, a detailed study of the influence of laser and the material parameters used for measuring the X-ray dose rate is crucial to assess appropriate safety measures. Legall et al. [9] measured the spectral X-ray emission and dose rates ($\dot{H}'(0.07)$) for different target materials using an ultrashort-pulse laser up to maximum peak intensity 2.6×10^{14} W/cm^2. They showed that the unwanted emission of X-ray radiation during ultrashort pulse laser processing of materials in air is commonly observed. These results were confirmed by Behrens et al. (2019) under similar experimental conditions [10]. Also, an analytical model was recently presented by Weber et al. (2019) to estimate the expected X-ray dose when laser machining with ultrashort-pulse lasers under industrial conditions up to an intensity 7×10^{14} W/cm^2 [4].

On the other hand, no work has been published for the X-ray emission during industrial machining, in the moderate intensity regime of 10^{14}–10^{16} W/cm^2 with high repetition rate lasers. One of the reasons is that until now there are no commonly available spectrometer with an immediate readout for measuring dose rates with high sensitivity. X-ray emission ranging from a few eV to about 10 keV occurs when processing most of the materials such as stainless steel, copper, various alloys, and ceramics [10]. This X-ray radiation can cause skin cancer hence, this emission range requires more attention. In this paper, we investigate the X-ray emission from different alloys, metals, and ceramics with two high repetition rate lasers. We employ a unique calibrated X-ray detector for measurements, offering good sensitivity within the 2–20 keV range. The experimental methods utilized for ascertaining dose rates and conducting spectral analysis of various alloys, metals, and ceramics are provided. We additionally compare the dosimeter OD-02 with the innovative spectrodosimeter Silix lambda (engineering office Prof. Dr.-Ing. Günter Dittmar) under uniform testing settings.

2. MATERIALS AND METHODS

2.1. Optical Setup

The studies utilize two distinct laser systems, with parameters detailed in Table 1. The laser beam is concentrated on the sample using a plano-convex lens/f-theta scanning objective with a focal length of 60 mm. The angle of incidence of the laser beam is orthogonal to the sample plane. The laser intensity varies from 10^{13}–10^{16} W/cm^2. The average power of the pulse is simultaneously measured at one arm from a beam splitter.

Table 1. Laser parameters used during micromachining.

Laser Parameter	KMLabs Dragon	Coherent Monaco
Wavelength	800 nm	1035 nm
Rep. rate	1 kHz	755 kHz
Pulse energy	1 mJ	72 μJ
Avg. power	1 W	54 W

Laser Parameter	KMLabs Dragon	Coherent Monaco
Pulse duration	35 fs	350 fs
Focus diameter ($1/e^2$)	12 ± 2 µm	22 ± 2 µm
Rayleigh length	0.2 mm	0.4 mm
Scanning speed	1.3 mm/s	1000 mm/s
Pulse overlap	10.7 µm	20.7 µm
Polarization	Linear	Linear
Beam profile	Gaussian	Gaussian
Angle of incidence	Perpendicular	Perpendicular
Scanning pattern	20 Spirals	20 Spirals
Scanning time	49 s	49 s

We utilize many metals for our measurements, including titanium, molybdenum, iron, copper, nickel, zinc, and tungsten. In addition, we process ceramics like ZrO_2, Si_3N_4, and Al_2O_3. Furthermore, brass, stainless steel, NiCuZn, and NiCrFe alloys have been exposure to strong laser radiation. The alloy % composition is available in Table 2. The sample is positioned with a motorized three-axis stage utilized for ultrafast machining in laboratories for the 1 kHz system and a scanner for the 755 kHz system (Figure 1). A spiral with a line spacing of 30 µm is inscribed to eliminate polarization dependence. The minimal internal diameter is 1 mm. The laser activation duration is 2.45 seconds each spiral, with a deactivation period of 4 milliseconds when returning to the starting location. The outer diameter was selected based on the repetition rate to ensure the overall time of the writing operation is approximately 49 seconds. The inter-pulse spacing is 1.3 µm for both lasers.

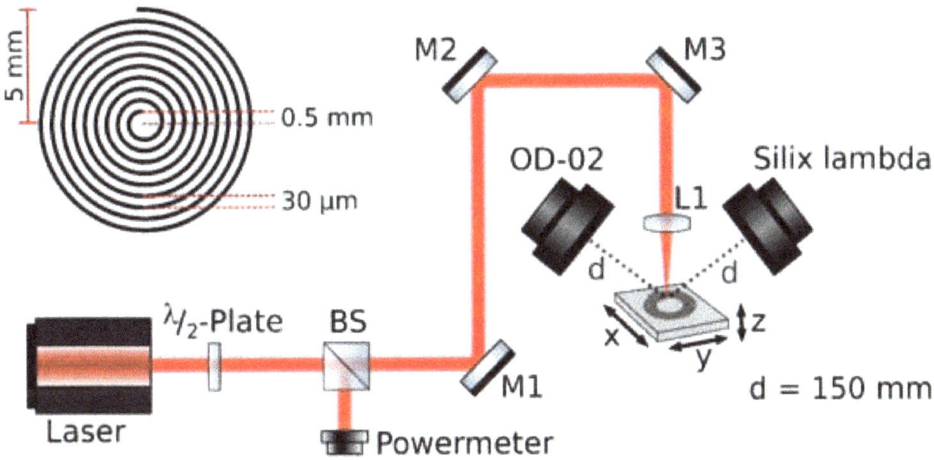

Figure 1. The schematic experimental setup is shown here. M1, M2, M3—Dielectric mirrors; L1—Plano-convex lens (f = 60 mm); BS—beam splitter. The schematic image of the spiral patterning on the material formed during the laser micromachining is shown in the left top. A power meter is used to verify the laser power used for the laser processing.

Table 2. List of alloys and its percentage of compositions used for the micromachining.

CuZn	Cu 58%	Zn 39%	Pb 3%
NiCuZn	Cu 47–64%	Ni 10–25%	Zn 15–42%
NiCrFe	Ni 72–76%	Cr 18–21%	Fe 5%
Steel 1.4310	Fe 70–77%	Cr 16–19%	Ni 6–9.5%

Two different detectors are used for the local dose rate ($\dot{H}'(0.07)$) measurements. A calibrated compact portable ion chamber dosimeter (STEP OD-02, PTW Freiburg, Freiburg im Breisgau, Germany) is used for skin dose measurements $\dot{H}'(0.07)$ at a distance of 150 mm from the processing point at an angle of 45° with respect to the laser line. At the same time, we use a calibrated spectrodosimeter Silix lambda, with the same measurement distance and angle, to obtain the X-ray spectral and dose rate information. The detailed description of this calibrated X-ray detector is given in the next section.

Figure 2 depicts the typical appearance of the material surface after processing of 20 spirals with the 755 kHz laser system with an irradiation of 4.2 ± 1014 W/cm2. It should be noted that during this experiment the laser parameters are not optimized for surface quality. Due to high pulse overlap and high average power heat the surface shows formation of resolidified material, indicating that heat accumulation occurred during the laser process. No predominant directional structures are observed in any sample.

(a) (b)

Figure 2. Shown is (a) the spiral after laser processing by the Coherent Monaco laser at 4.2 × 1014 W/cm2, (b) a detailed view from the structure.

2.2. Silix Lambda X-ray Spectrodosimeter

The Silix spectrodosimeter was specifically designed by the engineering firm of Prof. Dr.-Ing. Günter Dittmar for the measurement of X-ray radiation during material processing using ultrashort pulse lasers, providing a spectral resolution of 0.08

keV for evaluating X-ray output. The Silix spectrodosimeter is calibrated using a reference radiation source from the Physikalisch-Technische Bundesanstalt (PTB) in Braunschweig, Germany, through comparing measurements with standardized X-ray spectra. This spectroradiometer operates on the concept of the photoelectric effect to detect X-rays. The sensor is a silicon imaging device featuring up to 8 million pixels. Pulsed X-ray radiation produces electrical charge carriers in the silicon chip, which are analyzed by a rapid microcontroller. When an X-ray photon strikes an image pixel, a little current is produced. Higher intensity X-ray photons produce more photocurrent. Specialized signal processing isolates the photocurrent from interfering currents. A novel feature is the proprietary self-monitoring capability of the gadget to ensure the optimal performance of the Silix X-ray detector. The equivalent dose rates relevant for radiation protection are calculated internally in the device:

Directional dose equivalent rate for the skin $\dot{H}'(0.07)$.

Directional dose equivalent rate for the eyes $\dot{H}'(3)$.

Ambient dose equivalent rate $\dot{H}^*(10)$.

The energy spectrum of the photon irradiance is recorded in 235 channels ranging from 2–20 keV. It is stored and displayed as a curve with a unit of photons (keV cm^2 s). For measurement at very low irradiances (small dose rate), any number of individual spectra can be averaged. In this way, at photon energies between 10–2 keV—the most important range for laser-induced ionising radiation—dose rates as low as about 0.1 μSv/h can be displayed. In the range of up to 45°, the angular dependence of the measurement signal deviates less than 5% from the cosine function. The measuring sensor has an area of 0.1 cm^2 and thus enables homogeneous irradiation even at a very short distance of about 2 cm to the point of impact of the laser beam. The Silix lambda X-ray monitor, equipped with these qualities, is suitable for safety-critical monitoring activities in laser processing applications.

3. RESULTS AND DISCUSSION

The process by which X-rays are produced during ultrafast laser interactions with materials has been detailed in other sources [3,4,8]. Under experimental conditions, high-kinetic-energy electrons smash directly with electrons in the inner shell orbitals of atoms, leading to the emission of distinctive lines such as k-α and k-β, in addition to bremsstrahlung radiation. The X-ray photon flux generated by the laser-induced plasma is highly contingent upon the quantity of absorbed laser photons. The X-ray photon flow is contingent upon two factors: the initial interaction of the material with the laser and the following heating of plasma electrons [11]. Consequently, X-ray production is significantly influenced by laser and material characteristics. This study examines the X-ray emission produced during ultrashort pulse laser machining of materials frequently utilized in scientific laboratories and laser-based manufacturing sectors. The intensity is altered in the experiment by rotating a half-wave plate within the optical beam path. The outcomes of these trials are outlined in the subsequent sections.

3.1. Dose Rates

The maximum X-ray local dose rates measured for the Monaco (Coherent Inc., Santa Clara, CA, USA) and Dragon (KMLabs, Boulder, CO, USA) lasers in mSv/h as a function of the laser peak intensity using the OD-02 is shown in **Figure 3**. For the dose rate measurements, three ceramics (Al_2O_3, Si_3N_4, ZrO_2), four alloys (brass, steel, nickel brass, NiCrFe) and three metals (titanium, molybdenum, and tungsten) are used with the two laser systems.

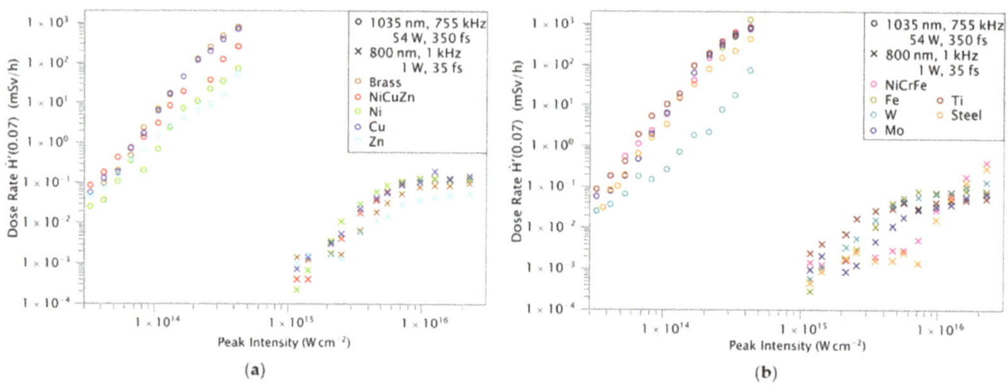

Figure 3. Measured dose rates $\dot{H}'(0.07)$ in dependence on the different target material and the incident laser peak intensity listed as: (a) Brass, and NiCuZn alloy and its composition metals Ni, Cu, and Zn are investigated using the ionization chamber dosimeter OD-02 at a distance of 150 mm from the interaction zone in air. The laser intensity was varied between 5×10^{13} and 10×5^{16} W/cm² by tuning the laser pulse energy for two different lasers. (b) NiCrFe and stainless steel alloys and its composition metals as well as W, Mo, and Ti are plotted in the same experimental conditions.

In the case of the Coherent Monaco laser system, the dose rate increases with the laser intensity for all materials. Thus, the dose rate reaches 1 Sv/h at the maximum laser intensity.

This corresponds to a maximum X-ray dose per laser pulse of about 0.37 nSv. At an intensity of 4.2×10^{14} W/cm², Iron and Molybdenum show the highest dose rate for metals. Among alloys, the highest dose rate is observed in NiCrFe and among ceramics in Si_3N_4. We notice a power dependence for the dose rate with the laser peak intensity $\dot{H}'(0.07) \propto I^5$.

The maximum dose rate for the KMLabs Dragon 1 kHz system is of 0.3 mSv/h which is significantly lower in comparison with the Coherent Monaco laser. This is about 22% of the dose per laser pulse (0.08 nSv) compared to the 755 kHz system. Despite the higher pulse energy of the KMLabs Dragon laser, we obtained a 4.6 × higher dose rate from the Coherent Monaco. Several parameters—such as central wavelength, pulse duration, spot size, and multi-pulse effects—contribute to this difference in X-ray dose rate. In particular, the bigger focal diameter of the Coherent Monaco laser leads to a 3.4 × larger illuminated area on the surface, while the longer wavelength allows for higher accelerations of the electrons in the plasma and thus contributes to the increase in dose rate. These two parameters could be primarily responsible for such difference in the dose rate measured [12]. Laser absorption to the material has a direct dependence over the wavelength of the laser. If the wavelength is low, the conversion efficiencies of laser photons to the plasma is reduced and it effectively reduces the X-ray dose rates [13]. The material properties have an influence on the conversion efficiency and thus on the resulting dose rate. Additionally, the high repetition rate of the Coherent Monaco laser system may contribute to the higher dose per pulse compared to the 1 kHz system. At high repetition rates heat accumulation in the material can reduce ablation thresholds [14] and increase material removal rates [15]. On the other hand, König et al. [16] have observed that hot material ejected by laser ablation of metals can be present up to several microseconds after the laser pulse. In [16] it is concluded that the material plume possibly consists of metal vapor and droplets which may lead to shielding effects at a repetition rate of some hundreds of kHz and beyond. Thus, it is likely that in our experiments with the 755 kHz laser system particle shielding is present and plasma shielding cannot be completely ruled out. Coupling a considerable amount of laser energy into a particle cloud or plasma of previous ablation events may contribute to an even stronger X-ray emission compared to low repetition rate laser ablation. When processing materials with the KMLabs Dragon laser, Copper has the highest dose rate for metals compared to the Coherent Monaco laser, while NiCrFe has the highest dose rate for alloys.

To understand the physical process of X-ray generation in laser materials processing, the fundamental mechanisms that result in X-ray generation during laser-plasma interaction must be studied. Two different mechanisms are needed to be considered for the generation of X-rays in the intensity range of 10^{12}–10^{16} W/cm^2, namely collisional absorption and collisional less absorption [17,18,19]. The plasma is heated via 'collisional' absorption (inverse bremsstrahlung), in which electrons disperse the kinetic energy absorbed from the laser radiation field through collisions with other electrons and ions in the plasma. Through this mechanism, hot electrons are generated and these electrons decelerate in the vicinity of ions and produce bremsstrahlung X-rays. These hot electrons produced in the plasma scale with intensity via inverse bremsstrahlung, and this ultimately results in higher X-ray emission and dose rates which is given by, $T_h \propto (I\lambda^2)^\alpha$ with T_h hot electron temperature, I, λ intensity and wavelength of the laser, respectively [20].

It is commonly accepted, that at intensities below 10^{13} W/cm^2 collisional absorption dominates the laser plasma interaction [17,18,19]. In this regime, the X-ray dose rate is power dependent on the intensity. However, for an intensity regime above 10^{15} W/cm^2, a mechanism where the laser electric field resonantly excites plasma waves, close to the surface of the over-dense plasma region, takes place. Here the laser field cannot propagate which is commonly referred to as 'collision-less' absorption regime and hence the dependency of the dose rate is weaker and almost saturates at a plateau. At this point the intensity depends on the dose rate as $(I\lambda^2)^{1/3}$ which confirms that resonance absorption dominates in this intensity regime.

When we compare the parameters of the used laser systems such as the wavelength, pulse duration, pulse energy and repetition rate, we observe a strong influence on the X-ray dose rate. At high laser pulse repetition rates in the hundreds of kHz range, the accumulated number of pulses over a time is very high. Hence, the heat accumulation on the target is increasing. This will support the plasma generation to produce more hot electrons and can explain the increase of X-ray doses at higher repetition rate.

In addition, the conversion efficiency of X-ray generation should be considered to estimate the dose rates, which is defined as the ratio of the energy emitted as X-rays to the energy of the incident laser pulse. This conversion efficiency depends critically on the target geometry in which the laser–plasma interaction takes place, material absorption coefficient, thermal conductivity and laser parameters—e.g., the laser pulse duration, wavelength, and intensity [11].

When we compare the X-ray emission from alloys with the emission from its composition metals, the dose rates of alloys can be increased or suppressed according to the dose rates of its composite metals. Take brass as an example, its dose rate scales similar to its composition metal of copper. On the contrary, the NiCuZn dose rate is suppressed due to its nickel composition. Similarly, in the case of the NiCrFe alloy, the dose rate follows that of iron and chromium rather than scaling with the Nickel dose rates. Therefore, the alloy dose rate is not completely relying on the dose rates of its metal compositions. However, these first observations are still preliminary and further work is needed in order to fully understand the dependencies on the materials.

The measured X-ray dose rates shown in **Figure 2** exceed the skin dose of 20 mSv per year already at an exposure time of 80 s [21]. The maximum dose rate $\dot{H}'(0.07)$ measured at the highest intensity is 899 mSv/h and 0.40 mSv/h for Coherent Monaco and KMLabs Dragon lasers, respectively. So, it is recommended to take necessary measures to ensure personal safety during the micromachining process. Also, we should find the working distance from the interaction zone with an acceptable dose rates and proper shielding for a safe working environment. The permissible occupational radiation exposure should be monitored before each experiment.

3.2. Spectra

This section presents the X-ray spectral distribution for all used materials at the maximum intensity of the KMLabs Dragon and Coherent Monaco laser systems. The Silix lambda spectrodosimeter is used for the spectral measurement and the X-ray detection range is spanned over 2–20 keV. The spectral photon sensitivity of the spectrodosimeter is corrected and calibrated for all 235 channels. In **Figure 4** and **Figure 5**, the spectra are averaged for 6 s and normalized to their maximum values. The X-ray photon counts are higher for the Coherent Monaco laser which is a high repetition rate laser.

The spectrum distribution of plasma emission comprises distinctive lines and a bremsstrahlung continuum, contingent upon the interaction of materials with laser energy. The bremsstrahlung emission is a continuous emission detected below the 5 keV X-ray energy range. In this intensity domain, multiphoton ionization (MPI) produces free electrons that acquire high kinetic energy, becoming 'hot' electrons by the resonant absorption of photons from the laser pulse. Bremsstrahlung emission of X-rays happens as these heated electrons decelerate. An augmentation in X-ray flux can consequently be attained by elevating the quantity of hot electrons, which are the principal contributors to bremsstrahlung [22]. The bremsstrahlung emission escalates with intensity.

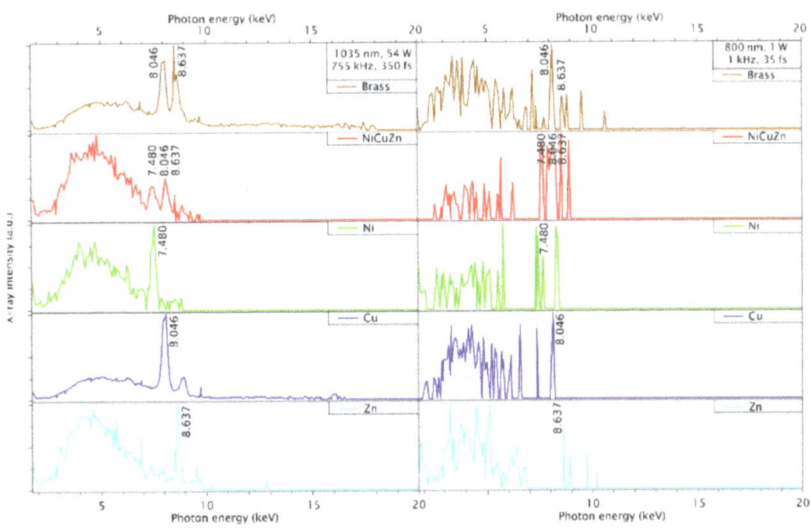

Figure 4. **X-ray spectra measured for the different target material and the incident laser peak intensity listed as brass and NiCuZn alloy and its composition metals Ni, Cu, and Zn are investigated using a Silix lambda detector at a distance of 150 mm from interaction zone in air at an intensity 4.2 × 10^{14} W/cm^2 (left) and 2.3 × 10^{16} W/cm^2 (right) respectively for the above mentioned two lasers.**

In high repetition rate lasers, bremsstrahlung and characteristic line emission occur with equal probability. The spectral emission of brass parallels its copper content; nevertheless, the spectral emission of NiCuZn aligns with the spectral distribution of nickel composition. Furthermore, the bremsstrahlung emission is augmented in the NiCuZn relative to brass, although it adheres to the spectrum emission of nickel. The spectrum emission of NiCrFe aligns with the spectral emissions of chromium and nickel, rather than those of iron. In comparison to the 1 kHz laser, the 755 kHz laser exhibits a more extensive bremsstrahlung emission. Bremsstrahlung emission or line emission predominates according on the substance. For example, comparing molybdenum with titanium, the dose rate of titanium consists mainly of the characteristic line emission at k-α ≅ 4.512 keV, while the dose rate of molybdenum consists mainly of bremsstrahlung. The characteristic line of molybdenum would be at 17.48 keV.

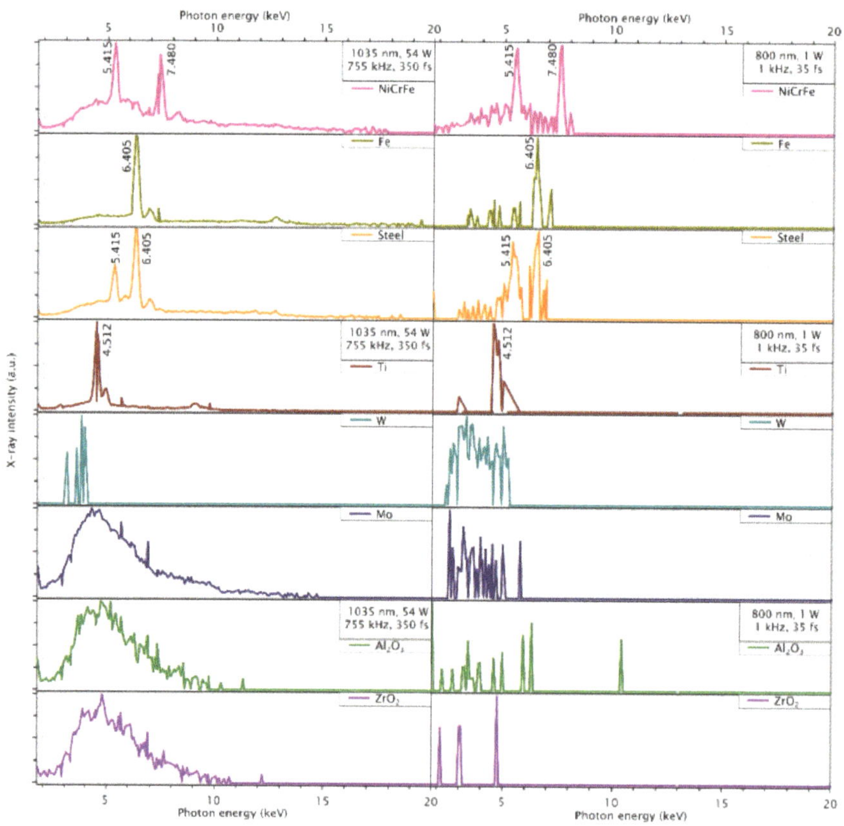

Figure 5. X-ray spectra measured for on the different target material and the incident laser peak intensity listed as NiCrFe and stainless-steel alloys and iron, titanium, tungsten, and molybdenum are investigated using the Silix lambda detector at a distance of 150 mm from the interaction zone in air at an intensity 4.2×10^{14} W/cm^2 (left), 2.3×10^{16} W/cm^2 (right) respectively for the above mentioned two lasers. The measured spectra of ceramics Al_2O_3 and ZrO_2 are plotted in the same experimental conditions.

In the case of ceramics, the X-ray emission of Si3N4 is significantly higher compared to other materials shown in Figure 6. It may be due to the nitrides mixed to the silicon. Hence more investigation is necessary to understand the X-ray emission and dose rates on the alloys and ceramics.

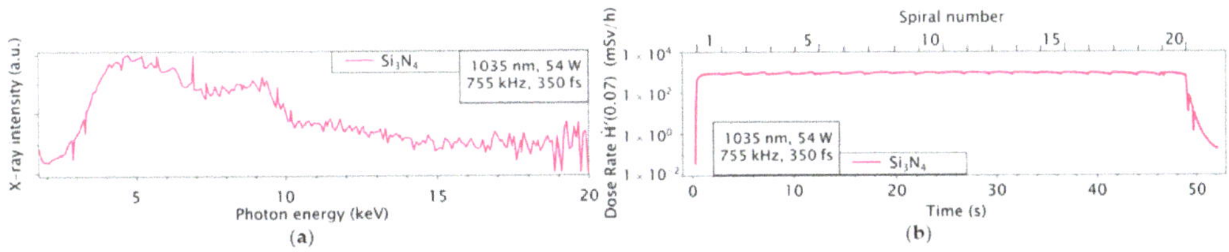

Figure 6. Shown in (a) is the spectrum of Si3N4 measured with the Silix lambda and in (b) is the corresponding dose rate over time measured with the OD-02.

The attenuation of X-ray doses by a shielding material is contingent upon the spectrum distribution of the incoming X-ray radiation; thus, a shielding factor can only be determined if the spectral photon flux is precisely known. Therefore, the spectrum distribution of the generated laser-X-ray radiation is essential for assessing a protection strategy [9]. The measured dose rates comprise a combination of bremsstrahlung and characteristic line emissions for all materials.

3.3. Discrepency in Dose Rates

This section will take a closer look at the dose rate over time since the laser spot is moving on the material in a spiral format during the material processing. Here, even small changes in laser intensity as well as the position of the focus undergo large effects on the generation of X-rays.

In **Figure** 7, we see the time-resolved measurement of the dose rate for stainless-steel with Coherent Monaco laser. The process takes about 49 s and consists of 20 spirals. The durations of the spirals are plotted at the top of the plot. The highest dose rate is observed when the laser focus is displaced by ~0.4 mm above the material surface (solid lines in **Figure** 7). In this situation, the X-ray emission stays at an almost constant level during the entire process duration. When the focus is set at the material surface (dashed lines) the X-ray emission drops after an initial high peak. Additionally, an oscillation is visible which correlates well with the time evolution of the spirals. The strong discrepancy induced by a small shift in focus position is only present at high laser irradiance. In the lower irradiance regime ($<10^{14}$ W/cm^2) the X-ray emission is strongest when the laser focus is set at the material surface, at an overall lower emission level.

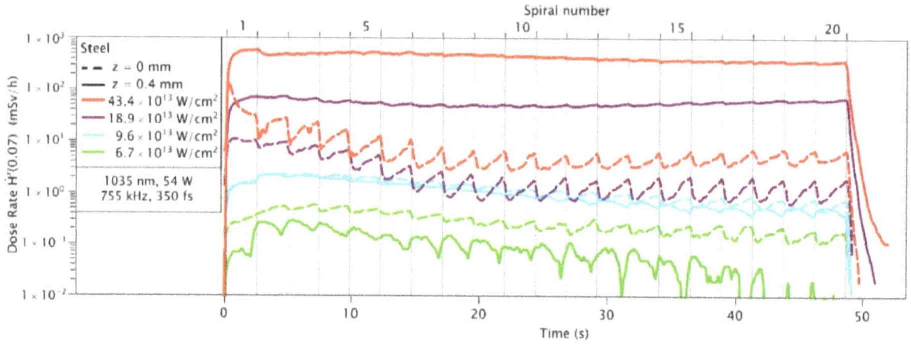

Figure 7. Dose rate Ḣ'(0.07) recorded during different laser processing steps as function of the processing time. The OD-02 dosimeter was placed at a distance of 150 mm from the interaction zone. Each of the individual processing steps is marked by a number on the top of the graph and separated by vertical lines. The laser intensity is 4.2 × 1014 W/cm2 at a repetition rate 755 kHz and the focal spot diameter 22 µm.

This observation supports the idea that particle and/or plasma shielding strongly affects the X-ray generation for high repetition rate processes at high laser intensities. In this scenario, a slight shift of the focus above the material surface may increase coupling efficiency of laser energy into the ablation plume of previous ablation events, and thus increases the amount of X-ray emission from the re-heated ablation plume. This effect is for high laser intensities most dominant as a large and long living plume can be present. At lower laser intensities the ablation plume is weaker and the effect disappears to a point where the ablation rate of a defocused laser beam is lower resulting in lower X-ray emission compared to the focused beam.

3.4. Comparison Silix and OD-02

The X-ray radiation during the micromachining is measured with two detection systems. For the dose rate measurements, the Silix lamda spectrodosimeter is used in addition to a highly sensitive ionisation chamber (OD-02). In addition to the dose rate, the Silix lambda simultaneously measures the spectrum in the X-ray energy range 2–20 keV. The measurement time for the dose rate of the Silix lambda spectrodosimeter is set to 6 s. For the dose rate measurements of the OD-02, the measurement time is 0.078 s. In order to compare both measurements, the measured dose rates shown in Figure 8 were averaged over the whole scanning time.

From the Figure 8, all the materials follow a similar trend. For a comparison of the two detectors, the maximum and average X-ray dose rates from the OD-02 detector are reported in Table 3, together with the X-ray dose rates from the Silix lambda spectrodosimeter for the two laser systems at the respective maximum intensity. The measurements of the two devices are in good agreement. The strong deviation of more than 70% in the case of Si3N4 can be explained by the spectrum, as the measured X-ray radiation exceeds the measurement range of the Silix lambda spectrodosimeter. However, the X-ray energy regime should be considered while comparing these two detectors. OD-02 is specified 6–100 keV and Silix lambda X-ray spectrometer is specified 2–20 keV. The deviations in the measurements can be explained by the different measuring ranges as well as the spectral radiation depending on laser parameters and the material.

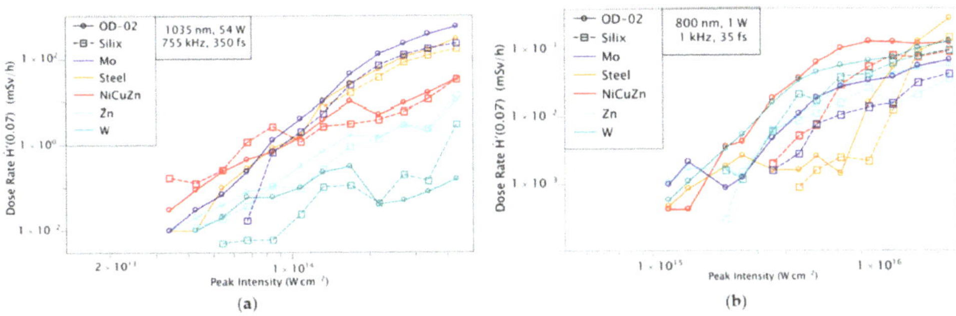

Figure 8. Comparison of the accumulated dose rates using Silix lambda and OD-02 detectors for an intensity range 5 × 10^{13} and 10 × 5^{16} W/cm^2 by tuning the laser pulse energy for Coherent Monaco (a) and KMLabs Dragon laser (b). The dotted line represents the dose measurement with the Silix and the solid line the dose measurement with the OD-02.

Table 3. Comparison of the measured dose rates from the Silix and the OD-02.

Dose Rate H'(0.07) mSv/h	KMLabs Dragon 2.26 × 10^{16} W/cm^2			Coherent Monaco 4.22 × 10^{14} W/cm^2		
	OD-02	OD-02 (Mean)	Silix Lambda	OD-02	OD-02 (Mean)	Silix Lambda
Brass	0.10	0.05	0.09	748	305	171
NiCuZn	0.13	0.05	0.09	248	33	33
Cu	0.15	0.11	0.12	678	229	212
Ni	0.13	0.06	0.08	70	0.77	5.46
Zn	0.06	0.04	0.04	53	10	14
NiCrFe	0.40	0.13	0.30	822	351	215
Steel	0.29	0.09	0.15	432	295	169
Fe	0.08	0.05	0.05	1220	453	237
Mo	0.07	0.06	0.04	745	610	231
Ti	0.05	0.04	0.05	793	166	159
W	0.13	0.12	0.10	72	0.16	2.98
Al_2O_3	0.0024	0.0011	-	21	20	10
ZrO_2	0.0031	0.0007	-	25	18	12
Si_3N_4		-	-	988	889	268

4. CONCISE SUMMARY

In conclusion, we have provided a comparative analysis of the emitted X-ray dose rate from laser-driven plasma across various target materials, including pure metals, diverse alloys, and ceramics, utilizing two distinct ultrafast laser systems. The X-ray dosage rate is significantly influenced by several laser parameters, such as wavelength, pulse duration, repetition rate, and pulse energy. For a laser intensity above 1013 W/cm^2, all target materials investigated, show a potential hazardous X-ray dose rate which increases with laser intensity. The application of mid-infrared laser wavelengths with multi-kHz repetition rates results in an enhanced dosage rate relative to conventional near-infrared kHz laser systems, even at reduced pulse energy. The pronounced dependence of the dosage rate on the laser focus position relative to the target surface suggests that an extraneous component, such as thermal accumulation or particle/plasma shielding, may influence the X-ray emission during material processing. Such effects are particularly pertinent in the context of multi-pulse irradiation with high repetition rate lasers. Accurate prediction of X-ray emission is a problem due to several aspects related to both the laser system and the target, which influence the generation process during laser-material interaction. To correlate the X-ray dose rate estimation with the spectral data from the laser-induced plasma, we employed a calibrated Silix lambda spectrodosimeter. This detector delivers a photon energy spectrum of X-ray radiation with a dosage rate, elucidating the source of X-ray emission below 20 keV. The dosage rates from the Silix lambda spectrodosimeter were compared with those from the widely utilized OD-02 dosimeter across all measurements, demonstrating strong concordance between the two detectors.

AT A GLANCE

Ultrashort pulse laser machining aims to enhance processing speeds by increasing average power and pulse repetition rates, resulting in elevated X-ray emission dose rates during laser-matter interaction. The X-ray energy range below 10 keV is infrequently examined using a quantitative methodology. We report observations utilizing a newly calibrated X-ray detector within the detection range of 2–20 keV, demonstrating the correlation between X-ray radiation dose rates and spectral emissions for various laser parameters from commonly employed metals, alloys, and ceramics in ultrafast laser machining. Our research encompasses the dosage rate dependence on multiple laser parameters present in ultrafast laser laboratories and industrial laser systems. The assessed X-ray dosage rates for high repetition rate lasers utilizing various materials significantly surpass legal thresholds in the absence of radiation shielding.

REFERENCES

1. Gower, M.C. Industrial applications of laser micromachining. *Opt. Express* **2000**, *7*, 56–67. [Google Scholar] [CrossRef] [PubMed]
2. Cheng, J.; Liu, C.-S.; Shang, S.; Liu, D.; Perrie, W.; Dearden, G.; Watkins, K. A review of ultrafast laser materials micromachining. *Opt. Laser Technol.* **2013**, *46*, 88–102. [Google Scholar] [CrossRef]
3. Thogersen, J.; Borowiec, A.; Haugen, H.K.; McNeill, F.E.; Stronach, I.M. X-ray emission from femtosecond laser micromachining. *Appl. Phys. A* **2001**, *73*, 361–363. [Google Scholar] [CrossRef]
4. Weber, R.; Giedl-Wagner, R.; Förster, D.J.; Pauli, A.; Graf, T.; Balmer, J.E. Expected x-ray dose rates resulting from industrial ultrafast laser applications. *Appl. Phys. A* **2019**, *125*, 1–12. [Google Scholar] [CrossRef] [Green Version]
5. Legall, H.; Bonse, J.; Krüger, J. Review of x-ray exposure and safety issues arising from ultra-short pulse laser material processing. *J. Radiol. Prot.* **2021**, *41*, R28. [Google Scholar] [CrossRef] [PubMed]
6. Bunte, J.; Barcikowski, S.; Püster, T.; Burmester, T.; Brose, M.; Ludwig, T. Secondary Hazards: Particle and X-Ray Emission. In *Femtosecond Technology for Technical and Medical Applications*; Dausinger, F., Lubatschowski, H., Lichtner, F., Eds.; Springer: Berlin, Germany, 2004; Volume 96, pp. 309–321. [Google Scholar]
7. Kephart, J.F.; Godwin, R.P.; McCall, G.H. Bremsstrahlung emission from laser—Produced plasmas. *Appl. Phys. Lett.* **1974**, *25*, 108–109. [Google Scholar] [CrossRef]
8. Legall, H.; Schwanke, C.; Pentzien, S.; Dittmar, G.; Bonse, J.; Krüger, J. X-ray emission as a potential hazard during ultrashort pulse laser material processing. *Appl. Phys. A* **2018**, *124*, 1–8. [Google Scholar] [CrossRef] [Green Version]
9. Legall, H.; Schwanke, C.; Bonse, J.; Krüger, J. X-ray radiation protection aspects during ultrashort laser processing. *J. Laser Appl.* **2020**, *32*, 022004. [Google Scholar] [CrossRef]
10. Behrens, R.; Pullner, B.; Reginatto, M. X-ray emission from materials processing lasers. *Radiat. Prot. Dosim.* **2019**, *183*, 361–374. [Google Scholar] [CrossRef]
11. Kruer, W.L. *The Physics of Laser Plasma Interactions*, 1st ed.; CRC Press: Boca Raton, FL, USA, 2019. [Google Scholar]
12. Weisshaupt, J.; Juvé, V.; Holtz, M.; Ku, S.; Woerner, M.; Elsaesser, T.; Ališauskas, S.; Pugžlys, A.; Baltuška, A. High-brightness table-top hard X-ray source driven by sub-100-femtosecond mid-infrared pulses. *Nat. Photonics* **2014**, *8*, 927–930. [Google Scholar] [CrossRef]
13. Schroeder, W.A.; Omenetto, F.G.; Borisov, A.B.; Longworth, J.W.; McPherson, A.; Jordan, C.; Boyer, K.; Kondo, K.; Rhodes, C.K. Pump laser wavelength-dependent control of the efficiency of kilovolt x-ray emission from atomic clusters. *J. Phys. B At. Mol. Opt. Phys.* **1998**, *31*, 5031–5051. [Google Scholar] [CrossRef]
14. Di Niso, F.; Gaudiuso, C.; Sibillano, T.; Mezzapesa, F.P.; Ancona, A.; Lugarà, P.M. Role of heat accumulation on the incubation effect in multi-shot laser ablation of stainless steel at high repetition rates. *Opt. Express* **2014**, *22*, 12200–12210. [Google Scholar] [CrossRef]
15. Vorobyev, A.Y.; Guo, C. Direct observation of enhanced residual thermal energy coupling to solids in femtosecond laser ablation. *Appl. Phys. Lett.* **2005**, *86*, 011916. [Google Scholar] [CrossRef]
16. König, J.; Nolte, S.; Tünnermann, A. Plasma evolution during metal ablation with ultrashort laser pulses. *Opt. Express* **2005**, *13*, 10597–10607. [Google Scholar] [CrossRef] [PubMed]
17. Chen, F.F. *Introduction to Plasma Physics*; Plenum Press: New York, NY, USA, 1974. [Google Scholar]
18. Attwood, D. *Soft X-rays and Extreme Ultraviolet Radiation: Principles and Applications*; Cambridge University Press: Cambridge, UK, 2000. [Google Scholar]
19. Eliezer, S. *The Interaction of High-Power Lasers with Plasmas*; CRC Press: Boca Raton, FL, USA, 2002. [Google Scholar]
20. Forslund, D.W.; Kindel, J.M.; Lee, K. Theory of hot-electron spectra at high laser intensity. *Phys. Rev. Lett.* **1977**, *39*, 284–288. [Google Scholar] [CrossRef]

21. German Federal Office for Radiation Protection. Limit Values in Radiation Protection. Available online: **https://www.bfs.de/EN/topics/ion/radiation-protection/limit-values/limit-values_node.html** (accessed on ٧ July ٢٠٢١).

22. Sankar, P.; Thomas, J.; Shashikala, H.D.; Philip, R. Enhanced bremsstrahlung X-ray emission from Ag nanoparticles irradiated by ultrashort laser pulses. *Opt. Mater.* **2019**, *92*, 30–35. **[Google Scholar] [CrossRef**

CHAPTER 9

MICROMACHINING OF INVAR UTILIZING 784 BEAMS WITH A 1.3 PS LASER SOURCE AT 515 NM

1. INTRODUCTION: AN OVERVIEW

Invar is an alloy composed of iron and nickel, characterized by exceptional strength, impact toughness, processability, and a low coefficient of thermal expansion. These qualities render invar highly appealing for diverse industrial applications, such as bi-metal applications, liquefied natural gas storage tanks, and shadow masks for the manufacturing of organic light-emitting diodes (OLED) [2]. High-resolution shadow masks are essential in the red, green, and blue evaporation processes of organic luminescent materials in the fabrication of OLED displays [3]. The geometry, dimensions, and general integrity of holes in the shadow mask are directly correlated with the quality of OLED pixels [4]. Chemical etching is currently a prevalent microfabrication technique for the manufacturing of OLED shadow masks [2]. Nonetheless, it is a complex multi-stage procedure typically involving coating, cleaning, exposure, and an etching process, lacking control over the taper angle. Furthermore, fabricating a pattern smaller than the material thickness is challenging due to the isotropic nature of chemical etching, hence hindering the attainment of the high resolution necessary for future virtual reality displays [5,6].

Laser micromachining offers a high-quality, single-step, and environmentally sustainable method that eliminates the need for chemicals. The investigation of laser microfabrication of shadow masks involved various laser systems, resulting in the creation of 85 μm channels in polymethyl methacrylate with a CO_2 laser, 250 μm wide channels in wax using a CO_2 laser, 250 μm wide lines in plastics and glass employing a Nd:YAG laser at the fundamental wavelength of 1064 nm, and 140 μm lines produced in a steel shadow mask utilizing the third harmonics of a Nd:YAG laser. To fabricate high-resolution shadow masks with microscopic features of a few tens of microns, the utilization of ultrashort pulsed lasers is necessary [8,11,12,13]. The investigation of Invar drilling using ultrashort laser systems revealed the capability to achieve a 200 μm diameter in Invar with a 785 nm Ti:sapphire laser, as well as hole dimensions under 30 μm, characterized by exceptional quality and a minimum heat-affected zone. Nonetheless, the primary limitation of ultrashort pulse laser micromachining is its processing speed, which constrains its extensive application in industry, particularly for the prevalent single-beam direct laser writing method. Furthermore, the advent of high-power ultrashort laser systems [14] renders high-quality invar micromachining highly wasteful, as a significant portion of the laser power remains unutilized. This is necessitated by the requirement to operate near the damage threshold to ensure superior quality and a limited heat-affected zone. Consequently, innovative methods for expedited large-scale processing are necessary. The most promising rapid large-scale methods encompass polygonal scanning systems [15,16], direct laser interference patterning [17,18], and multi-beam scanning techniques [19]. Utilizing a diffractive optical element (DOE) in conjunction with a galvanometer scanner and high-power ultrashort laser systems for beam splitting is a viable approach to achieving industry standards in high-speed processing of extensive areas. The DOE beam-splitter allocates the incident laser beam intensity into the intended far-field configuration, typically a one-dimensional or two-dimensional array of beams. Each beam possesses identical characteristics to the original beam, with the exception of pulse energy and propagation angle [20]. Recently, commercially available multibeam systems, which achieve high throughputs by dividing beams into 10–100 beamlets, can drill holes with a diameter of 20 μm in 20 μm thick metal foils [21,22].This study seeks to showcase a method that efficiently produces microstructures by presenting a prototype of a Meopta beam-splitting, focussing, and monitoring module. The module segments an initial high-quality beam into over 700 beamlets for the parallel processing of a thin invar foil. By employing the module in conjunction with the HiLASE high-energy pulse ultrashort laser system and a galvanometer scanner, it became feasible to precisely modify the dimensions and shape of the generated microholes appropriate for OLED display fabrication.

2. MATERIALS AND METHODS

Invar (FeNi36) foils provided by Goodfellow GmbH (Friedberg, Germany) with a thickness of 20 μm and dimensions of 50 × 50 mm were used as received (Ra~0.07 μm). The foils were treated by Perla laser system from HiLASE (Dolni Brezany, Czech Republic), equipped with second harmonic generation module emitting at wavelength of 515 nm, with pulse duration of 1.3 ps, beam quality factor M2 better than 1.5, repetition rate of 1 kHz and pulse energy up to 10 mJ **[14,23]**. The generated beam (**Figure 1**c) was guided through the galvanometer scanner without F-theta lens into the optical module developed by Meopta-optika, s.r.o., consisting of a DOE beamsplitter, 100 mm focusing lens and a polarizing beamsplitter for pattern observation on the sample surface. The module is responsible for focusing and splitting the initial beam by means of diffractive optics into a matrix of 28 × 28 beamlets with homogeneous distribution across the whole pattern and suppression of zero order diffraction. The module allows the separation distance between beamlets to be adjusted by changing the distance between the DOE and focusing lens. Beam splitter and camera allows to monitor the intensity, shape and focal position of each beamlet. The beamsplitter passes linearly polarized beam on to the sample surface. The reflected light from the sample with random polarization is then able to reach the camera. The separation distance between each beamlet on the sample plane was 150 μm. The sample was placed vertically on to the vacuum holder with an optional connection to the cooling water circuit. The whole setup is depicted in **Figure 1**.

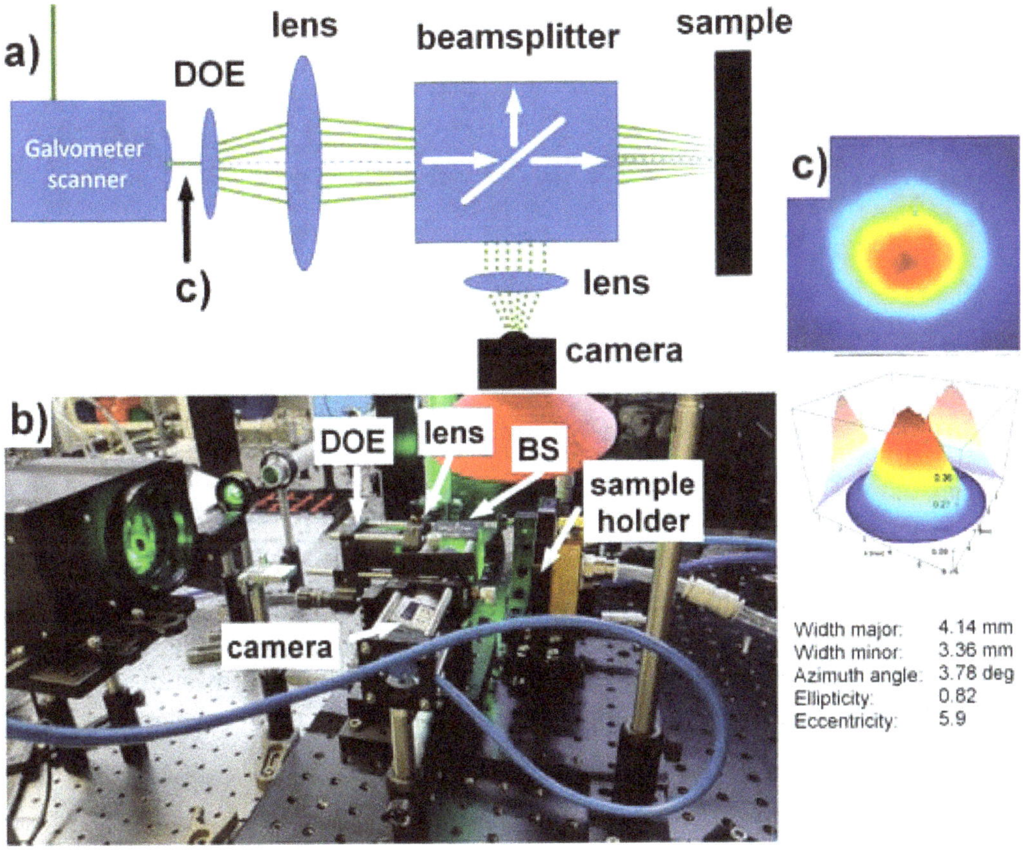

Figure 1. Schematics (a) and photo (b) of the experimental setup with a detail of input beam (c).

Topography of selected surfaces was analyzed with a laser scanning confocal microscope Olympus OLS 5000. The pattern shape and beamlets intensity distribution were captured by a Basler ace acA4024-29um camera with pixel size of 1.85 μm.

3. RESULTS AND DISCUSSION

The influence of misalignments in the experimental setup on the uniformity of the multi-spot pattern, the permissible deviation angles during beam deflections affecting the pattern configuration, and the laser and processing parameters for optimal cutting and drilling have been examined in a series of experiments. Initially, the pattern homogeneity required adjustment to provide uniform intensity across each beamlet by angling the beam-splitting component. The pattern shape was viewed in real-time using a camera within the Meopta beam-splitting module during alignment, as illustrated in Figure 2.

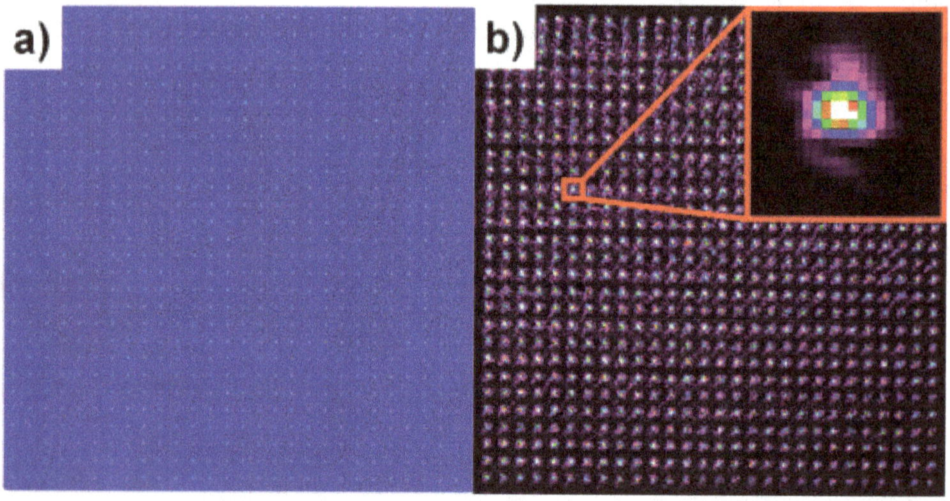

Figure 2. Final pattern shape on camera in a sample plane (a) and the same pattern reflected from the sample surface (b).

As shown in **Figure 2**, the pattern shape reflected from the sample surface exhibits undesired speckles, which can be further improved by mirror polishing of the sample. In addition to pattern homogeneity, the working plane was easily found by observing the pattern on camera as the sharp beam profile of each beamlet (inset of **Figure 2b**) can be observed only when the sample is located exactly in the focal plane.

In the next step, the sample was irradiated with 100,000 consecutive laser pulses with energy of 0.6 mJ which corresponds to the fluence of 0.28 J/cm2 for each beamlet. The first experiments revealed extensive heat accumulation connected with a large heat-affected zone and interconnection of microholes into one hole in the center of the pattern due to extensive heating and melting (Figure 3a). Consequently, the samples were placed on a water-cooled holder to minimize the effect of heat accumulation, as depicted in Figure 3b.

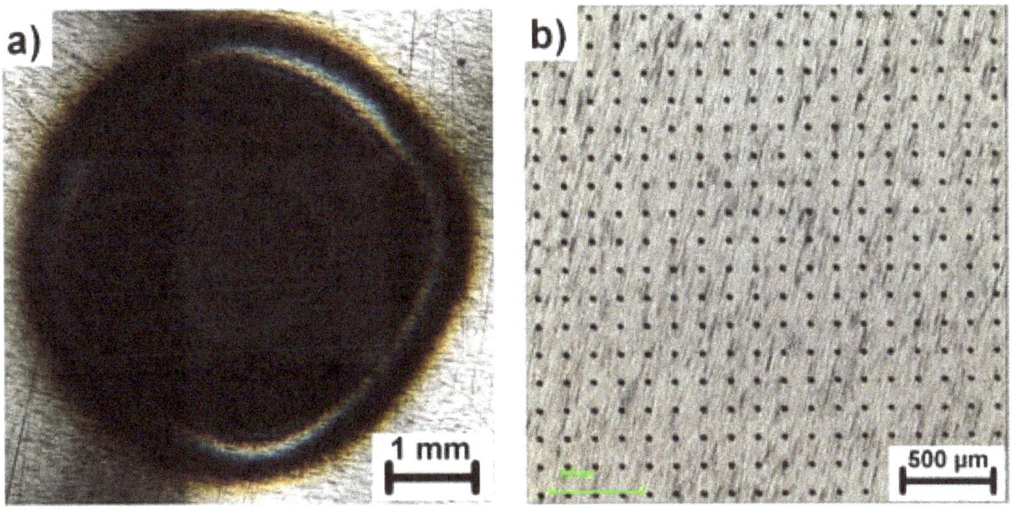

Figure 3. Overview of the heat affected zone on the sample processed without water-cooled holder (a) compared to the sample processed with the water-cooled holder (b). Both samples were processed with the same fluence of 0.28 J/cm2 and 100,000 pulses.

To determine an optimal processing window for high-quality multi-beam drilling, different combinations of laser parameters were analyzed. First, the ablation threshold for the simultaneous drilling of 784 holes was estimated. **Figure 4**a presents a microhole ablated close to the ablation threshold with 100,000 pulses and pulse energy of 0.3 mJ in the initial beam (beamlet fluence of 0.15 J/cm^2). No ablation was observed for pulse energies below 0.3 mJ, hence it is considered as the ablation threshold.

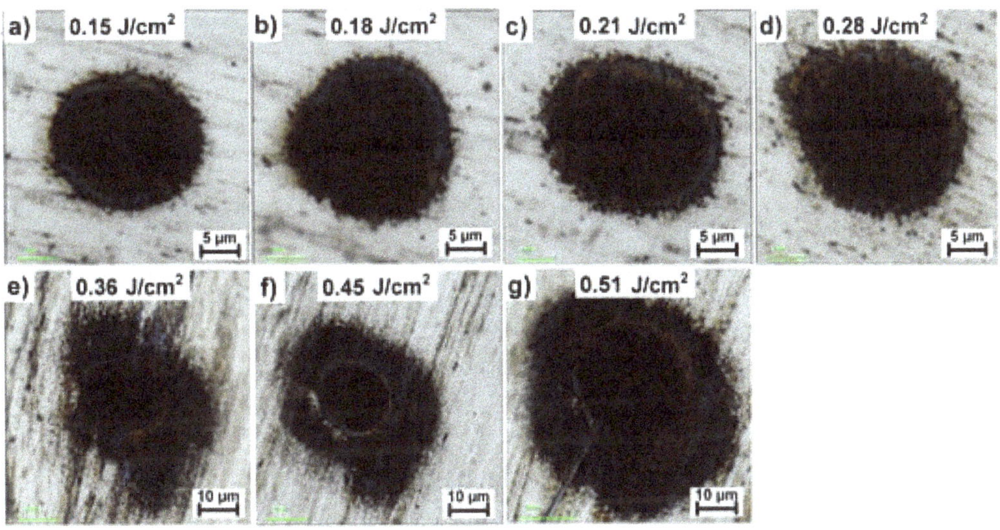

Figure 4. Evolution of microhole quality and heat affected zone with the increase in fluence. Sample is irradiated with 100,000 consecutive laser pulses. (a–d) gentle ablation regime, (e–g) high fluence regime.

The extension of the heat-affected zone and overall hole quality have been examined using different pulse energies in the range of 0.3 mJ to 1.2 mJ which correspond to the beamlet fluence of 0.15 J/cm^2 to 0.51 J/cm^2, as shown in Figure 4.

In line with the results presented in Figure 4, two ablation regimes have been identified. The highest quality microholes have been reached in a gentle ablation regime for fluences up to two-times the ablation threshold, i.e., up to 0.28 J/cm2. In this case, the extension of the heat-affected zone was found to be less than 3 μm. By increasing the fluence over this value, the quality of microholes significantly deteriorated due to the extensive heat-affected zone. Moreover, as shown in Figure 5a, the removal rate (ablated volume per time and power) for microholes drilled in a higher fluence regime over 0.28 J/cm2 is much smaller compared to lower fluences.

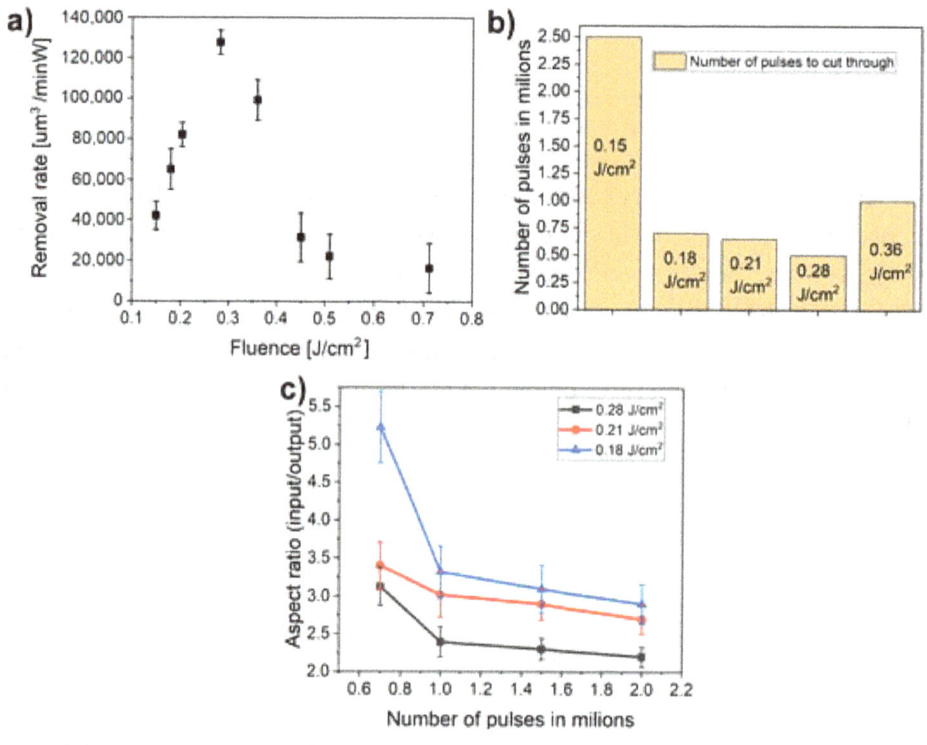

Figure 5. (a) Removal rate as a function of fluence; (b) Number of pulses necessary to cut through the whole sample; (c) Evolution of the aspect ratio as a function of number of pulses.

This phenomenon may be explained by the decrease in beam quality connected with high pulse energies which may overcome the tolerances of a diffractive beamsplitter and thus affect the intensity distribution within the beamlet. This issue will be further addressed in the following research.

On the other hand, the removal rate is sharply increasing with the fluence in the gentle ablation regime reaching the peak around 0.28 J/cm2, which was identified as the optimal fluence for effective drilling.

Furthermore, the number of pulses required for piercing the sample have been determined (Figure 5b). With the threshold fluence of 0.15 J/cm2 at least 2.5 million pulses were required to pierce through the 20 µm thick invar foil, thus removing less than 0.01 nm per pulse. By increasing the fluence above the ablation threshold, the sample was pierced with less than 700,000 pulses for fluences between 0.18 J/cm2 and 0.28 J/cm2. The lowest number of pulses required for piercing the foil was in the case of 0.28 J/cm2 with only 500,000 pulses, which is in accordance with the optimal removal rate value. For the higher fluence regime, the sample was pierced only in the case of 0.36 J/cm2 after exposition with 1 million pulses. Higher fluences did not lead to cutting through the invar foil even after 3 million pulses. Therefore, the optimal processing window for effective micromachining of invar foil has been identified as 0.18 J/cm^2 to 0.28 J/cm^2.

Additionally, the effect of laser parameters on the microhole diameter input/output aspect ratio have been studied. As demonstrated in **Figure** 5c, the aspect ratio decreases with higher fluences as well as with increasing number of pulses. Similarly, the shape of the exit diameter also improves with a higher number of applied pulses **(Figure 6).**

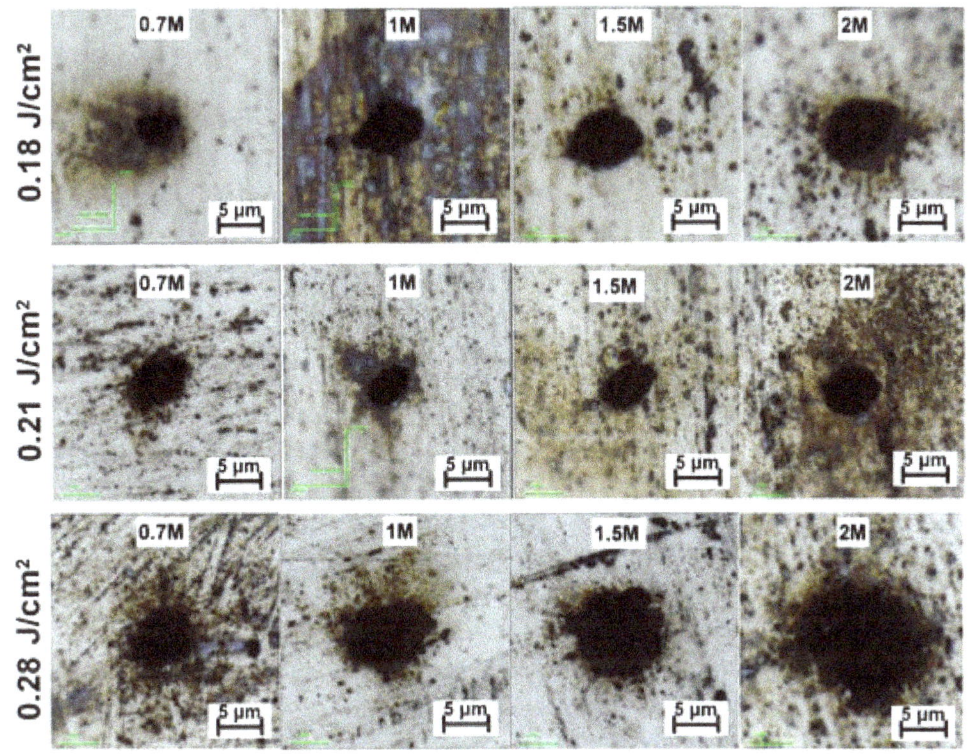

Figure 6. Geometry evolution of exit hole diameters with increasing number of pulses from 0.7 million to 2 million pulses.

Generally, the exit microhole diameter is larger and less elliptical for higher fluences and pulse counts, as demonstrated in **Figure 6.** The best aspect ratio and circularity were reached for the sample irradiated with 2 million pulses and 0.28 J/cm^2 per beamlet.

Following up on the optimal parameters, a galvanometer scanner was deployed to deflect the beam in a small area between microholes to fabricate different geometries on the invar foil. Since the separation distance between produced microholes is 150 μm, the square with a side width of 100 μm was chosen for the cutting experiment. To improve cutting efficiency, the input cut width was increased by cutting 15 squares with the same center position and decreasing diameter per 3 μm for each square. The cutting speed was adjusted to 1 mm/s to ensure high enough overlapping of ~95% with the laser repetition rate of 1 kHz. The cutting results are depicted in **Figure 7.**

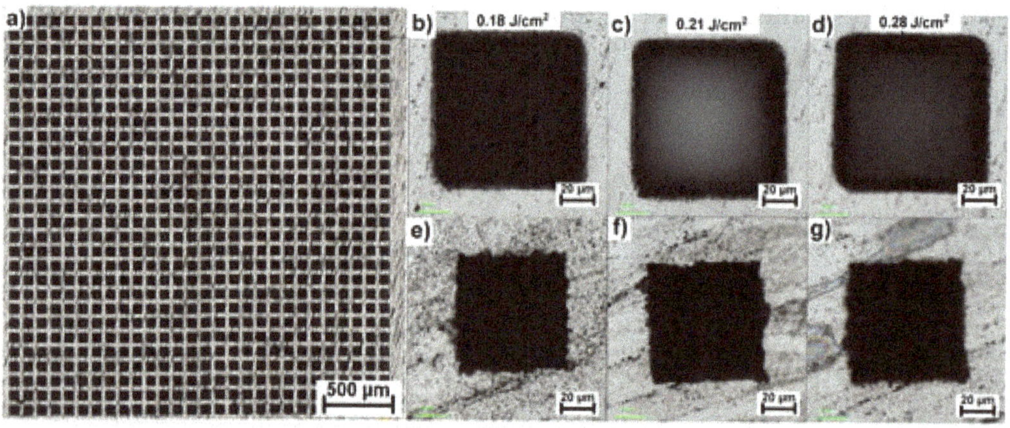

Figure 7. (a) Overview of a pattern cut with a combination of 28 × 28 beamsplitter and galvanometer scanner; (b–d) details of a square cuts fabricated with different fluences inside the optimal processing window and 400 overscans and their corresponding exit sides (e–g).

As demonstrated in Figure 7, it was possible to fabricate square geometries and cut through the invar foil with all 784 beams at once. The complete penetration was reached for 700, 440 and 250 overscans for the fluences of 0.18 J/cm^2, 0.21 J/cm^2, and 0.28 J/cm^2, respectively. The uniformity of square shape and their interspacing was assessed on 20 randomly chosen squares throughout the pattern, verifying identical values with a standard variation of less than 3 μm in all instances.

4. CONCISE SUMMARY

The application of a Meopta beam-splitting, focusing, and monitoring module in conjunction with the HiLASE high-energy pulse laser system, which emits ultrashort pulses at 515 nm, facilitated the effective cutting and drilling of invar foil concurrently employing 784 beamlets organized in a 28 × 28 matrix. The characteristics of the optical module provide rapid alignment of the setup to achieve a uniform intensity distribution throughout the whole pattern, along with focus adjustment. The ablation threshold, appropriate fluence levels, and pulse counts have been established, identifying the ideal processing window for high-quality drilling and cutting with a heat-affected zone below 3 μm as 0.18 J/cm^2 to 0.28 J/cm^2. The integration of a galvanometer scanning head with a beam-splitting module facilitated the multi-beam manufacturing of square-shaped microholes with variable size. This solution can enhance throughput by sevenfold relative to contemporary commercial multi-beam systems and over 700 times compared to single beam methods, indicating substantial potential for improved fabrication speed and efficiency in the production of invar shadow masks.

AT A GLANCE

To meet the specifications for high-resolution organic light-emitting diode (OLED) displays, accurate and high-quality micrometer-scale patterns must be produced within metal shadow masks. Invar has been chosen for this purpose because to its distinctive characteristics, particularly its low coefficient of thermal expansion. This study will present a unique, cost-effective approach for multi-beam micromachining of invar. The integration of a Meopta beam-splitting, focusing, and monitoring module with a galvanometer scanner and HiLASE high-energy pulse laser system, which emits ultrashort pulses at 515 nm, enables the simultaneous drilling and cutting of invar foil using 784 beams with exceptional precision and minimal thermal effects and heat-affected zones, thereby markedly enhancing throughput and efficiency.

REFERENCES

1. Ono, F.; Hamatani, Y.; Mukumoto, Y.; Komatsu, S.; Ishikawa, N.; Chimi, Y.; Iwase, A.; Kambara, T.; Mueller, C.; Neumann, R. Modification of Fe–Ni Invar alloys by high-energy ion beams. *Nucl. Instrum. Methods Phys. Res. Sect. B Beam Interac. Mater. At.* **2003**, *206*, 295–298. **[Google Scholar] [CrossRef]**
2. Kim, C. Production of shadow-mask-improved technology. *J. Mater. Process. Technol.* **2002**, *127*, 409–418. **[Google Scholar] [CrossRef]**
3. Shin, H.Y.; Suh, M.C. A study on full color organic light emitting diodes with blue common layer under the patterned emission layer. *Org. Electron.* **2014**, *15*, 2932–2941. **[Google Scholar] [CrossRef]**
4. Chung, I.-Y.; Kim, J.-D.; Kang, K.-H. Ablation drilling of invar alloy using ultrashort pulsed laser. *Int. J. Precis. Eng. Manuf.* **2009**, *10*, 11–16. **[Google Scholar] [CrossRef]**
5. Kim, S.H.; Choi, S.G.; Choi, W.K.; Yang, B.Y.; Lee, E.S. Pulse electrochemical machining on Invar alloy: Optical microscopic/SEM and non-contact 3D measurement study of surface analyses. *Appl. Surf. Sci.* **2014**, *314*, 822–831. **[Google Scholar] [CrossRef]**
6. Choi, W.; Kim, H.Y.; Jeon, J.W.; Chang, W.S.; Cho, S.H. Vibration-assisted femtosecond Laser drilling with controllable taper angles for AMOLED fine metal mask fabrication. *Materials* **2017**, *10*, 212. **[Google Scholar] [CrossRef] [Green Version]**
7. Klank, H.; Kutter, J.P.; Geschke, O. CO_2-laser micromachining and back-end processing for rapid production of PMMA-based microfluidic systems. *Lab Chip* **2002**, *2*, 242–246. **[Google Scholar] [CrossRef]**
8. Fan, Y.; Li, H.; Yi, Y.; Foulds, I.G. Laser micromachined wax-covered plastic paper as both sputter deposition shadow masks and deep-ultraviolet patterning masks for polymethylmethacrylate-based

microfluidic systems. *J. Micro/Nanolithogr. MEMS MOEMS* **2013**, *12*, 049701. **[Google Scholar]** **[CrossRef]** **[Green Version]**

9. Tahir, B.A.; Ahmed, R.; Ashiq, M.G.; Ahmed, A.; Saeed, M.A. Cutting of nonmetallic materials using Nd: YAG laser beam. *Chin. Phys. B* **2012**, *21*, 044201. **[Google Scholar]** **[CrossRef]**
10. Shiu, P.; Knopf, G.; Ostojic, M.; Nikumb, S. Rapid fabrication of tooling for microfluidic devices via laser micromachining and hot embossing. *J. Micromech. Microeng.* **2008**, *18*, 025012. **[Google Scholar]** **[CrossRef]** **[Green Version]**
11. Rizvi, N.H.; Apte, P. Developments in laser micro-machining techniques. *J. Mater. Process. Technol.* **2002**, *127*, 206–210. **[Google Scholar]** **[CrossRef]**
12. Kumar, A.; Gupta, A.; Kant, R.; Akhtar, S.N.; Tiwari, N.; Ramkumar, J.; Bhattacharya, S. Optimization of laser machining process for the preparation of photomasks, and its application to microsystems fabrication. *J. Micro/Nanolithogr. MEMS MOEMS* **2013**, *12*, 041203. **[Google Scholar]** **[CrossRef]**
13. Heo, J.; Min, H.; Lee, M. Laser micromachining of permalloy for fine metal mask. *Int. J. Precis. Eng. Manuf. Green Technol.* **2015**, *2*, 225–230. **[Google Scholar]** **[CrossRef]** **[Green Version]**
14. Smrž, M.; Novák, O.; Mužík, J.; Turčičová, H.; Chyla, M.; Nagisetty, S.S.; Vyvlečka, M.; Roškot, L.; Miura, T.; Černohorská, J.; et al. Advances in high-power, ultrashort pulse DPSSL technologies at HiLASE. *Appl. Sci.* **2017**, *7*, 1016. **[Google Scholar]** **[CrossRef]** **[Green Version]**
15. Mincuzzi, G.; Gemini, L.; Faucon, M.; Kling, R. Extending ultra-short pulse laser texturing over large area. *Appl. Surf. Sci.* **2016**, *386*, 65–71. **[Google Scholar]** **[CrossRef]**
16. De Loor, R. Polygon scanner system for ultra short pulsed laser micro-machining applications. *Phys. Proced.* **2013**, *41*, 544–551. **[Google Scholar]** **[CrossRef]** **[Green Version]**
17. Voisiat, B.; Gedvilas, M.; Indrišiūnas, S.; Račiukaitis, G. Picosecond-laser 4-beam-interference ablation as a flexible tool for thin film microstructuring. *Phys. Proced.* **2011**, *12*, 116–124. **[Google Scholar]** **[CrossRef]** **[Green Version]**
18. Hauschwitz, P.; Jagdheesh, R.; Alamri, S.; Rostohar, D.; Kunze, T.; Brajer, J.; Kopeček, J.; Mocek, T. Fabrication of functional superhydrophobic surfaces on carbon fibre reinforced plastics by IR and UV direct laser interference patterning. *Appl. Surf. Sci.* **2019**, *508*, 144817. **[Google Scholar]** **[CrossRef]**
19. Gillner, A.; Finger, J.; Gretzki, P.; Niessen, M.; Bartels, T.; Reininghaus, M. High power laser processing with ultrafast and multi-parallel beams. *J. Laser Micro Nanoeng.* **2019**, *14*, 129–137. **[Google Scholar]**
20. Katz, S.; Kaplan, N.; Grossinger, I. Using Diffractive Optical Elements: DOEs for beam shaping–fundamentals and applications. *Opt. Photonik* **2018**, *13*, 83–86. **[Google Scholar]** **[CrossRef]** **[Green Version]**
21. Holtkamp, J.; Eifel, S.; Ryll, J. Material Processing with Ultrashort-pulsed Lasers: Potentials—Requirements—Approaches. *Laser Tech. J.* **2014**, *11*, 36–39. **[Google Scholar]** **[CrossRef]**
22. Holtkamp, J. Pulsar Photonics: Innovative Beam Forming Concepts, Swissphotonics. Available online: **https://swissphotonics.net/libraries.files/holtkamp** (accessed on 1 July 2020).
23. Duda, M.; Novák, O.; Chyla, M.; Smrž, M.; Mocek, T. Balancing the conversion efficiency and beam quality of second harmonic generation of a two-picosecond Yb: YAG thin-disk laser. *Laser Phys.* **2020**, *30*, 025405. **[Google Scholar]** **[CrossRef]** **[Green Version]**

CHAPTER 10

USING TWO LASER BEAMS TO PROCESS THIN FILMS OF SEMICONDUCTORS BY ABSORBING EXCITED STATES

1. INTRODUCTION: AN OVERVIEW

In 1994, Hell et al. theoretically elucidated stimulated emission depletion (STED) microscopy, which was experimentally validated a few years later and currently permits the resolution of structures less than 10 nm [1,2,3]. A Gaussian laser beam excites carriers, which are then returned to their ground state by a secondary donut-shaped STED beam with a wavelength that is unique to the material. Despite the diffraction limitations on the focusability of both laser beams, the resultant fluorescent volume is markedly reduced to the zero-intensity center of the STED beam [4]. This method is now employed for building various photopolymers with feature sizes in the low double-digit nanometer range [5]. In this instance, polymerization in the external, high-intensity portion of the annular beam is inhibited by accelerated depletion of the electrons necessary for polymerization or by the activation of inhibitors. The approach is restricted to particular photopolymers and designated light sources. The structures can alternatively be translated to different kinds of materials using replica casting. Nonetheless, the direct sub-diffraction structuring of semiconductors and dielectrics using the STED technology would enhance flexibility, diminish process complexity, and decrease total production expenses [6,7]. In this regard, the methodology suggested by Engel et al. [8] is highly promising for optically active materials such as direct semiconductors. In summary, akin to STED microscopy, a diffraction-limited Gaussian laser beam with adequate photon energy EP is employed to excite electrons into the conduction band (CB). A second temporally and spatially aligned donut-shaped beam with appropriate energy potential in the vicinity of the optical band gap energy depletes the excited states and facilitates the transport of electrons back to the valence band by stimulated emission. The depletion happens solely in the high-intensity annular region of the beam, while the excited electrons in the zero-intensity core remain unaffected. The excited core region can then absorb a third Gaussian laser beam by excited state absorption (ESA) at a wavelength that is otherwise transmitted in non-excited areas. The extra energy from this ESA-based process, derived from mechanisms like intra-band absorption, avalanche effects, or Coulomb explosion, may be appropriate for sub-diffraction structuring, despite the single beams being constrained by diffraction-limited focusing. This study concentrates on ESA-based ablation, a crucial component of the proposed sub-diffraction direct laser writing technique, as indicated in prior research [9] and illustrated in Figure 1a [8].

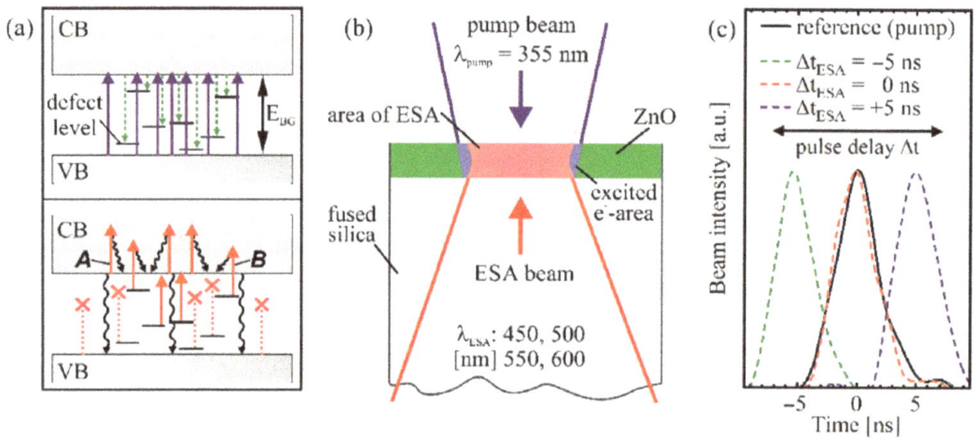

Figure 1. (a) Schematic illustration of the theoretical excited state absorption (ESA)-based ablation process with excitation (top) and a subsequent ESA (bottom), indicated by solid arrows. The electrons, excited prior by the pump beam with $E_P > E_{BG}$, rapidly relax radiative or non-radiative (dashed arrows) and partially migrate to defect levels located within the band gap. The subsequent ESA beams with $E_P < E_{BG}$ are absorbed by either intra-band *A* or defect level → conduction band (CB) *B* transitions, both enabled only by the prior pump excitation. Waved and dashed arrows indicate non-radiative and forbidden transitions, respectively. (b) Sample cross section of 1 mm thick fused silica substrate coated with a 200 nm thin ZnO layer. The beam directions are indicated by arrows. (c) Illustration of selected pulse delays Δt = -5, 0, +5 ns (dashed lines) and their temporal overlap with the pump pulse (solid line).

Numerous research have already employed ESA-based microfabrication by a hybrid laser technique on broad bandgap semiconductors, including fused silica, sapphire, and quartz [7,10,11,12,13]. The non-thermal ablation using electron excitation above the vacuum energy enhances ablation quality and reduces surface debris generation. In dual-beam processes, Excimer and Raman lasers with UV-range wavelengths, along with (ultra-)short pulse lasers, are predominantly employed to enhance absorption by photodissociation and alterations in transient absorption [7]. This process is also applicable to lower bandgap semiconductors like gallium nitride, lithium niobate, and silicon carbide [13,14,15,16,17]. In this case, the single laser beams already exceed EBG, but the excitation beyond vacuum energy for improved non-thermal ablation is only achieved by the dual-beam setup. Similarly, an increased absorption of a second, otherwise transmitted laser beam with EP < EBG, without exceeding the vacuum energy enables energy and cost efficient processing as already demonstrated for silicon and silicon carbide [18,19]. The key factor here is that commercially available, compact, and less expansive lasers in the visible and infrared spectrum can be used [18].

This dual-beam ablation approach employed Zinc(II) oxide (ZnO) as a direct bandgap semiconductor. ZnO is frequently utilized in advanced applications such as optoelectronics, metamaterials, and photonics, particularly as thin films on low-absorbing substrates like fused silica, making it a promising candidate for ESA and potential STED-based ablation in conjunction with ESA. Consequently, we concentrated on selective, high-quality thin film ablation utilizing optimized processing settings and cost effectiveness. The dual-beam setup's appropriateness for thin film structuring is evaluated by examining the ablation process in relation to laser peak fluences, temporal delays, and the wavelengths of the two beams.

2. MATERIALS AND METHODS

2.1. Sample Preparation and Characterization

Fused silica substrates (Infrasil 301, Heraeus, Hanau, Germany) with a thickness of 1 mm were coated by RF-magnetron sputtering of 200 nm thin ZnO layers using a sputtering time of 37 min at a pressure of 0.295 Pa, a gas flow (argon with 2% oxygen) of 6 cm^3/min, and a power of 150 W. The coated and uncoated samples were optically characterized by UV/VIS transmittance spectroscopy (MCS/100-3, J&M Analytik, Essingen, Germany).

The photoluminescence emission spectra of the coated samples were measured at room temperature with a HeCd laser (IK3202R-D, Kimmon Koha, Tokyo, Japan) at a laser wavelength $\lambda = 325$ nm and a continuous wave power of 17 mW. The fluorescence was detected at an angle of 45° by an optical spectrometer (Maya2000 Pro, Ocean Insight, Rostock, Germany). A 350 nm longpass filter (XUL0350, Asahi Spectra Co., Tokyo, Japan) was utilized in front of the detector to restrict the detection range above E_{BG}, where only weak emission is expected [22].

2.2. Laser Processing Setup

The dual-beam setup consists of two Nd:YAG lasers (SLI-10, Amplitude Systems) with a fundamental wavelength $\lambda = 1064$ nm. The third harmonic of one of them was used as pump beam with $\lambda_{pump} = 355$ nm and a pulse duration of $\tau_{pump} = 4.7$ ns. The second Nd:YAG laser equipped with an additional optical parametric oscillator provides selected ESA beam wavelengths λ_{ESA} = 550 ,500 ,450, and 600 nm with $\tau_{ESA} = 4.8$ ns.

The beams were collinearly focused on the sample surface from opposite sides. Consequently, the ESA beams must pass the fused silica substrate (**Figure 1b**). For synchronization and temporal peak-to-peak delay, adjustments of both beams at the sample surface, a digital delay generator (DG535, Stanford Research Systems) and a silicon photodetector (ET-2000, EOT) were used. A digital oscilloscope (XDS3302 Plus, OWON Technology Inc., Zhangzhou, China) with a rise time < 1.2 ns monitored the temporal pulse shape (**Figure 1c**).

The ablation process was investigated using single laser pulses of the respective pump and ESA beams in order to exclude accumulation and incubation effects that would lead to an altered absorption. These include steady sample heating, contamination of the surface by resolidified ZnO, or the formation of deep level defects and color centers [**11,23,24**]. To ensure optimum laser operation with highest output stability, the laser was operated at its standard repetition rate of 10 Hz during the measurements. The separate pulses were then selected by mechanical shutters, synchronized with the delay generator already used for laser operations. Furthermore, the lasers were operated at their optimum output energy E_{pump} = 160 mJ and E_{ESA} = 29–27 mJ and attenuated by neutral density filters (FW2AND and NDC-100C-4M, Thorlabs Inc., Newton, MA, USA). To further reduce the impact of pulse energy fluctuations and sample inhomogeneities, the results of 5 independent ablation spots fabricated with identical processing parameters were averaged for all performed single- and dual-beam ablation measurements.

2.3. Generation and Analysis of Ablation Spots

The ablation behavior of the ZnO layer for the single laser beams was characterized by optical microscopy (VH-Z100, Keyence, Osaka, Japan) and white light interferometry (CCI HD, Ametek Inc., Berwyn, IL, USA) as a function of their respective pump-pulse energy $E_{pump} \leq 13$ μJ and ESA-pulse energy E_{ESA}. The latter was labeled according to the utilized ESA-wavelengths λ_{ESA} as $E_{450} \leq 19$ μJ, $E_{500} \leq 12$ μJ, $E_{550} \leq 25$ μJ, $E_{600} \leq 31$ μJ. The single-beam ablation threshold energy E_{th} and the beam diameter $2\omega_f$ ($1/e^2$-diameter) were determined using the method proposed by Liu [**25**]. For this purpose, the squared ablation diameters D^2 were plotted in semilog plot versus the pulse energy E. According to

$$D^2 = 2\omega_f \cdot \ln\left(\frac{E}{E_{th}}\right), \tag{1}$$

the intersection of the linear fit at $D^2 = 0$ and its slope determines E_{th} and $2\omega_f$, respectively. Using these parameters, the laser peak fluence at the ablation threshold was calculated by $F_{th} = 2E_{th}/\pi\omega_f^2$. Hereinafter, all fluence values refer to laser peak fluences.

For fluences below F_{th} in order to prevent single-beam ablation, ESA-based laser processing with a combination of pump and ESA beam was analyzed. The impact of the different parameters for optimal ablation quality and efficiency was characterized by varying the ESA beam fluence F_{ESA} ($F_{450} \leq 1.6$ J/cm², $F_{500} \leq 4.5$ J/cm², $F_{550} \leq 5.7$ J/cm², $F_{600} \leq 5.3$ J/cm²) below their individual F_{th} at a constant pulse delay of Δt = 5 ns and several pump beam fluences at F_{pump} = 0.2 J/cm0.1 ,² J/cm², and 0.02 J/cm². Using the highest pump beam fluence F_{pump} = 0.2 J/cm² and the same values of F_{ESA}, Δt was varied from −10 to 500 ns to investigate the influence of the pulse delay. Here, a negative temporal delay (Δt < 0) implies that the ESA beam precedes the pump pulse (**Figure 1c**).

3. RESULTS AND DISCUSSION

3.1. Optical Characterization of the ZnO Thin Film

The UV/VIS transmittance spectrum of the uncoated fused silica substrate T_S (**Figure 2a**) shows no absorption for the applied λ and thus no interaction with the radiation. The deviation to T = 1 is caused by reflectance losses at the interfaces [**26**]. The transmittance of the 200 nm thin ZnO layer on the substrate T_{ZnO} shows a single steep absorption edge at λ ≈ 380 nm with a direct optical band gap at E_{BG} = 3.26 eV [**20,27**], determined by a Tauc-plot [**28**]:

$$\alpha E_P = G \cdot (E_P - E_{BG})^R, \quad (2)$$

where G is a constant that depends on the transition probability. The parameter R is determined by the optical transmittance type and is R = 0.5 for directly allowed optical transmissions [**29,30**]. Plotting αE_P over E_P and using a linear fit of the straight portion of the absorption edge, the intersection at $\alpha E_P = 0$ determines E_{BG} (**Figure 2b**). The results demonstrate optimal conditions for a strong fundamental single photon absorption of the pump beam (λ_{pump} = 355 nm). The optical penetration depth is determined by $d_p = 1/\alpha$ with the absorption coefficient $\alpha = \ln(T^{-1})/d_L$. d_p = 62 nm and corresponds by definition to the depth at which the beam intensity has dropped to 1/e. This value is smaller than the layer thickness d_L and thus almost all of the pulse energy is used to promote electrons to the CB, which are available for the subsequent ESA-based ablation process.

By analyzing the sinusoidal interference fringes in the weak and medium absorbing region of the T_{ZnO}-spectrum by the envelope method of Swanepoel [**26**], the interference free transmittance T_E was derived. Here, the T_{ZnO}-maxima and minima were used to fit an upper T_M and lower T_m envelope function, respectively. Both envelopes were computer-generated using the program OriginPro 2020b (Origin-Lab Corp.) [**30**]. At shorter wavelengths in the strong absorbing region above E_{BG}, the envelope functions converge to a single curve, merging with the measured values of

T_{ZnO}. The interference free absorption coefficient was calculated from the determined T_E values (**Table 1**) [31].

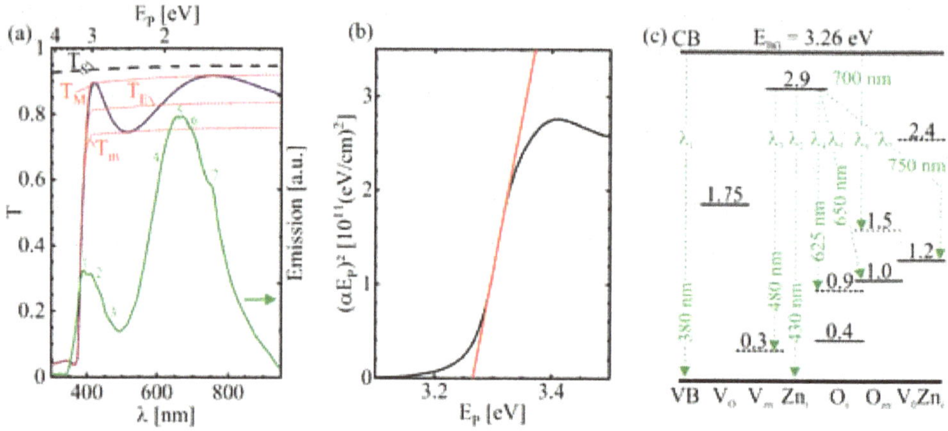

Figure 2. (a) UV/VIS transmittance spectra of 1 mm thin uncoated fused silica substrate (- - -) and fused silica coated with a 200 nm thin ZnO-layer (—). The dotted lines labeled T_M, T_m, and T_E are the fits of the upper and lower envelope function as well as their geometric mean, respectively [26]. In the emission spectrum of the ZnO film (—), λ = 325 nm was used for excitation. Selected emission peaks assigned to specific transitions are numbered. (b) Tauc plot [28] of T_{ZnO} (—). The linear fit (—) at the straight portion of the band edge allows to determine E_{BG} by the intersection at $\alpha E_P = 0$. (c) Schematic of the location of intrinsic point defect levels within the band gap by full-potential linear Muffin-tin orbital method calculations according to Xu et al. [32]. The numbers on the lines indicate ΔE_P of the defect states compared to the energy of the valence band (VB)-maximum with filled (—), partially filled (---), and empty (- - -) states. The positions of the energy levels are true to scale with respect to E_{BG} = 3.26 eV. Selected transitions are labelled with dashed arrows and a number corresponding to the emission wavelengths in (a).

Table 1. Optical parameters (T_E, n, α, d_p) at the utilized λ calculated from the transmittance data by the method of Swanepoel (Figure 2a) [26], as well as ablation parameters for single-beam ablation (E_{th}, $2\omega_f$, F_{th}) determined by the method of Liu (Figure 3a) [25].

λ [nm]	355	450	500	550	600
T_E	0.04	0.82	0.82	0.83	0.83
n	-	2.07	2.05	2.04	2.03
α [1/cm]	1.6×10^5	9.7×10^3	9.5×10^3	9.4×10^3	9.2×10^3
d_p [nm]	62	1027	1047	1065	1088
$2\omega_f$ [μm]	28.4	28.5	20.2	23.6	19.8
E_{th} [μJ]	1.6	6.2	7.8	12.9	8.5
F_{th} [J/cm²]	0.49	1.96	4.86	5.90	5.51

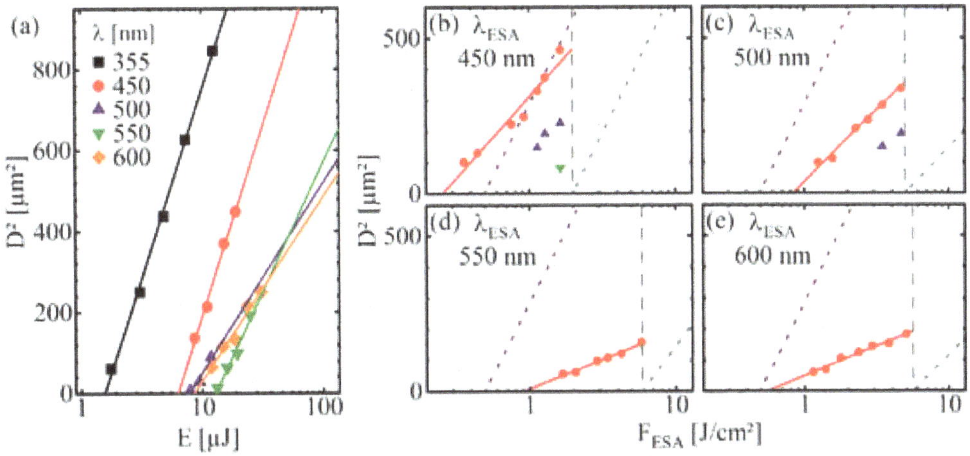

Figure 3. (a) Semilog plots of D^2 over E with the corresponding linear fits, to determine the ablation parameters E_{th}, $2\omega_p$, and F_{th} of single-beam absorption by the method of Liu [25]. The exact values at the utilized λ are listed in Table 1. (b–e) Semilog plots of D^2 over F_{ESA} for ESA based ablation at different λ_{ESA} with (b) λ = 450 nm, (c) λ = 500 nm, (d) λ = 550 nm, and (e) λ = 600 nm. For excitation, different F_{pump} values of 0.2 J/cm² (-●-), 0.1 J/cm² (▲), and 0.02 J/cm² (▼) were used. F_{ESA} was varied below their respective F_{th} (Table 1), indicated by the vertical dashed line. The values for F_{pump} = 0.2 J/cm² were linearly fit to D^2 = 0. The diagonal dotted lines highlight the F_{th} fits of λ = 355 nm (---) and the respective λ_{ESA} (---) plotted in (a).

Furthermore, the amplitude and oscillation of the interference fringes contain information about the ZnO-layer thickness d_L and the refraction index n, respectively. The refraction index of the substrate was derived from the transmittance of the uncoated substrate. The interference free refraction index in the weak absorbing region was calculated using the envelope functions as demonstrated by Swanepoel [26]. By using λ and n of adjacent fringe maxima or minima and the order number, d_L was calculated to be 203 nm. The measured and calculated parameters for the utilized λ are listed in **Table 1** [26,31].

Below the band gap energy E_{BG}, n and α steadily increase and the small variation of λ below E_{BG} indicates the comparably weak absorption of the ESA wavelengths without additional pump excitation. The difference between T_E and T_S is caused primarily by the deviating n, as the upper envelope function that intersects the fringe maxima almost coincides with T_S [26]. With increasing absorption at E_{BG}, α increases rapidly.

The sinusoidal interference fringes in the spectrum are caused by reflection losses due to interference effects of the radiation between the air–film and film–substrate interfaces [33]. This indicates homogeneous, smooth, and low scattering surfaces [34]. Thus, only at sufficiently high beam fluences, an independent ESA beam ablation is expected by either a multi-photon absorption or processes including various intrinsic point defects. The latter include deep level defects inside the ZnO layer or shallow donors that are located typically at the surface [**20,22,35,36**]. As illustrated in **Figure** 2c, the location of the intrinsic defect levels were calculated by Xu et al. with a full-potential linear Muffin-tin orbital method [**32**]. These include vacant zinc (V_{Zn}), vacant oxygen (V_0), interstitial zinc (Zn_i), interstitial oxygen (O_i), substituted oxygen at zinc positions (O_{Zn}), and complexes of V_0 and Zn_i (V_0Zn_i).

The emission spectrum in Figure 2a shows a weaker emission in the UV-range from 360–500 nm (close to E_{BG}), compared to λ = 550–800 nm. The sharp band emission at λ_1 = 390–380 nm overlaps with the band edge and is caused by recombination of photo-induced charge carriers through exciton–exciton collision processes (Figure 2c) [21,22,37]. The specific emissions at λ_2 = 430 nm and λ_3 = 480 nm are attributed to $Zn_i \rightarrow$ VB and $Zn_i \rightarrow V_{Zn}$ transitions, respectively [22]. Their weak emission indicates low roughness and homogeneous surface quality, in line with the findings in the T_{ZnO} spectrum as these transitions occur primarily at the surface [22]. The strongest emission in the visible region corresponds to deep level defects in the bulk of the layer [22,32,36]. Here, the broad emission is comprised of several transitions at λ_4 = 625 nm, λ_5 = 650 nm, λ_6 = 700 nm, and λ_7 = 750 nm, induced by $Zn_i \rightarrow O_i$, $Zn_i \rightarrow O_{Zn}$, CB $\rightarrow O_{Zn}$, and $Zn_i \rightarrow V_0Zn_i$ [22].

3.2. Single-Beam Ablation

It becomes evident from Table 1 that F_{th} of the ESA beams is about one order of magnitude larger than the pump beam. The pump beam with λ_{pump} = 355 nm is absorbed fundamentally by single photon absorption, indicated by the linear increase in D^2 (at the semilog scale in Figure 3a) and the transmittance in Figure 2a ($E_{pump} > E_{BG}$) [38,39]. In contrast, the ESA beams require either populated defect levels that absorb the laser wavelengths and promote electrons to CB [35], or a multi photon process ($E_{ESA} < E_{BG}$) [7]. However, such nonlinear processes would cause an exponential deviation from the linear trend [38] and is therefore unlikely. Thus, the ESA at defect levels is expected as the primary absorption mechanism. The steeper slope angle of the pump beam indicates a larger beam diameter $2\omega_f$ = 28 μm [40]. It is comparable or larger than the ESA beams (Table 1) and therefore it provides ideal conditions for optimal utilization of the pulse energy for the ESA process.

The behavior at λ_{ESA} = 450 nm deviates slightly from the other wavelengths, where F_{th} is smaller and $2\omega_f$ is larger. This suggests a partially different excitation process due to the high E_P closest to E_{BG}. This might be explained by the excitation of VB-electrons directly to defect levels close to CB. Moreover, defect levels could be able to absorb the beam and excite electrons directly to CB (Figure 2c), which would lower F_{th} [32,35]. As the defect concentration is expected to be highest at the ZnO–fused silica interface, most of the energy would be absorbed here. At sufficient F_{ESA}, the abrupt absorption at this interface can cause an explosion-like ablation of the whole layer without the necessity to heat the entire ZnO inside the laser spot. For single-beam ablation of the ESA-beams, the resulting spots are mostly inhomogeneous at low F_{ESA} (Figure 4b,g) and the further increase of F_{ESA} leads to substrate damage (Figure 4a,f). Here, an inhomogeneous ablation with ablation depths d_a of several microns were observed in the dark areas of the ablation spot. The intense absorption at the fused silica interface exceeds the substrate ablation threshold and limits the F_{ESA}-range significantly. Thus, for the F_{th} calculations, F_{ESA} was decreased and the spots at higher F_{ESA} exceeding the damage threshold of the fused silica substrate were excluded from analysis, as the measured D^2 are highly inconsistent and do not represent the ablation characteristic of the ZnO-layer.

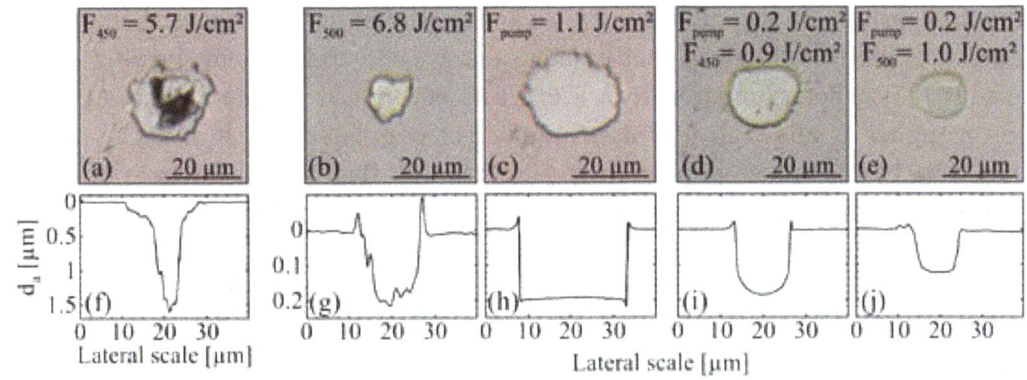

Figure 4. Microscopic images of single-beam ablation spots with (a) F_{450} = 5.7 J/cm², (b) F_{500} = 6.8 J/cm², and (c) F_{pump} = 1.1 J/cm², as well as ESA-based dual-beam ablation spots at (d) Δt = 5 ns with F_{pump} = 0.2 J/cm² + F_{450} = 0.9 J/cm² and (e) Δt = 10 ns with F_{pump} = 0.2 J/cm² + F_{500} = 1.0 J/cm². (f–j) White light interferometric depth profiles of spots (a–e).

3.3. Dual-Beam Ablation

3.3.1. Influence of the ESA Beam Fluence

At first, F_{ESA} was varied (Figure 3b–e) and the values for optimum ablation conditions were set to a small positive temporal delay of Δt = 5 ns and F_{pump} = 0.2 J/cm². In the case of an ablation process solely based on ESA, the pump beam induces the transient absorption of the spatially and temporally superimposed ESA beam (Figure 1). This requires to prevent surface modification of the individual beams by ablation [8]. Therefore, all measurements were performed using laser peak fluences below their respective F_{th} (Table 1). This is indicated in Figure 3b–e where all values are below the vertical dashed lines, originating at F_{th} of the respective ESA-beams. Additionally, to compare the dual-beam process with single-beam ablation, the diagonal dashed lines show the linear fits of the F_{th} calculations, determined by the fits of E_{th} in Figure 3a.

Decreasing F_{ESA} results in a decreased D^2 as less additional energy is available to further excite the already excited electrons. This limits the ablation to an area closer to the beam center, where the intensity of the Gaussian beam profile is still sufficient for ablation. Here, the utilizable fluence range for ESA-based ablation extends about one order of magnitude down to $F_{ESA} \approx 0.3\ J/cm^2$ for $\lambda = 450$ nm. Slightly below these values, the ablation spots are characterized by their non-circular shape. By further decreasing F_{ESA}, inhomogeneously distributed substructures were observed within the irradiated area. These surface modifications at low laser fluences are caused by locally enhanced carrier densities that are generated by either an inhomogeneous laser beam profile or locally enhanced ablation at dust, scratches, or crystal defects [41]. It has to be noted that the analysis only includes ablated structures that resulted in distinct and well pronounced ablation spots, excluding the aforementioned value-ranges.

The comparison of the different ESA wavelengths reveals a similar behavior of the beams with the highest photon energy at $\lambda = 450$ and 500 nm showing highly increased D^2 when compared to $\lambda = 550$ and 600 nm at similar fluences (Figure 3b–e). Higher E_P allow an increased number of populated defect levels to absorb these ESA beams and to populate the CB by an ESA (Figure 2c). It has to be noted that a single photon absorption of VB-electrons directly to the CB is not possible at λ_{ESA} with $E_P < E_{BG}$. In addition, a multiphoton absorption is negligible at the low applied fluences. Furthermore, a transition to another defect level is unlikely due to the specific required wavelength. Thus, besides the intra-band transition of CB-electrons, a transition from defect levels close to the VB into the CB is the most likely absorption mechanism. The existence of a considerable amount of these defect levels is confirmed by the characteristic fluorescence (Figure 2a), where the highest number of emitted photons is attributed to these defect level transitions [42,43,44].

Thus, with the highest E_P, using $\lambda_{ESA} = 450$ nm is optimal to generate the biggest ablation spots at the lowest F_{ESA} (Figure 3b). Here, at $F_{pump} = 0.2\ J/cm^2$ (red markers), even less additional F_{ESA} was necessary than the single pump beam requires for ablation. This is indicated by values above the purple line (linear F_{th}-fit of Figure 3a).

To analyze the effect of F_{pump} on the ESA-based ablation, $\Delta t = 5$ ns at the identical values of F_{ESA} for each wavelength were used. As seen in Figure 3d,e, below $F_{pump} = 0.2\ J/cm^2$ no distinct ablation spots can be observed for $\lambda = 550$ and 600 nm. The utilizable F_{ESA} range reduces drastically with decreasing F_{pump}. Without a sufficient excited electron concentration, the ESA-based ablation process stops abruptly. Only for $\lambda = 450$ nm at $F_{pump} = 0.02\ J/cm^2$ distinct ablation spots can be observed (Figure 3b,c).

3.3.2. Ablation Efficiency

When compared to single-beam ablation, the ESA-based process allows to decrease the amount of laser energy required for ablation (Table 1). This improvement was evaluated based on the method of Liu [25]. For this purpose, D^2 of the ESA-based ablation spots in Figure 3b–e (for $F_{pump} = 0.2\ J/cm^2$ at $\Delta t = 5$ ns) was plotted in a semilog plot as a function of F_{ESA}. The linear extrapolation allows to calculate the effective ESA-beam diameter $2\omega_f^{eff}$ (slope of the fit) and the effective ESA-beam threshold peak fluence F_{th}^{eff} (fit intersect at $D^2 = 0$) (Table 2).

Table 2. Calculation of dual-beam ablation parameters ($2\omega_f^{eff}$, F_{th}^{eff}) based on the method of Liu [25] and dual-beam efficiency parameters (F_E, F_{ET}) determined from Equations (3) and (4).

λ [nm]	450	500	550	600
$2\omega_f^{eff}$ [µm]	21.4	20.1	12.8	13
F_{th}^{eff} [J/cm²]	0.26	0.83	0.90	0.55
F_E	0.86	2.03	2.16	1.46
F_{ET}	0.39	0.50	0.48	0.43

Here, the values of $2\omega_f^{eff}$ for λ = 450 and 500 nm are close to their respective $2\omega_f$ of single-beam ablation. Consequently, almost the entire area of the fluence profile exceeds the ablation threshold and contributes to the ESA-based ablation. This suggests an effective utilization of the pulse energy. For λ = 550 and 600 nm, the values deviate significantly by a factor of two, which becomes evident by the decreased slope of the fits in Figure 3d,e.

Similarly to F_{th} that defines the minimum required fluence for single-beam ablation, F_{th}^{eff} defines the minimum F_{ESA} at given F_{pump} that is required for ESA-based dual-beam ablation. These values decreased by a factor of 10–5, but are still proportional to F_{th}. With the highest E_p, λ_{ESA} = 450 nm can utilize the energy, induced by the pump, most optimally. Here, F_{th}^{eff} = 0.26 J/cm is even below F_{th} = 0.49 J/cm² of the single pump beam, indicated by values above the purple line, as described earlier (Figure 3b).

To evaluate the improvements of the ESA-based process, compared to single-beam ablation, fluence efficiency

$$F_E = \frac{F_{th}^{eff} + F_{pump}}{F_{th}^{pump}} \quad (3)$$

and total fluence efficiency as the sum of the normalized single-beam fluences

$$F_{ET} = \frac{F_{th}^{eff}}{F_{th}^{ESA}} + \frac{F_{pump}}{F_{th}^{pump}} \quad (4)$$

were defined (Table 2). For F_E-values below one, the combined fluences of the ESA- and the pump-beam are still below F_{th}^{pump} for single-beam ablation ($F_E = 1$, if $F_{th}^{eff} + F_{pump} = F_{th}^{pump}$). As already mentioned, only λ = 450 nm at Δt = 5 ns and F_{pump} = 0.2 J/cm² reached F_E = 0.86. Similarly, replacing the single-beam threshold fluence in Equation (3) to F_{th} of the ESA-beams, allows the comparison of the efficiency improvement with the ESA-beams. In order to exclude an ordinary accumulation effect of the pump and ESA beam, F_{ET} was defined. For $F_{ET} < 1$, an additional induced transient absorption must be considered as part of the ablation process (Table 2).

3.3.3. Ablation Quality

Figure 4a,b show selected single-beam ablation spots for the utilized ESA wavelengths. At high F_{ESA}, the absorption of defect levels at the fused silica–ZnO interface leads to substrate damage (Figure 4a,f). At F_{ESA} close to F_{th}, the ablation spots become increasingly inhomogeneous (Figure 4b). As described earlier, the further decrease of F_{ESA} results in only partially interconnected spots inside the beam. In comparison, the pump beam hits the ZnO–air interface first (Figure 1b) and the fundamental absorption leads to an almost circular and homogeneous single-beam ablation of the entire ZnO layer (Figure 4c). When reaching the ZnO–substrate interface, the pump beam is already highly attenuated preventing abrupt heating and damaging of the substrate (Figure 4h). The maximum d_a observed in the depth profiles (Figure 4g–j) are in good agreement with the calculated layer thickness d_L = 203 nm.

Compared to single-beam ablation, the spot diameters are decreased by the ESA-based ablation process, indicated by decreased values of $2\omega_f^{eff}$ when compared to $2\omega_f$ (Table 1 and Table 2). The ablation area is restricted by the superposition of both pulses. This limits the ablation to the mostly uniform high-intensive center of the pulses and results in an enhanced ablation quality with circular ablation spots. In comparison to single pump beam ablation, the absorbed energy and induced heat at the ZnO–air interface is reduced and homogeneously distributed throughout the entire ZnO-layer. This results in a reduced melt formation at the edges protruding above the layer surface (Figure 4g–j), as well as decreasing the surface contamination by resolidified ZnO-droplets around the ablation spots (Figure 4d,e). By increasing Δt, these effects are reduced even further, as the heat dissipates and an ESA at defect levels dominates the ablation process (Figure 4i,j).

The optimization of the parameters allows the complete removal of the ZnO layer without damaging the fused silica substrate (Figure 4d,i) and a precise control of d_a (Figure 4e,j). The reduced F_{ESA} prevents the fused silica–ZnO interface from abruptly heating to the point of substrate damage (Figure 4a,f). As the fundamentally absorbed pump beam is focused on the ZnO–air surface, the initial abrupt heating together with a high concentration of excited carriers diffuse gradually and homogeneously through the ZnO-layer (d_p = 62 nm). Optimal ablation conditions were found at $\Delta t \approx$ 5 ns. The ZnO layer is already excited by the pump beam, in particular at the ZnO–substrate interface, where the defects are located primarily. It thus requires only a low excess of additionally absorbed ESA photons and energy to ablate the ZnO layer from the fused silica substrate.

3.3.4. Influence of Temporal Pulse Delay

To determine the importance of defect levels compared to absorption solely based on intra-band transition, the temporal delay between the pulses was varied for different F_{ESA} at the fixed optimum of F_{pump} = 0.2 J/cm (Figure 5). At the highest F_{ESA}, for λ = 450 and 500 nm (Figure 5a,b), delays above 300 ns still resulted in a measurable ablation. Compared to λ = 550 and 600 nm (Figure 5c,d), this delay is much larger with Δt mostly below 50 ns. As the fluorescent lifetime of ZnO band gap electrons is well below 10 ns [42,44], an ablation at these high Δt cannot be explained solely by intra-band transition-based absorption. Thus, the defect levels with their increased fluorescent lifetime serve as the main driving force [35]. First, the pump beam promotes electrons into CB (Figure 1a). From there, besides relaxation back to the VB, electrons transfer rapidly to deep defect levels, which are suitable as absorption centers for the subsequent ESA beam (Figure 2c). The decrease of D^2 and the Δt-range is almost proportional to F_{ESA}. Only F_{ESA}-values close to their respective F_{th} deviate from this trend. Especially for λ = 550 and 600 nm, these values increase overproportional at the highest F_{ESA}, where the ESA-beams can almost ablate the ZnO solely by single-beam ablation.

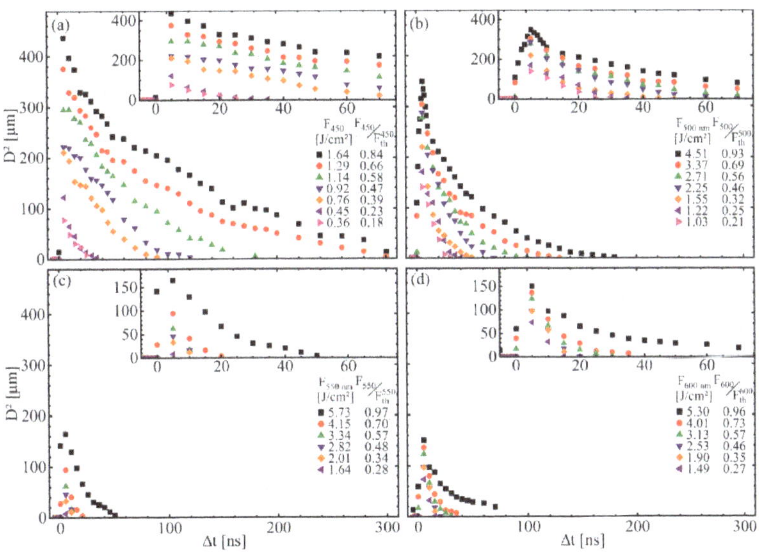

Figure 5. D2 plotted versus Δt to determine the optimal delay between the pulses. For excitation, Fpump = 0.2 J/cm² was used at (a) λESA = 450 nm, (b) λESA = 500 nm, (c) λESA = 550 nm, (d) λESA = 600 nm. FESA was varied in the identical increments as for Figure 3b (color coded with FESA and the FESA/Fth-ratio). In the top-right corners, the section of short Δt is extended.

Figure 5c,d reveals deviations from this trend, especially for λ = 550 and 600 nm, where the values increase overproportionately at the highest values of F_{ESA}. These fluence values are close the respective ablation thresholds, which almost allow ablation of the ZnO layer solely by the ESA beam.

At low temporal delays ($\Delta t < 10$ ns), the impact of λ_{ESA} is less pronounced (Figure 5). Here, the effect of an intra-band transition and sample heating dominates as the beams still overlap. For $\lambda = 550$ and 600 nm, an increasing Δt leads to a rapidly decreasing D^2 because the defect level absorption emerging as the dominant process is only utilized effectively at $\lambda = 450$ and 500 nm. As shown in Figure 1c, at $\Delta t = +5$ and -5 ns, the temporal overlap is similar, but the resulting D^2 differ significantly, as prior excitation of electrons by the pump beam is required for ESA.

Exemplary, for $\lambda = 500$ nm at $F_{ESA} = 4.51$ J/cm², Δt was varied in one ns-steps between $\Delta t = -5$ to +10 ns to analyze the area of temporal beam overlap in more detail (Figure 5b). At the optimum conditions around $\Delta t \approx 5$ ns, the highest number of prior excited electrons is available for an ESA (Figure 1c). At $\Delta t = 0$ ns, where both beams overlap optimally, the excited electrons are only partially used for the ESA, as parts of the radiation excite carriers after the highest intensity of the ESA beam reaches the sample. At negative temporal delays and further decreasing beam overlap, only minor ablation is observed for F_{ESA} close to F_{th}. Here, the number of excited electrons is insufficient to reach the required ablation threshold. Increasing Δt above the optimum also leads to a decreasing D^2-values. However, the rate is much lower than for negative delays.

At $\Delta t < 50$ ns, the heat generated by the pump excitation must be considered, as it effects the ablation process considerably, particularly close to the temporal beam overlap (Figure 1c). Here, increased sample temperatures act as an offset for the ESA, as less additionally absorbed energy is required for ablation. Thus, a wider area of the ESA beam surpasses F_{th}, resulting in increased D^2. $F_{pump} = 0.2$ J/cm², well below $F_{th} = 0.49$ J/cm² was used to reduce sample heating.

For polycrystalline ZnO-layers, thermal conductivities $k_{th} < 10$ W/m·K are found in literature, which is well below $k_{th} \sim 100$–30 W/m·K for epitaxially grown or bulk ZnO [45,46,47,48,49]. The thermal diffusion length is determined by $\mu = 2(\alpha_{th} \cdot \Delta t)^{0.5}$ with $\alpha_{th} = k_{th}/c_V$ where c_V is the volumetric heat capacity with about $10 \cdot 2.8$–2.4^6 J/m³·K [47,49]. With $d_p = 62$ nm, most of the heat is created by the pump beam at the ZnO–air interface. It takes about 5 ns to reach the ZnO–fused silica interface, where the highest defect concentration and thus, the strongest ESA is expected. This is in good agreement with the ablation maximum measured at a delay of about 5 ns.

Furthermore, the rapid D^2-decrease in the delay range 20–50 ns (Figure 5) can be attributed to heat dissipation and correlates with literature values [45,50]. As $2\omega_f$ is about 100 times larger than d_L and μ for the time scale in consideration, one-dimensional heat conduction can be assumed [47,51,52]. Here, the heat should be mostly dissipated. Consequently, the defect level absorption is now the dominant factor for ESA, and it is no longer superimposed by pump beam heating. At higher Δt, the ablation slowly decreases due to the finite fluorescent lifetime of the defect level electrons that slowly lowers the available excited electron density for the ESA [43,53].

4. CONCISE SUMMARY

In the present study, we investigated a dual-beam laser process based on ESA by applying several wavelengths (λ_{pump} = 355 nm, λ_{ESA} = 600 ,550 ,500 ,450 nm) at various F_{pump} and F_{ESA}. The ESA-beams with the highest photon energies produced the largest ablation spots at the lowest beam fluences. A small temporal delay of +5 ns is optimal for the dual-beam ablation process, as it utilizes a combination of heat and excited carriers for increased absorption. Extended time delays lead to heat dispersion and rapidly reduce the ablation diameters. The absorption of the ESA beam is exclusively dictated by the concentration of excited carriers. The quality of the ablation spots was enhanced by minimizing debris and melt production, resulting in a reduction in the necessary laser energy. Additionally, a methodology for assessing the efficiency enhancement based on ESA was created. Future investigations into the integration of STED and ESA concerning STED direct writing processes were proposed based on these findings.

AT A GLANCE

We introduce a distinctive dual laser beam processing method utilizing excited state absorption to structure 200 nm thin zinc oxide films deposited on fused silica substrates. The integration of two pulsed nanosecond laser beams with distinct photon energies—one beneath and one exceeding the zinc oxide band gap energy—facilitates precise, efficient, and uniform ablation of the films without causing substrate damage. A comprehensive notion of energy transfer and excitation mechanisms during irradiation was created based on structuring experiments that considered laser wavelength, pulse fluence, and pulse delay of both laser beams. It offers an extensive comprehension of the thermal and electrical mechanisms involved in ablation. An efficiency model was constructed to measure the improvements of the dual-beam procedure relative to single-beam ablation.

REFERENCES

1. Hell, S.W.; Wichmann, J. Breaking the diffraction resolution limit by stimulated emission: Stimulated-emission-depletion fluorescence microscopy. *Opt. Lett.* **1994**, *19*, 780–782. **[Google Scholar] [CrossRef] [PubMed]**
2. Klar, T.A.; Hell, S.W. Subdiffraction resolution in far-field fluorescence microscopy. *Opt. Lett.* **1999**, *24*, 954–956. **[Google Scholar] [CrossRef]**
3. Rittweger, E.; Han, K.Y.; Irvine, S.E.; Eggeling, C.; Hell, S.W. STED microscopy reveals crystal colour centres with nanometric resolution. *Nat. Photon.* **2009**, 3, 144–147. **[Google Scholar] [CrossRef]**
4. Gottfert, F.; Wurm, C.A.; Mueller, V.; Berning, S.; Cordes, V.C.; Honigmann, A.; Hell, S.W. Coaligned dual-channel STED nanoscopy and molecular diffusion analysis at 20 nm resolution. *Biophys. J.* **2013**, *105*, L01–L03. **[Google Scholar] [CrossRef] [Green Version]**

5. Malinauskas, M.; Farsari, M.; Piskarskas, A.; Juodkazis, S. Ultrafast laser nanostructuring of photopolymers: A decade of advances. *Phys. Rep.* **2013**, *533*, 1–31. [Google Scholar] [CrossRef]

6. Frolich, A.; Fischer, J.; Zebrowski, T.; Busch, K.; Wegener, M. Titania woodpiles with complete three-dimensional photonic bandgaps in the visible. *Adv. Mater.* **2013**, *25*, 3588–3592. [Google Scholar] [CrossRef] [PubMed]

7. Sugioka, K.; Meunier, M.; Piqué, A. *Laser Precision Microfabrication*; Springer: Berlin, Germany, 2010; Volume 135. [Google Scholar]

8. Engel, S.; Wenisch, C.; Gräf, S.; Müller, F. Sub-diffraction direct laser writing by a combination of STED and ESA. In *Laser-Based Micro-and Nanoprocessing XIV*; International Society for Optics and Photonics, SPIE LASE: San Francisco, CA, USA, 2020; p. 112680D. [Google Scholar]

9. Wenisch, C.; Engel, S.; Gräf, S.; Müller, F.A. Fundamentals of a new sub-diffraction direct laser writing method by a combination of stimulated emission depletion and excited state absorption. *J. Laser Micro Nanoeng.* **2020**, *15*, 169–173. [Google Scholar]

10. Théberge, F.; Chin, S. Enhanced ablation of silica by the superposition of femtosecond and nanosecond laser pulses. *Appl. Phys. A* **2005**, *80*, 1505–1510. [Google Scholar] [CrossRef]

11. Zhang, J.; Sugioka, K.; Takahashi, T.; Toyoda, K.; Midorikawa, K. Dual-beam ablation of fused silica by multiwavelength excitation process using KrF excimer and F2 lasers. *Appl. Phys. A* **2000**, *71*, 23–26. [Google Scholar]

12. Obata, K.; Sugioka, K.; Akane, T.; Midorikawa, K.; Aoki, N.; Toyoda, K. Efficient refractive-index modification of fused silica by a resonance-photoionization-like process using F 2 and KrF excimer lasers. *Opt. Lett.* **2002**, *27*, 330–332. [Google Scholar] [CrossRef]

13. Sugioka, K.; Wada, S.; Tashiro, H.; Toyoda, K. Ablation of wide band-gap materials by multi-wavelength irradiation using a VUV Raman laser. *Appl. Surf. Sci.* **1997**, *109*, 179–183. [Google Scholar] [CrossRef]

14. Zhang, J.; Sugioka, K.; Wada, S.; Tashiro, H.; Toyoda, K. Direct photoetching of single crystal SiC by VUV-266 nm multiwavelength laser ablation. *Appl. Phys. A* **1997**, *64*, 367–371. [Google Scholar] [CrossRef]

15. Zhang, J.; Sugioka, K.; Wada, S.; Tashiro, H.; Toyoda, K.; Midorikawa, K. Precise microfabrication of wide band gap semiconductors (SiC and GaN) by VUV–UV multiwavelength laser ablation. *Appl. Surf. Sci.* **1998**, *127*, 793–799. [Google Scholar] [CrossRef]

16. Akane, T.; Sugioka, K.; Hammura, K.; Aoyagi, Y.; Midorikawa, K.; Obata, K.; Toyoda, K.; Nomura, S. GaN ablation etching by simultaneous irradiation with F 2 laser and KrF excimer laser. *J. Vac. Sci. Tech. B Microelectron. Nanometer Struct. Proces. Meas. Phenom.* **2001**, *19*, 1388–1391. [Google Scholar] [CrossRef]

17. Obata, K.; Sugioka, K.; Midorikawa, K.; Inamura, T.; Takai, H. Deep etching of epitaxial gallium nitride film by multiwavelength excitation process using F 2 and KrF excimer lasers. *Appl. Phys. A* **2006**, *82*, 479–483. [Google Scholar] [CrossRef]

18. Zoppel, S.; Zehetner, J.; Reider, G.A. Two color laser ablation: Enhanced yield, improved machining. *Appl. Surf. Sci.* **2007**, *253*, 7692–7695. [Google Scholar] [CrossRef]

19. Zoppel, S.; Merz, R.; Zehetner, J.; Reider, G.A. Enhancement of laser ablation yield by two color excitation. *Appl. Phys. A* **2005**, *81*, 847–850. [Google Scholar] [CrossRef]

20. Djurišić, A.; Ng, A.M.C.; Chen, X. ZnO nanostructures for optoelectronics: Material properties and device applications. *Prog. Quantum Electron.* **2010**, *34*, 191–259. [Google Scholar] [CrossRef]

21. Ogata, K.; Sakurai, K.; Fujita, S.; Fujita, S.; Matsushige, K. Effects of thermal annealing of ZnO layers grown by MBE. *J. Cryst. Growth* **2000**, *214*, 312–315. [Google Scholar] [CrossRef]

22. Marković, S.; Simatović, I.S.; Ahmetović, S.; Veselinović, L.; Stojadinović, S.; Rac, V.; Škapin, S.D.; Bogdanović, D.B.; Častvan, I.J.; Uskoković, D. Surfactant-assisted microwave processing of ZnO particles: A simple way for designing the surface-to-bulk defect ratio and improving photo (electro) catalytic properties. *RSC Adv.* **2019**, *9*, 17165–17178. [Google Scholar] [CrossRef] [Green Version]

23. Wang, X.; Lim, G.; Zheng, H.; Ng, F.; Liu, W.; Chua, S. Femtosecond pulse laser ablation of sapphire in ambient air. *Appl. Surf. Sci.* **2004**, *228*, 221–226. [Google Scholar] [CrossRef]

24. Rudolph, P.; Bonse, J.; Krüger, J.; Kautek, W. Femtosecond-and nanosecond-pulse laser ablation of bariumalumoborosilicate glass. *Appl. Phys. A* **1999**, *69*, S763–S766. **[Google Scholar] [CrossRef]**

25. Liu, J. Simple technique for measurements of pulsed Gaussian-beam spot sizes. *Opt. Lett.* **1982**, *7*, 196–198. **[Google Scholar] [CrossRef]**

26. Swanepoel, R. Determination of the thickness and optical constants of amorphous silicon. *J. Phys. E Sci. Instrum.* **1983**, *16*, 1214. **[Google Scholar] [CrossRef]**

27. Sernelius, B.E.; Berggren, K.-F.; Jin, Z.-C.; Hamberg, I.; Granqvist, C.G. Band-gap tailoring of ZnO by means of heavy Al doping. *Phys. Rev. B* **1988**, *37*, 10244. **[Google Scholar] [CrossRef] [Green Version]**

28. Tauc, J.; Grigorovici, R.; Vancu, A. Optical properties and electronic structure of amorphous germanium. *Phys. Stat. Solidi B* **1966**, *15*, 627–637. **[Google Scholar] [CrossRef]**

29. Farrag, A.A.-G.; Balboul, M.R. Nano ZnO thin films synthesis by sol–gel spin coating method as a transparent layer for solar cell applications. *J. Sol. Gel Sci. Technol.* **2017**, *82*, 269–279. **[Google Scholar] [CrossRef]**

30. Shaaban, E.; Yahia, I.; El-Metwally, E. Validity of Swanepoel's method for calculating the optical constants of thick films. *Acta Phys. Pol. Ser. A Gen. Phys.* **2012**, *121*, 628. **[Google Scholar] [CrossRef]**

31. Xue, S.; Zu, X.; Zhou, W.; Deng, H.; Xiang, X.; Zhang, L.; Deng, H. Effects of post-thermal annealing on the optical constants of ZnO thin film. *J. Alloys Compd.* **2008**, *448*, 21–26. **[Google Scholar] [CrossRef]**

32. Xu, P.; Sun, Y.; Shi, C.; Xu, F.; Pan, H. The electronic structure and spectral properties of ZnO and its defects. *Nucl. Instrum. Methods Phys. Res. Sec. B Beam Interact. Mater. Atoms* **2003**, *199*, 286–290. **[Google Scholar] [CrossRef]**

33. Wang, T.; Zheng, S.; Hao, W.; Wang, C. Studies on photocatalytic activity and transmittance spectra of TiO2 thin films prepared by rf magnetron sputtering method. *Surf. Coat. Technol.* **2002**, *155*, 141–145. **[Google Scholar] [CrossRef]**

34. Caglar, M.; Caglar, Y.; Ilican, S. The determination of the thickness and optical constants of the ZnO crystalline thin film by using envelope method. *J. Optoelectron. Adv. Mater.* **2006**, *8*, 1410. **[Google Scholar]**

35. Han, N.S.; Shim, H.S.; Seo, J.H.; Kim, S.Y.; Park, S.M.; Song, J.K. Defect states of ZnO nanoparticles: Discrimination by time-resolved photoluminescence spectroscopy. *J. Appl. Phys.* **2010**, *107*, 084306. **[Google Scholar] [CrossRef] [Green Version]**

36. Lin, B.; Fu, Z.; Jia, Y. Green luminescent center in undoped zinc oxide films deposited on silicon substrates. *Appl. Phys. Lett.* **2001**, *79*, 943–945. **[Google Scholar] [CrossRef]**

37. Johnson, J.C.; Knutsen, K.P.; Yan, H.; Law, M.; Zhang, Y.; Yang, P.; Saykally, R.J. Ultrafast carrier dynamics in single ZnO nanowire and nanoribbon lasers. *Nano Lett.* **2004**, *4*, 197–204. **[Google Scholar] [CrossRef]**

38. Sugioka, K.; Midorikawa, K. VUV-UV multiwavelength excitation process for high-quality ablation of fused silica. In *Damage to VUV, EUV, and X-Ray Optics IV; and EUV and X-Ray Optics: Synergy between Laboratory and Space III*; International Society for Optics and Photonics, SPIE OPTICS + OPTOELECTRONICS: Prague, Czech Republic, 2013; p. 877704. **[Google Scholar]**

39. Mardare, D.; Tasca, M.; Delibas, M.; Rusu, G. On the structural properties and optical transmittance of TiO2 rf sputtered thin films. *Appl. Surf. Sci.* **2000**, *156*, 200–206. **[Google Scholar] [CrossRef]**

40. Martin, S.; Hertwig, A.; Lenzner, M.; Krüger, J.; Kautek, W. Spot-size dependence of the ablation threshold in dielectrics for femtosecond laser pulses. *Appl. Phys. A* **2003**, *77*, 883–884. **[Google Scholar] [CrossRef]**

41. Bonse, J.; Baudach, S.; Krüger, J.; Kautek, W.; Lenzner, M. Femtosecond laser ablation of silicon–modification thresholds and morphology. *Appl. Phys. A* **2002**, *74*, 19–25. **[Google Scholar] [CrossRef]**

42. Layek, A.; Manna, B.; Chowdhury, A. Carrier recombination dynamics through defect states of ZnO nanocrystals: From nanoparticles to nanorods. *Chem. Phys. Lett.* **2012**, *539*, 133–138. **[Google Scholar] [CrossRef]**

43. Tang, A.-H.; Mei, Z.-X.; Hou, Y.-N.; Du, X.-L. Photodynamics of GaZn–VZn complex defect in Ga doped ZnO. *Chin. Phys. B* **2018**, *27*, 117802. **[Google Scholar] [CrossRef]**

44. Özgür, Ü.; Alivov, Y.I.; Liu, C.; Teke, A.; Reshchikov, M.; Doğan, S.; Avrutin, V.; Cho, S.-J.; Morkoç, A.H. A comprehensive review of ZnO materials and devices. J. Appl. Phys. 2005, 98, 11. [Google Scholar] [CrossRef] [Green Version]

45. Huang, Z.X.; Tang, Z.A.; Yu, J.; Bai, S. Thermal conductivity of nanoscale polycrystalline ZnO thin films. *Phys. B Condens. Matter* **2011**, *406*, 818–823. **[Google Scholar] [CrossRef]**

46. Akhir, R.M.; Abd Wahab, Z. Thermal diffusivity studies of ZnO-CuO at high temperatures. *J. Teknologi* **2015**, *76*, 19–23. **[Google Scholar]**

47. Yamashita, Y.; Honda, K.; Yagi, T.; Jia, J.; Taketoshi, N.; Shigesato, Y. Thermal conductivity of hetero-epitaxial ZnO thin films on c-and r-plane sapphire substrates: Thickness and grain size effect. *J. Appl. Phys.* **2019**, *125*, 035101. **[Google Scholar] [CrossRef]**

48. Alvarez-Quintana, J.; Martínez, E.; Pérez-Tijerina, E.; Pérez-García, S.; Rodríguez-Viejo, J. Temperature dependent thermal conductivity of polycrystalline ZnO films. *J. Appl. Phys.* **2010**, *107*, 063713. **[Google Scholar] [CrossRef]**

49. Xu, Y.; Goto, M.; Kato, R.; Tanaka, Y.; Kagawa, Y. Thermal conductivity of ZnO thin film produced by reactive sputtering. *J. Appl. Phys.* **2012**, *111*, 084320. **[Google Scholar] [CrossRef]**

50. Nedyalkov, N.; Koleva, M.; Nikov, R.; Atanasov, P.; Nakajima, Y.; Takami, A.; Shibata, A.; Terakawa, M. Laser nanostructuring of ZnO thin films. *Appl. Surf. Sci.* **2016**, *374*, 172–176. **[Google Scholar] [CrossRef]**

51. Diez, M.; Ametowobla, M.; Graf, T. Time-resolved reflectivity and temperature measurements during laser irradiation of crystalline silicon. *J. Laser Micro Nanoeng.* **2017**, *12*, 230–234. **[Google Scholar]**

52. Jiang, P.; Qian, X.; Yang, R. Time-domain thermoreflectance (TDTR) measurements of anisotropic thermal conductivity using a variable spot size approach. *Rev. Sci. Instrum.* **2017**, *88*, 074901. **[Google Scholar] [CrossRef] [PubMed] [Green Version]**

53. Zhong, Y.; Djurisic, A.B.; Hsu, Y.F.; Wong, K.S.; Brauer, G.; Ling, C.C.; Chan, W.K. Exceptionally long exciton photoluminescence lifetime in ZnO tetrapods. *J. Phys. Chem. C* **2008**, *112*, 16286–16295. **[Google Scholar] [CrossRef]**

CHAPTER 11

RE-DEPOSITION PATTERNS OF BACKWARD FLUX DURING MULTI-SPOT LASER ABLATION OF STAINLESS STEEL UTILIZING PICOSECOND AND FEMTOSECOND PULSES IN AIR

1. INTRODUCTION: AN OVERVIEW

The engagement of intense ultrafast laser radiation with atoms, gases, and solid materials has resulted in significant scientific breakthroughs, including High Harmonic Generation for attosecond spectroscopy[1,2], filamentation for atmospheric probing[3,4], two-photon microscopy[5,6], and intraocular fs-LASIK in ophthalmic surgery. Specifically, laser ablation (LA) of materials utilizing picosecond (ps) and sub-picosecond (sub-ps) pulse durations has enhanced materials analysis in fields such as Laser Ablation Inductively Coupled Mass Spectrometry (LA-ICP-MS) [8,9,10]and Laser-Induced Breakdown Spectroscopy (LIBS)[11,12]. With temporal pulse lengths <10 ps in ultrafast laser ablation, the instantaneous energy deposition reduces plasma absorption while minimizing melt and thermal diffusion during the pulse and yield high precision material removal [13,14]. This allows for easy generation of plasmonic, interferometric and hierarchical microstructures on, for example, stainless steel [15,16,17]. In metals, which have absorption coefficients typically $\alpha \sim 10^6$ cm^{-1}, a thin layer with thickness d ~10–30 nm is converted to a plasma at a solid density which expands well after the pulse is absorbed [18,19].

In a vacuum, laser ablation is characterized as a free expansion[20,21,22,23], whereas the laser plume formed in the presence of an ambient gas is far more intricate, encompassing durations ranging from picoseconds to nanoseconds to microseconds. Following electron heating during the pulse, the lattice typically heats over many picoseconds due to electron-phonon interaction, elevating the surface temperature much above the evaporation threshold. This results in fast surface electron emission due to thermionic [24] or multi-photon absorption [25,26] and a remarkable early stage "air plasma" can develop during the first 150 ps [27,28], with electron density $n_e \sim 10^{20}$ cm^{-3} and a transient electric field developing above the surface [29].

On the nanosecond timescale in air, the plasma plume expands rapidly and initially longitudinally (at speed v ~ 10^6 cm s^{-1}), colliding with the high density of surrounding gas molecules, ($N_{1atm} \sim 10 \times 2.6^{19}$ cm^{-3}), confining the plasma to the surface region while creating a blast (shock) wave which moves ahead of the plume [30,31,32,33,34]. As the longitudinal expansion decelerates rapidly, the plasma expands laterally after a few ×100 ns ($v_{lat} \sim 10^5$ cm s^{-1}) and can form a vortex structure with plasma plume concentrated at the ablation spot edge [35]. Transient electron temperatures in the plasma can reach $T_e >> 10^4$ K [24], then plasma cooling (over a few hundreds of ns to µs) occurs due to plasma expansion, electron-ion recombination and collisions with the background gas molecules. When thermal equilibrium is achieved in the plasma (~100 ns), this is the regime for LIBS and material analysis with fluorescence from highly excited atoms/ions [33,34]. As the plasma cools further, molecular species are formed at later times (~2–50 µs) due to atomic collisions, recombination and oxidation occur after the shock wave collapses [36]. Aerosols and nanoclusters are generated during a nucleation-condensation process with nanoparticle agglomerates re-deposited symmetrically by backward flux around a single spot [35,37]. While time-resolved spectral information using gated ICCDs can help elucidate the plasma dynamics, time-resolved scattering can detect nanoparticle formation and re-deposition on the microsecond timescale [35].

Debris re-deposition was first investigated during UV, excimer micro-structuring of polymers (e.g., Polyimide, PI), responsible for the developing cone structure with debris field dependent on spot shape, inferring strong horizontal forces close to the substrate surface [38]. Apparent field rotations with non-spherical spots were numerically modelled as a purely gas dynamic effect [39]. The debris radius was also shown to follow blast wave theory during excimer ablation of polyimide in air and inert gases [40]. Time-delayed (20–80 µs delay) weak emission detected at the plume periphery consisted mainly of continuum radiation consistent with thermal radiation from solid particles [20]. The difference in ablation geometries observed with a linear array of closely spaced ablation spots inferred stronger plume interactions in the middle [41]. Monte Carlo simulation of the expansion of a copper plasma in the presence of a background gas (Ar) shows compression of the ambient gas atoms by the plume (snowplough effect) for pressures >50 Pa and narrowing of the plume with increasing pressure while predicting backward plume motion and vortical flow at the plume periphery [42]. In the case of excimer ablation of stainless steel in air, (F = 10 J cm^{-2}/308 nm/25 ns) time-resolved ICCD images of the ablation plume up to 10 µs delay show this developing vortical structure after 1 µs and fully developed at 10 µs delay with a high density of nanoparticles deposited at the periphery [35].

Time-resolved light scattering during ablation also demonstrated that debris re-deposition occurred over a time delay of 1–60 μs after the pulse.

Polymers (such as PI, a photoresist) have strong absorption coefficients $\alpha > 10^4$ cm^{-1} in the UV leading to electronic excitation and "cold" ablation via direct bond breaking [43]. In the air, a plasma plume, carbonization (thermal component) and re-deposition around the ablation spot occurs but can be minimized using, for example, H_2 as a reactive, ambient gas [44]. Low ablation thresholds $F_{th} < 50$ mJ cm^{-2} and precise etch rates from 25 nm/pulse are observed [45]. On steel, UV (248 nm/25 ns) ablation re-deposition generates iron oxide nanoparticles with a wide distribution $\varphi = 2–20$ nm while at 500 fs/248 nm, this narrows to $\varphi = 3 \pm 2$ nm [46]. The use of NIR, ns pulses result in the incorporation of nitrogen, as well into the re-deposited layer [47].

Recently, there has been a focus on temporal plume expansion utilizing bursts of ultrashort pulses, alongside ultrafast lasers with high repetition rates, driven by the necessity to enhance material ablation rates through improved comprehension of plume expansion and plasma absorption by subsequent pulses. Recent observations of the interaction dynamics of shock waves during femtosecond laser multi-spot ablation at a water/air interface revealed clashing supersonic airflows through transient reflection [51,52]. The authors indicate the possibility of generating targeted ablation plumes pertinent to pulsed laser thin film deposition [53,54]. This concept has been examined using the observed backward re-deposition patterns arising from multi-spot ultrafast laser ablation on stainless steel, emphasizing the interactions of the resultant plasma plume and shock wave. Symmetric jets, characterized by debris expelled perpendicularly to the spot axis, are reported for the first time, and we offer a preliminary elucidation of the underlying physics of these observations. The results may significantly influence the critical domain of Laser-Induced Forward Transfer of nanoparticle thin films at ambient pressure [55,56].

2. MATERIALS AND METHODS

Experiments were conducted using two distinct laser and optical systems located at the Universities of Liverpool and Mittweida, with temporal pulse lengths of 10 ps and 600 fs, respectively. Figure 1 illustrates a schematic representation of the optical system in Liverpool. The output from a Nd:VAN seeded Regenerative amplifier (High-Q IC-355-800 ps, 10 ps/1064 nm Photonic Solutions Ltd., Edinburgh, UK) was attenuated, expanded (×3), then directed at low AOI to a phase-only Spatial Light Modulator (SLM, Hamamatsu-10468-03, Hamamatsu Photonics, Hamamatsu, Japan) for generating multi-spot patterns and focussed to a sample mounted on a three-axis stage able to bring the substrate surface to the focal plane. Computer Generated Holograms (CGHs) controlling spot numbers and separation on the substrate were generated using Inverse Fourier Transforms in the Hamamatsu software, while a pick-off optic (not shown) could direct spot patterns to a CCD camera prior to multi-spot ablation. This allows precise electronic control of spot separation with no mechanical movements. A 4f optical system ($f_1 = f_2 = 400$ mm) relayed the modulated beam from the SLM (addressed with CGHs) to the input aperture of a galvo system which directed the beams to the target surface, focussed by an f-theta lens (f = 100 mm). A fast-mechanical shutter (Thorlabs SH05, Thorlabs Ltd., Ely, UK) allowed the pulse number on target to be varied, synchronised to the scanning software (SCAPS GmbH). The expanding plumes were imaged to a time-resolved spectrometer (Andor Shamrock, model SR303i with intensified CCD, iStar 734, 2 ns gate width, Andor Technology Ltd., Belfast, UK) and synchronised from the 5 kHz Regenerative amplifier.

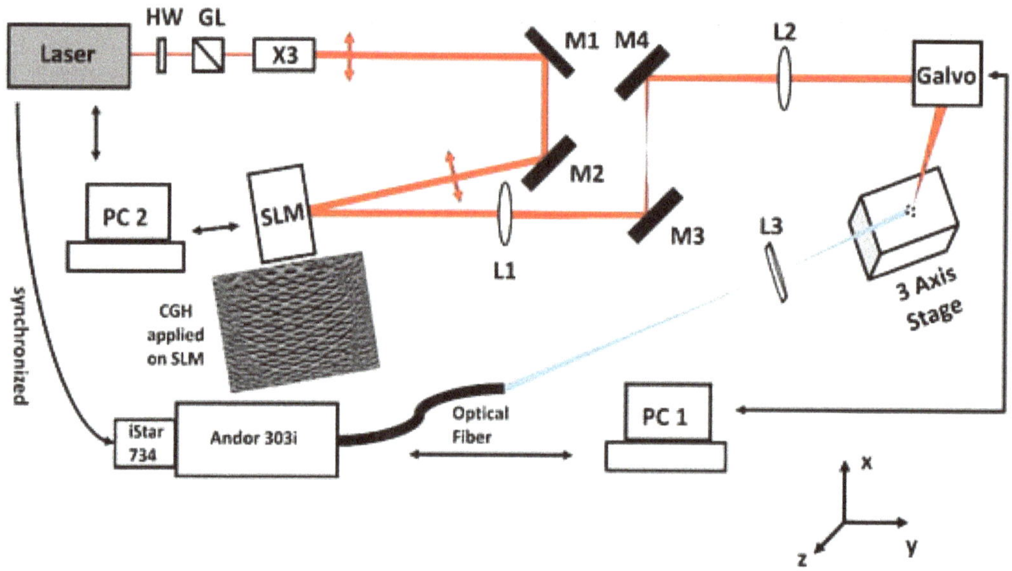

Figure 1. Optical set-up for 10 ps/5 kHz multi-spot ablation of stainless steel in the air. An attenuated and expanded beam from a Nd:Van Regen amplifier is directed to the phase-only SLM at low AOI then imaged via a 4f optical system (f = 400 mm) to the galvo input aperture. Appropriate phase-only CGH's could create arbitrary spot geometries and separations at the steel surface. A time-resolved spectrometer with triggered ICCD (synchronised to Regen amplifier) allowed measurement of the spectral plasma dynamics by focusing the plasma emission on a fibre coupler.

Figure 2 shows a schematic of the optical system for 600 fs/5 kHz double spot ablation (Mittweida). The pump laser (FX-Series, edgewave GmbH, 1030 nm/600 fs, Würselen, Germany) output was attenuated by HWP/polarising beam splitter BS. Afterwards, a second HWP/PBS combination produce two optical lines (1,2) with adjustable power distribution. Line 1 path length was altered by translating mirrors M (1.1) and M (1.2) to synchronise the ablation spots temporally. This synchronisation is proven by the ablation interference patterns with Line 2 at the sample surface. The ablation spot separation was adjusted by slightly tilting mirror M (1.4). Both lines were focussed onto the sample surface by an f = 100 mm optic L2. At the focal plane, the laser spot radii are w_{86} = 15 μm. A spot monitor (CCD) on the level of the material surface to enable a precise temporal and geometrical adjustment. A probe laser (Cavilux, Fa. Cavitar Ltd., 688 nm/13 ns, Tampere, Finland) electronically synchronised to the pump laser allowed pump-probe shadowgraphy of expanding plasma plumes. The shadowgraphs were recorded with a 14-bit cooled CCD camera (pco.1600, PCO AG, Kelheim, Germany).

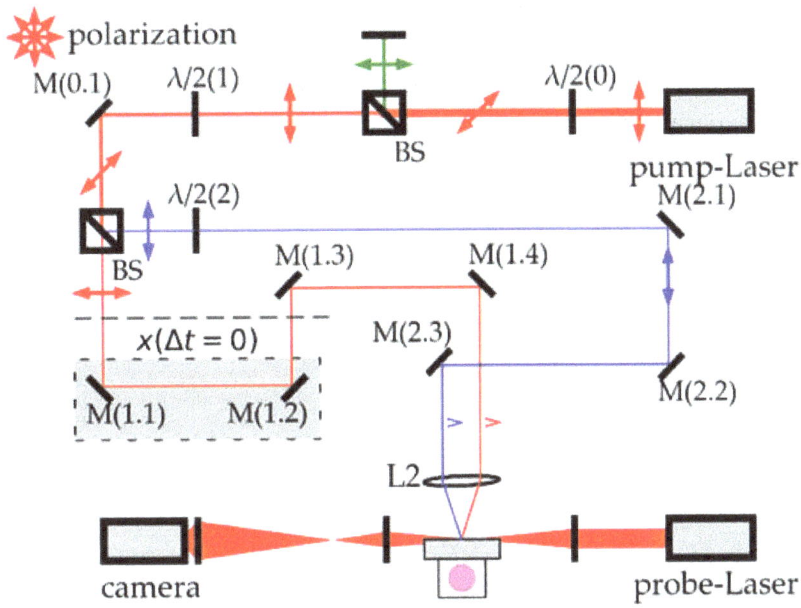

Figure 2. Scheme of the optical set-up for double spot ablation with simultaneous shadowgraphy of stainless steel in the air. The pump laser, probe laser and axis and shadowgraph capture were electrically synchronised. The attenuated and split pump beams are focussed and temporally overlapped at the sample surface by one optic lens L2. The temporal (by moving M (1.1) and M (1.2)) and geometrical adjustment (by tilting M (1.4)) can be monitored with a spot monitor in the focal plane.

3. RESULTS

3.1. 2-Spot Ablation with 10 ps Pulses at 1064 nm/5 kHz Repetition Rate

The substrate used was ANSI 304 stainless steel and optically polished to a roughness Ra ~50 nm. The focussed single beam diameter was measured from the observed ablation crater diameters with increasing pulse energy [57] and found to be φ = 22.2 ± 0.2 μm. No significant variation with multi-spot geometry was found. Single-pulse ablation threshold (N = 1) was measured to be F_{th} = 0.01 ± 0.29 J cm^{-2} decreasing with pulse number, and incubation coefficient measured to be S = 0.01 ± 0.85 in excellent agreement with the literature [58]. Figure 3 shows the re-deposition patterns observed during two spot ablation with 10 ps laser ablation of stainless steel in air while varying separation and pulse number at fluence F = 0.90 J cm^{-2}, (a) 200 pulses, (b) 400 pulses (c), 800 pulses, all 75 μm separation, (d)–(f) 95 μm separation, (g)–(i) and 145 μm separation, respectively. It required multi-hundred pulse exposure to observe debris for good optical contrast. The ablation debris is concentrated between the spots at d = 75 μm separation with a width comparable to the ablation spot diameter, while at 95 μm separation this narrows to around a 10 μm wide "filament". The concentration of this directed re-deposition, observed at low fluence, is an interesting phenomenon, indicating transient forces during the plume expansion and collisions with the air, resulting in some ablation debris acquiring momentum components preferentially directed along the axis between the spots. There is evidence also that at the midplane, some debris is expanding normal to the axis. As spot separation increased to 145 μm at this fluence, there was negligible interaction between the plumes. The debris radius near spots R_d ~ 30 μm.

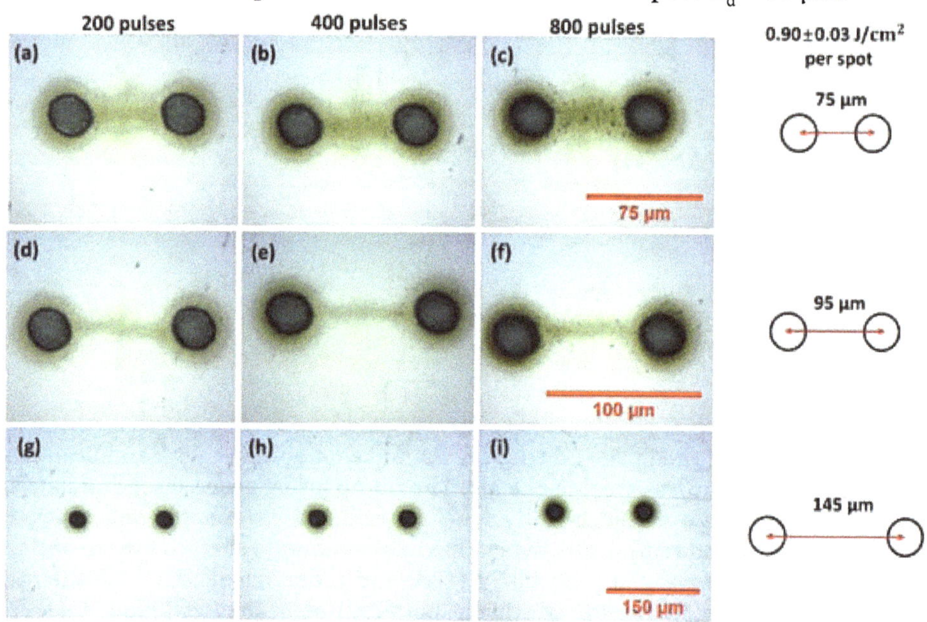

Figure 3. Optical images of observed two spot re-deposition during 10 ps laser ablation of stainless steel in air while varying spot separation and pulse number at F = 0.90 J cm-2, (a) 200 pulses, (b) 400 pulses, (c) 800 pulses, all 75 μm separation, (d–f) 95 μm separation, (g–i) 145 μm separation respectively. Debris is concentrated in a filament between the spots at 75 μm, thinning to a line about 10 μm wide at 95 μm spot separation. No interaction between the spots is observed at 145 μm separation at this low fluence.

The effect on debris re-deposition of increasing fluence to F = 4.51 J cm^{-2} on stainless steel is shown in the optical images of Figure 4a–i. At this higher fluence in Figure 4a–c, we observe diverging debris jets ejected normal to the spot axis. This extends to a radial distance of R_{jet} ~200 μm, well beyond that around each spot R_d ~ 30 μm, inferring that an energetic process may be involved. In Figure 4b,c, with increased exposure, removal of debris from the surface between the spots (in the form of two slightly curved lines) supports the view that strong shock wave interactions between the colliding plumes during ablation may be responsible, clear in Figure 4c,f. These effects essentially disappear at the highest, 145 μm separation with a return to the concentration of material between the spots Figure 4g–i, similar to the patterns at low fluence (Figure 3). The minute ablation points along the axis result from low-energy ghost beams generated during multi-pulse exposure, whereas the upper spot represents the residual zero order. Their existence facilitates the detection of local physical impacts of shock waves and air movements influencing debris mobility.

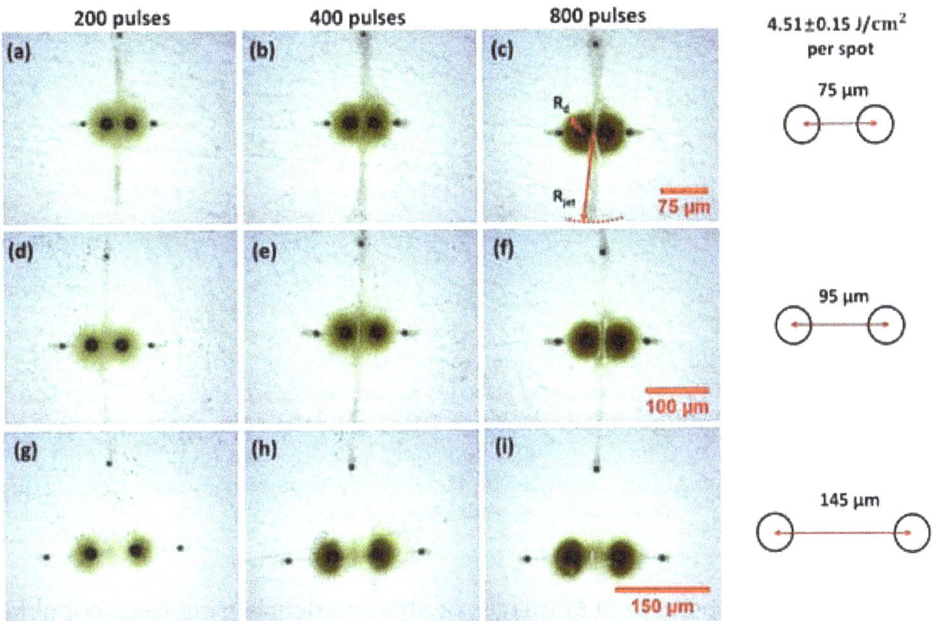

Figure 4. Optical images of observed two spot re-deposition during 10 ps laser ablation of stainless steel in air while varying spot separation and pulse number at fluence F = 4.51 J cm-2, (a) 200 pulses, (b) 400 pulses (c) 800 pulses, all 75 μm spot separation, (d–f) 95 μm separation, (g–i) 145 μm separation. At this higher fluence in (a–c), we observe diverging debris jets ejected normal to the spot axis while in (c,f), removal of surface debris by shock waves is apparent. These effects essentially disappear at the highest separation with a return to the concentration of material between the spots (g–i). The tiny ablation spots along the axis are due to low energy ghost beams while the top spot remains zero order.

3.2. Ablation Rates and Debris Radii (10 ps Pulses)

The measured ablation volume/pulse of single and two spot geometries with pulse number N is shown in Figure 5a for a fluence F = 2.9 J cm^{-2} (E_p = 6 μJ/pulse). These results confirm that ablation rates are essentially independent of spot geometry and spot separation when N ≥ 200, whether single or double spot, within experimental error. Hence, the proximity of the spots does not affect ablation rates significantly. The ablation volume/pulse V ~ 5.7 μm^3/pulse corresponding to a mass ablation M

~ 0.05 ng/pulse. Crater volumes and geometries were measured with a white light interferometer, Wyko NT3300. The single spot debris radius and jet radius with pulse energy and exposure are shown in Figure 5b on an Ln-Ln plot. This confirms that the debris radii follow a power-law $R \propto E^{0.47}$ for a single spot while the much higher Jet debris radii, $R_{jet} \propto E^{0.41}$. This relationship approximates that predicted by blast wave theory [59], however, higher than the expected $R \propto E^{1/3}$ which has also been observed by other authors [60].

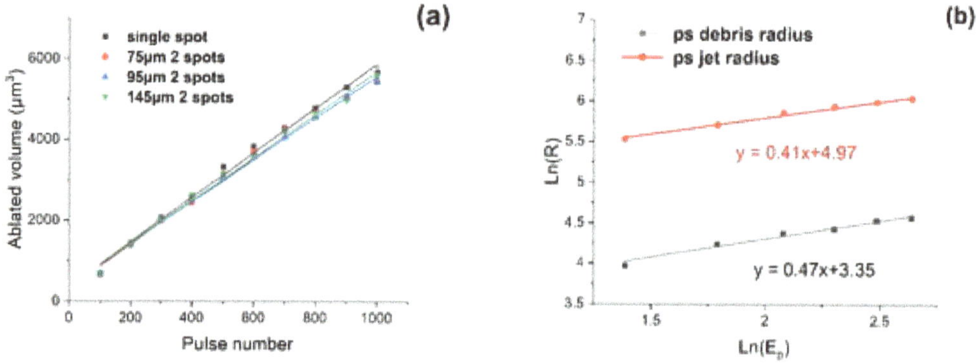

Figure 5. (a) Measured ablation volume versus pulse number for single and double spot ablation at d = 75 µm, 95 µm and 145 µm separation with fluence F = 2.9 J cm^{-2}/spot (Ep = 6 µJ/spot). Spot geometry does not affect ablation volume for $N \geq 200$, within experimental error, (b) Observed single spot debris radius and jet radius with pulse energy and exposure on an Ln-Ln plot. The radii follow a power-law $R \propto E^{0.47}$ for debris and $R_{jet} \propto E^{0.41}$ respectively.

3.3. SEM Imaging of the Debris Fields, Two Spot Patterns

Figure 6a–d shows a series of SEM images of ablation debris from two spot ablation of stainless steel in the air with 10 ps pulses at low fluence F = 0.9 J cm^{-2}, N = 800 pulses and spot separation d = 75 µm. The concentration of debris between spots is clear in Figure 6a, with evidence of some material ejected normal to the spot axis in the centre. At low fluence, there are two main components to the debris—particle agglomerates appear on the collision plane (and around the spots), while solid spherical nanoparticles with diameters 50–150 nm appear at the spot edge, Figure 6b. These are likely formed during collisions of the expanding plumes and condensed from the stagnation region at the midplane. Figure 6c (2000×) shows the particle agglomerates near the centre at higher magnification while Figure 6d (35,000×) confirms that the particle agglomerate consists of both solid np's and agglomerated chains of fine np's. There is a significant change in the nature of the debris generated at higher fluence, F = 4.8 J cm^{-2}, as shown in the SEM images of Figure 6e–h. Deposition now produces a thick deposit near the ablation spots while the shock waves lift material from the surface between the spots Figure 6e and deposit well away from the spots, Figure 6f. Figure 6g,h, with increasing magnification show that this jet debris consists of np chain agglomerates and is almost devoid of solid nanoparticles.

This is consistent with higher surface temperature achieved during ablation, well above the evaporation temperature, T_{ev} = 2861 K. The high concentration of the np chain agglomerate debris supports the assertion that this material is formed during strong stagnation of the plume collisions in the midplane.

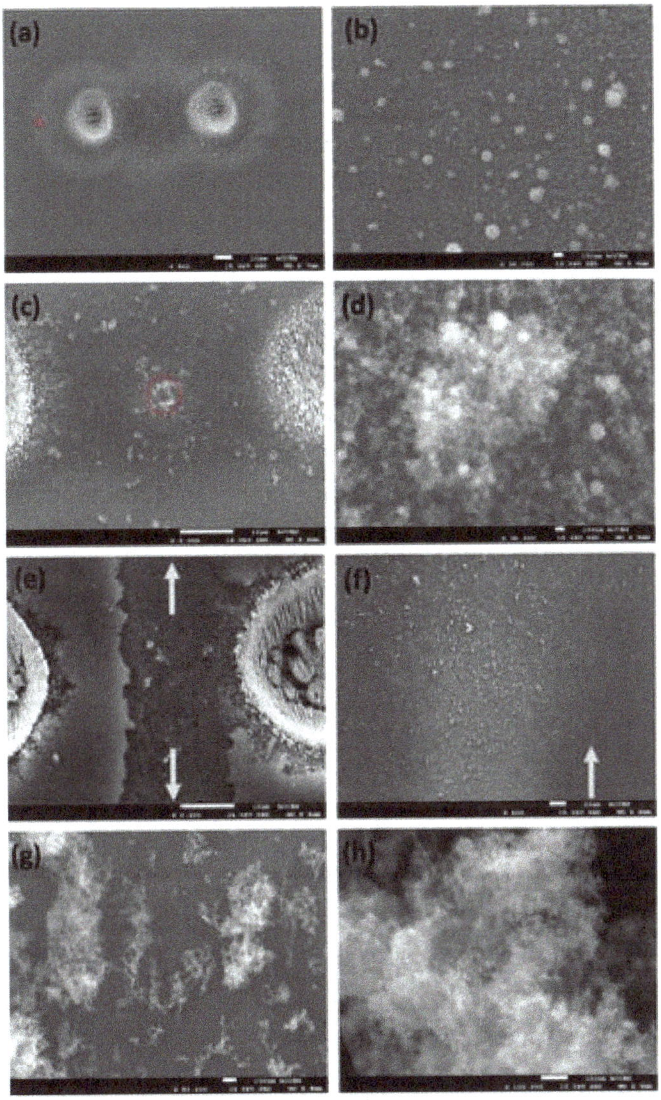

Figure 6. SEM images of the ablation debris from two spot ablation of stainless steel at low fluence (F = 0.9 J cm−2 (a–d) and high fluence F = 4.8 J cm−2 (e–h), N = 800 pulses, spot separation d = 75 μm. (a) image of the whole debris field, (b) 50–150 nm diameter np's observed at spot periphery in (a) red ring, (c) particle agglomerate at the centre, (d) high magnification image of (c) showing round np's and np agglomerate, (e) High fluence re-deposition pattern showing material lifted directly from the surface by the energetic colliding plumes and shock waves, (f) jet debris ejected normal to spots, (g) high magnification image of jet np chain agglomerate, (h) highest, 100,000× magnification of jet np chain agglomerate which still shows little structure.

3.4. Time-Resolved Plasma Emission Spectra

Figure 7 shows the two spots (E = 20 μJ/spot, d = 75 μm) time-resolved (95–0 ns) plasma emission spectrum of stainless steel in air over Δλ = 395 nm–415 nm. Gate width was set at 5 ns, gate delay interval 2 ns and data accumulated over 50 spectra. The substrate was scanned at 2 mm/s while an f = 125 mm bi-convex lens imaged the whole plasma emission to a fibre coupler (NA ~0.2, Figure 1) then to the spectrometer (Andor Shamrock 303i, 50 μm slit, 1800 L/mm grating) and cooled ICCD camera. The ICCD was triggered externally from the Laser Regenerative amplifier. Continuum dominated the spectra at early times, likely black body radiation from the hot plasma near the surface [21], while spectral line intensities rise sharply, then decrease along with the background continuum as the ablation plasma cools. Spectral line widths also decrease with time as electron density decreases, reducing Stark broadening [61]. The spectral lines in this region have been identified as due mainly to excited neutral atoms of Fe I, Cr I and Mn I: Fe I: 395.667 nm, Cr I: 396.368 nm, Cr I: 396.974 nm, Mn I: 397.708 nm, Fe I: 398.396 nm, Mn I: 399.161 nm, Fe I/Fe II blended line: 403.130 nm, Mn I: 403.307 nm, Fe I: 404.581 nm, Fe I: 406.359 nm, Fe I: 407.581 nm (centre), Fe I: 411.854 nm, Fe I: 413.290 nm, Fe I/Fe II/Fe III blended line: 414.26 nm. Single spot (20 μJ/pulse) spectra were very similar with lower intensities [62].

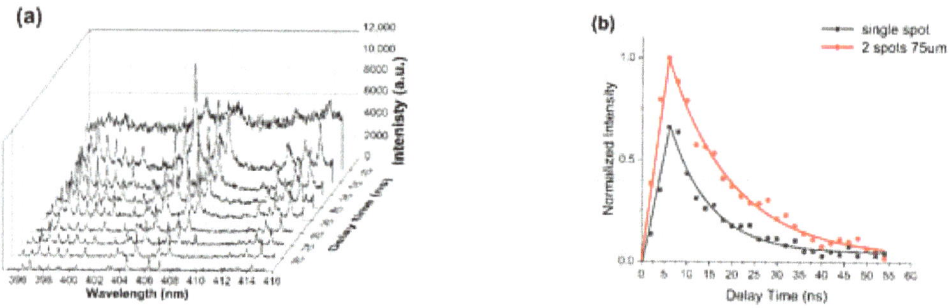

Figure 7. (a) Two spots (E = 20 μJ/spot, *d* = 75 μm,) time-resolved plasma emission spectrum (Δλ = 395–415 nm) of stainless steel over time delay 0–95 ns. Continuum dominates at early times while atomic lines appear as the plasma cools. The three intense lines near the centre are due to Fe I: 404.581 nm, Fe I: 406.359 nm, Fe I: 407.581 nm, (b) Integrated intensity (Δλ = 395 nm–415 nm) with time for single and double spot (*d* = 75 μm) showing exponential decrease with plasma lifetimes τ1/e ~ 0.5 ± 9.2 ns (single) and 0.7 ± 13.9 ns (double) respectively. The fits are exponential.

Figure 7b shows the time-integrated plasma emission (Δλ = 395–415 nm) for single and double spot (d = 75 μm) confirming that the plasma lifetimes $\tau_{1/e}$ ~ 1.0 ± 9.2 ns and 0.7 ± 13.9 ns, respectively, hence double spot plasma lifetime increasing over the single spot. These lifetimes are similar to those observed with 20 μJ, 150 fs laser ablation of Al in the air [21]. The short plasma lifetimes are a consequence of both the low pulse energies and rapid collisional cooling of the dense ambient air.

Plasma excitation temperature for single spot ablation (E = 20 μJ) has been estimated by the well-known Boltzmann method [11] from the Fe I line intensities I_{mn}, transition probabilities (gA_{mn}, g degeneracy) and upper energy levels, E_m. A plot of $Ln(\lambda I_{mn}/gA_{mn})$ versus E_m yielded a linear plot inferring T_e ~7500 K for single spot near 40 ns delay time. We also estimated the electron density from the Stark broadening of the Fe I line at λ = 404.58 nm yielding $N_e \sim 10^{18}$ cm^{-3}.

By removing the ICCD from the spectrometer and placing this at the image plane of the focus lens, (f = 75 mm, M ~ 4), the time-resolved plasma plumes were observed, Figure 8. Gate width here is 5 ns, energy/spot = 20 μJ and spot separation d = 75 μm. With spots normal to the optic axis, we can observe plasma expansion, collision and stagnation after 15 ns. When the plumes are imaged parallel to the optic axis and 15–10 ns delay, we see some interesting structure at right angles to the spot axis which may be connected to the jets. The lateral plume expansion velocity can be estimated to be $v_{\perp} \sim 10 \times 3.5^3$ ms^{-1}, decelerating after 15 ns, while the elliptical plume shape which develops supports lateral plasma expansion [35,63].

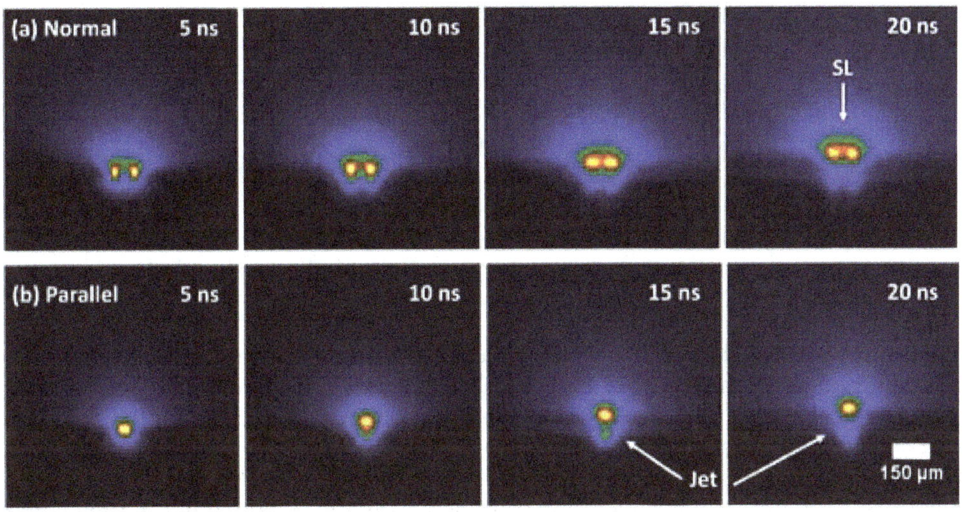

Figure 8. Time-resolved imaging of single and double spot ablation plume emission (20 μJ/spot, *d* = 75 μm) of stainless steel with time delays 20–5 ns. ICCD gate width was 5 ns (a) 2-spot normal to optic axis showing plume collisions and stagnation at 20 ns delay (b) two spots parallel to optic axis with an indication of jet structure 15–20 ns.

3.5. Two Spot Ablation with 600 fs/5 kHz Temporal Pulses at 1030 nm

The effect of increasing peak intensity by over 1 order of magnitude was investigated with the experimental system of Figure 2, while maintaining the same repetition rate of 5 kHz. Results are shown in Figure 9 for pulse numbers 500 ,200 and 1000 at fluence F = 1.41 J cm^{-2} (peak intensity I = 10×2.4^{12} W cm^{-2}). Again, we observe symmetric debris ejection or jets at right angles out to a radius of $R_{jet} \sim 150$ μm with spot separation d = 75 μm. As spot separation increases, we return to the directed, filamentary re-deposition (d = 100 μm) between the spots, and just apparent at d = 150 μm, N = 1000), similar to patterns observed with 10 ps pulses, Figure 3a–f.

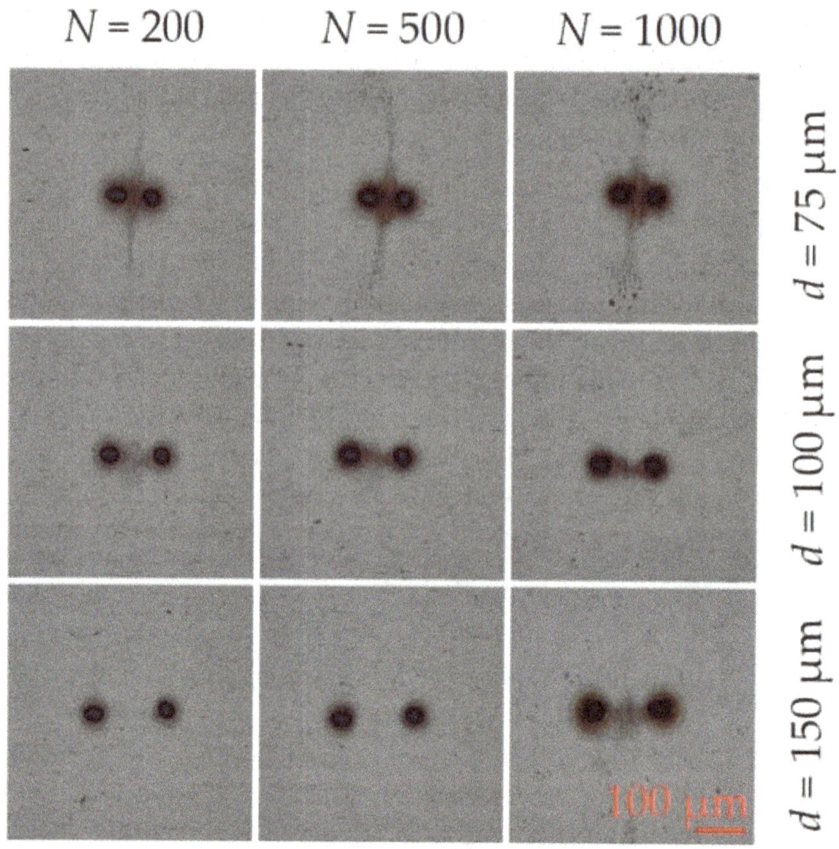

Figure 9. Resulting debris re-deposition and jet formation after 600 fs laser double pulse ablation with fluence F_0 = 1.41 J cm^{-2} (E_p = 5 µJ/pulse). Optical images of stainless steel surfaces after applying for various pulse numbers (1000 ,500 ,200) at different spot separations (50 µm, 100 µm and 150 µm). Debris jets are visible at 75 µm at the investigated *N*. At larger separations, the debris is concentrated in a filament between the spots (*d* = 100 µm) and becomes apparent at *d* = 150 µm separation, *N* = 1000. Fluence here is 1.5 times higher than with 10 ps pulses (Figure 3).

The results of high fluence two spot ablation with fluence F = 5.7 J cm^{-2} (20 µJ/spot, I = 10 × 9.6^{12} W cm^{-2}) are shown in Figure 10. where plume interactions create strongly diverging jets at spot separation d = 75 µm. At d = 100 µm separation, jets are more collimated with re-deposition jet radius R_{jet} > 300 µm. There is a slight tilt in the spot axis here relative to the horizontal. With N = 500 and 1000 at d = 100 µm, shock wave and plume interactions also remove debris from the surface (compare Figure 4c,f), evidence of the quasi-stationary shock waves [51,52]. At d = 150 µm material is again concentrated between the spots as interactions weaken. Peak intensity therefore appears to play a minor role in the debris re-deposition during the plume interactions from 600 fs to 10 ps pulse length on stainless steel in the air.

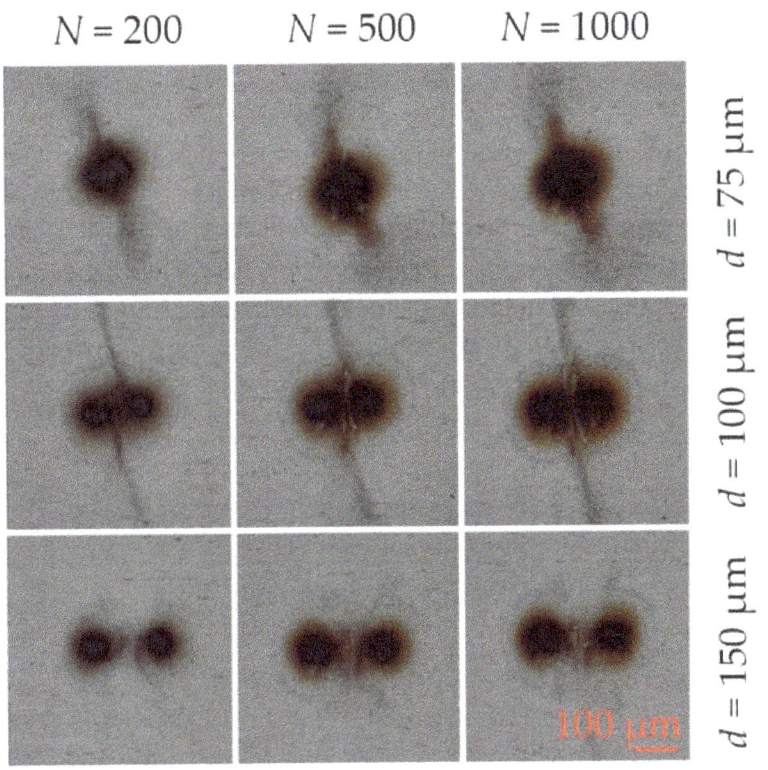

Figure 10. Resulting debris re-deposition and jet formation after 600 fs laser double spot ablation with fluence F_0 = 5.7 J cm^{-2} (E_p = 20 µJ/pulse). Within the investigated pulse numbers *N* (1000 ,500 ,200) and pulse separations *d* of 50 µm, 100 µm and 150 µm, the optical images reveal an interaction between the two spots. Strongly diverging debris jets at 75 µm and collimated jets at *d* = 100 µm are visible. At the largest separation *d* = 150 µm, where the interactions are weaker, the material is again concentrated between spots.

Figure 11a,b show SEM images of the diverging symmetric jet debris from 600 fs double spot (d = 100 µm) ablation of stainless steel in air (20 µJ/spot, F = 4.5 J cm^{-2}, N = 1000). This material likely consists of np chain and is concentrated at the jet ends, almost 400 µm from the spots.

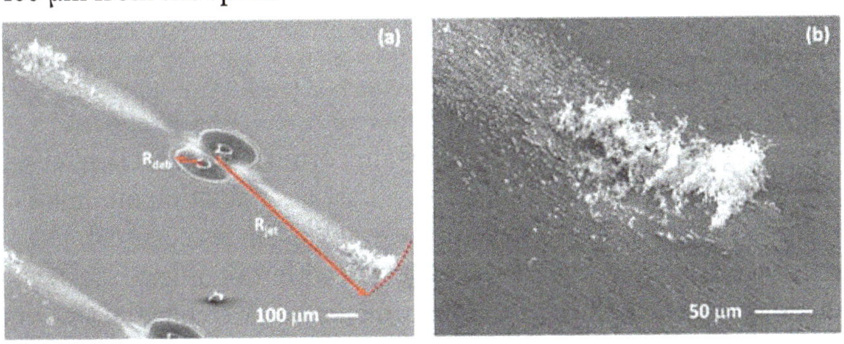

Figure 11. Tilted SEM images of the jets after double spot femtosecond ablation of s. steel in the air (E_p = 20 µJ/pulse, *N* = 1000, *d* = 100 µm). (a) shows the symmetric

diverging debris jets (np chain agglomerates) with an accretion of material at the jet ends, almost 400 μm from the double spot centre. (b) higher magnification of concentrated jet debris near the jet end. The debris radii near the spots and jet radii are indicated (R_{deb} and R_{jet} respectively).

The debris and jet radii generated with 600 fs pulses are shown in **Figure 12.** For comparison, the data from 10 ps two spot ablation. Logarithmic fits are included, and the exponents vary from n = 0.33 to 0.47. The fs and ps debris radii are close exponent while the fs jet radii appear to have the lowest, n = 0.33. The higher radii for given energy with 10 ps pulses is due to the smaller spot separation of 75 μm, while this was 100 μm with 600 fs pulses. There may also be differences due to uncertainties in the estimation of the radii.

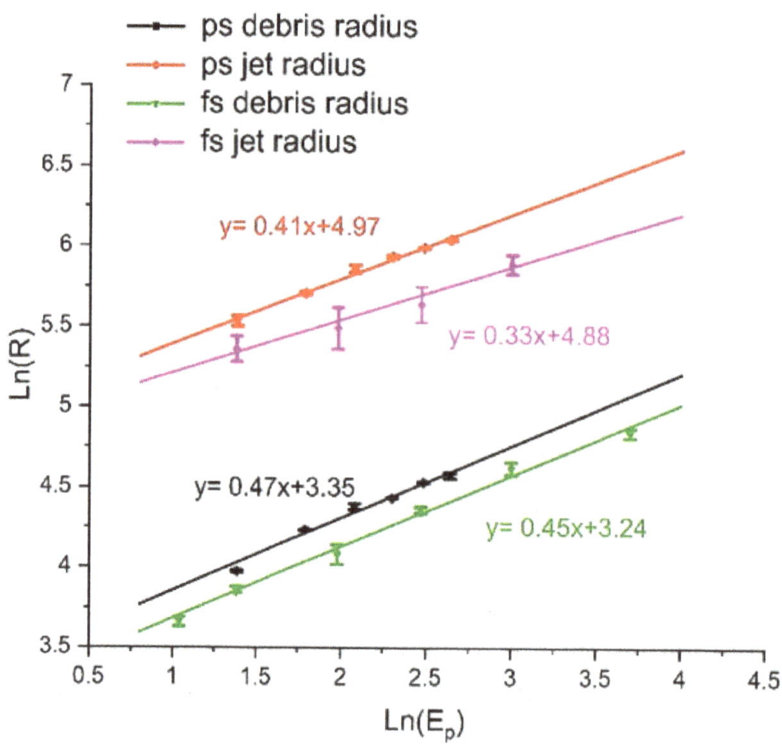

Figure 12. Ln-Ln plots of the two spots debris and jet radii ([R] = μm) measured for 600 fs and 10 ps ablation. The logarithmic fits ([E_p] = μJ) are shown to be reasonably close in exponent except for femtosecond ablation which is lower. The spot separations were 100 μm (600 fs) and 75 μm (10 ps), respectively. Error bars represent 1 σ.

The reproducibility of the results reported is supported by the excellent observed fits to the Ln-Ln plots with indicated errors. Ultrafast laser ablation minimises thermal diffusion and melt leading to a deterministic evaporation process [19]. The observed nanoparticle agglomerated jets can be deflected with a significant airflow over the substrate. However, when the surrounding ambient air is stable, the jets emanate at right angles and have the same length, **Figure 11a.**

3.6. Time-Resolved Shadowgraphy (600 fs/5 kHz).

Ablation plumes and their development after ablation were imaged with the pump-probe experimental set-up shown in Figure 2. The probe beam (λ = 688 nm, τ_H = 13 ns) was electronically synchronised to the pump beam (λ = 1030 nm). Time-resolved measurements are often carried out with a single pulse exposure–but here, after simultaneous multi-pulse, multi-spot exposure, we can see the developing plumes and their interactions. The connection axis of two spots is aligned perpendicular to the imaging plane. The total delay time regarding the arrival of the first pulse pair with 5 kHz repetition rate is τ = 1002 µs, hence a delay time of 2 µs after the last, N = 6th pulse, Figure 13. There is a strong plasma plume overlap confined to the surface at d = 100 µm separation with diverging plume above the spots containing solid (dark) micron size particles, strongly absorbing/scattering. The previous pulses have formed these. These also appear at 150 µm spot separation but disappear entirely at d = 200 µm, where the plasma plumes near the surface are distinct and plume absorption much more uniform. This particle is likely the np chain agglomerate observed on the surface, shown in the SEM images of Figure 11. which appear in the jets due to the strong plume interactions. Note in the shadowgraphs a visible expanding spherical shock wave from the last pulse pair overlaps the material previously ejected. The shock wave speed, from the time delay, v_s ~375 m/s just above the speed of sound while the lateral plasma expansion has slowed significantly to around v ≤ 50 m/s after 2 µs delay.

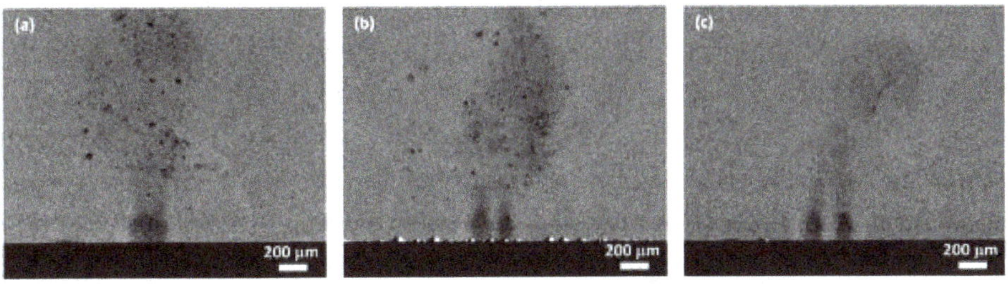

Figure 13. Time-resolved shadowgraphs of the interacting plumes with two spot ablation in the air after *N* = 6 and delay time of τ = 2 µs after ablation and τ (total) = 1002 µs. These images demonstrate that large condensed absorbing particles are apparent in the debris above the spots when the separation is smallest and where the plume collisions are strongest, (a) *d* = 100 µm, (b) *d* = 150 µm, (c) *d* = 200 µm where these particles have all but disappeared. Note that the spherical shock waves from the last pulse overlap the material ejected beforehand.

4. DISCUSSION

Laser ablation on metal in the ambient atmosphere can be likened to a mechanical detonation, creating an ablation plume expanding supersonically against the background gas, causing a shock wave. As the plume does work by expanding against atmospheric pressure, the expansion velocity decreases with time, and the shock wave radius is given by Taylor's blast wave theory [59],

$$R=(\gamma)(E/\rho 0)^{1/5t2/5} \quad (1)$$

where S(γ) ~1 is a function only of the air specific heat γ (~1.4), E is the energy released, ρ_0 is the undisturbed background gas density and t is the time after ablation. The effect is to force most of the air within the shock front into a thin shell just inside the front, compressing and heating the air. The initial pressure driving the front $p_{max} \gg p_0$, (1 atm). When $p_{max} \sim p_0$, the self-similar solution of Equation (1) is no longer valid. In this case, the radius for this upper limit is given by [20],

$$R=(E/p o)^{1/3} \quad (2)$$

The ablation plasma plume follows behind the shock front and a contact discontinuity can be observed inside the shock front [33]. For the range 2 μJ ≤ E ≤ 20 μJ, single spot debris radii range from 120–40 μm while R calculated from Equation (2) yields 270 μm ≤ R ≤ 580 μm. The debris radius can be related to the shock radius through $R_D \sim fR$ where f < 1 and here, f ~ 0.15–0.2, similar to that observed by other authors [60]. From the time-resolved plasma emission (10 ps), backward re-deposition with single spot ablation likely starts about ~100 ns after ablation and from the pump-probe observations, (600 fs) continues over 10 μs and longer. The debris power laws yield $R_{deb} \propto E^{0.3-0.5}$ for both single spot and jets but the much larger jet radii if applied to $R_D \sim fR$ yields f ~ 0.7–0.8 using Equation (2) inferring that an additional energetic process is involved during the plasma plume interactions.

Multi-spot ablation in vacuum between two independent and closely spaced seed plasmas collide due to the lateral plasma expansion. With a low density and high relative velocity, plasmas tend to interpenetrate, relevant in collisionless astrophysical plasmas, leading to ion reflection and particle acceleration [64]. However, with higher density and a low relative velocity, the plasmas rapidly decelerate at the collision plane, forming a stagnation layer (SL). Accretion and compression of the material within the SL leads to a local increase in density and temperature. The degree of stagnation can be described by a collisionality parameter $\xi = d/\lambda_{ii}$, where d is the distance between the two plasmas and λ_{ii} is the ion-other ion mean free path, given by, [65,66]

$$\lambda_{ii\,(1-2)} = \frac{4\pi\varepsilon_0^2 m_i^2 v_{12}^4}{q^4 Z^4 n_i \ln(\Lambda_{1-2})} \quad (3)$$

where ε_0 is the permittivity of free space, m_i is the ion mass, $v12$ is the relative ion flow velocity (prior to impact), q is the elementary charge, Z is the average ionisation state, n_i is the plasma density at the collision plane and ln (Λ_{1-2}) is the Coulomb Logarithm, a slowly varying function, with a value O (20–10) [67]. The parameter ξ is very sensitive to the relative plasma velocity term $v12^4$ while only linearly dependent on separation d. If used in ambient air, this description is more complex than in vacuum due to the presence of shock waves and rapid plasma deceleration due to collisions with the air. From time-resolved plasma emission at 15 ns, Figure 8. We estimated that $v_{\perp} \sim 10 \times 3.5^3$ m s^{-1} so that $v12 = 2v_{\perp} \sim 10 \times 7^3$ m s^{-1}. Inserting this value into Equation (3), we find $\lambda_{ii} \sim 3.1$ µm so that the plasma collision region is limited to the micron scale with collisionality parameter $\xi = d/\lambda_{ii} = 75$ µm/3.1 µm ~ 24.

In a plasma, the Debye length is the characteristic length over which electrons and ions can be separated, and in an ideal plasma (e.g., astrophysical) has many particles per Debye sphere N_D given by

$$N_D = N_e\left(\frac{4\pi}{3}\right)\lambda_D^3 \gg 1 \quad (4)$$

The classical plasma parameter $g = 1/N_e\lambda_D^3 << 1$ so that collective effects dominate the plasma. In laser ablation, we can calculate g for each plasma plume knowing N_e and λ_D which is given by [67] $\lambda_D = 743\,(T_e)^{1/2}\,(N_e)^{-1/2}$ where the electron temperature T_e is in eV while N_e is in cm^{-3}. At delay time $\tau = 50$ ns, we measured T_e (10 ps, $E_p = 20$ µJ) from a Boltzmann plot to be ~7500 K (1 eV = 11,600 K) hence plasma temperature $T_e \sim 0.65$ eV while $N_e \sim N_i \sim 10^{18}$ cm^{-3} from measured Stark broadening of the Fe I line at $\lambda = 404.58$ nm. The Debye length in each plasma plume is then $\lambda_D \sim {}^{7\text{-}}10 \times 6.0$ cm (6 nm) while the particle number in a Debye sphere is, from Equation (4), $N_D = 0.92$. The collision parameter $g = 1/N_e\lambda_D^3 = 4.6$ hence the plasma plumes are each highly collisional as expected. We can also estimate the Coulomb Logarithm [67], $\text{Ln}\Lambda = 9N_D/Z = 8.3$ assuming $Z \sim 1$ is the average ionisation state, supported by the time-resolved spectroscopy, Figure 7a.

Recently, the interaction of two colliding Al plasmas (in the air) and their shock waves were observed in two spot nanosecond laser ablation using shadowgraphy, schlieren images and interferometry [68]. From refractive index profiles, expanding shock fronts and their reflections were observed after collision along with compression and stagnation and of the air behind the shock fronts. With pulse fluence F ~ 27 J cm^{-2}, focus intensity I ~ 1.6 GWcm^{-2} and 1 mm spot separation, compressed air density reached $n_0 \sim 10 \times 5^{20}$ cm^{-3} behind the interacting shocks while plasma density $n_i \sim 5.10^{18}$ cm^{-3}. The stagnation was described as "soft" [69] with a plasma hill developing over times scales from 0.7–3 µs.

More recently, stationary transient straight shockwaves (on a 10 ns timescale) were detected during multi-spot fs ablation of water in the air creating supersonic air flows which collided [51,52]. Fluence per spot F ~ 18 J cm^{-2} with peak intensities I ~ 10^{14} W cm^{-2} and spot separations from d ~ 20–14 µm. The length of the shock waves correlated with the local flow velocity at a specific radial position from the sources and was noted when the relative speed of shock wave propagation surpassed the speed of sound in air. The symmetric architecture of the observed transient shock waves with two spots closely resembles the patterns of debris eliminated between double spot ablation at higher fluence in this study, as illustrated in Figures 4, 6, and 11, utilizing picosecond and femtosecond pulse durations. This suggests that material removal across locations during high fluence ablation may reflect shock wave interactions.

5. CONCISE SUMMARY

The jets appearing at higher fluence (and small spot separation) could be created as follows using a simplified 2D approach. When the high-pressure shock waves meet at the collision plane, the pressure due to each shock wave $p_{max} = 0.155\ E/R^3$ ($\gamma = 1.4$) [59]. If we set $E \sim E_p = 20\ \mu J$ and $R = 50\ \mu m$, $p_{max} = 10 \times 2.48^7$ Pa ~248 atm. The air between these shocks will be highly compressed by this "piston" as they meet and the curved expanding shocks could compress, accelerate and expel the air symmetrically in a diverging jet in both directions normal to the axis, Figure 14a. The plasmas following then stagnate and ion Coulomb repulsion off-axis at low impact parameters converts their axial momentum to transverse momentum assuming elastic collisions. The observed material lifted from the surface leaves patterns very similar to the transient stationary shock waves observed during multi-spot induced supersonic air flows at a water/air interface [51,52].

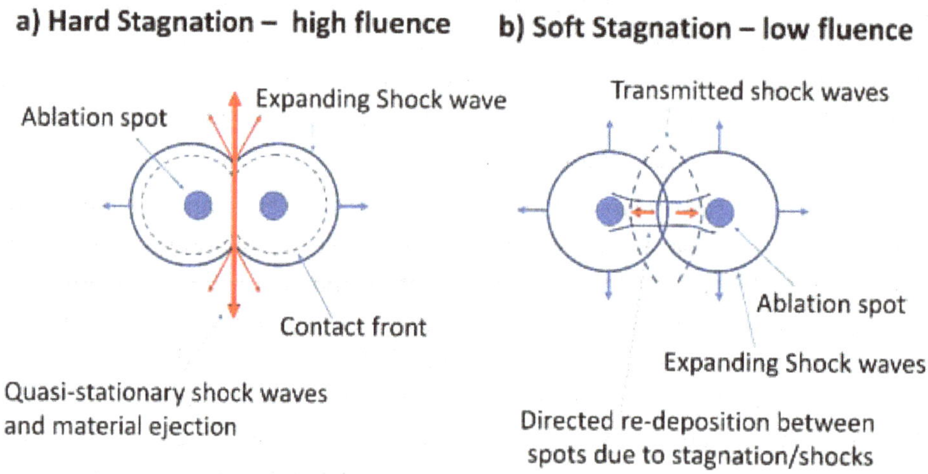

Figure 14. Schematic diagrams summarising current view of two spot ablation at high and low ablation (a) at high fluence, the air between the shocks is compressed by a "piston" as they meet and the curved shocks will accelerate and expel the air symmetrically in a diverging jet in both directions normal to the axis. The plasmas then stagnate and Coulomb interactions off-axis at low impact parameters convert their axial momentum to transverse momentum, (b) at low fluence and larger separations, shock wave pressures are much reduced, plasma density decreases, reducing stagnation to "soft", hence allowing plasma interpenetration and directed deposition between the spots.

At low fluence and larger separations, shock wave pressures are much reduced, plasma density decreases, reducing stagnation to "soft", hence allowing plasma interpenetration between the spots. Weak shock waves pass through each other and likely reflect back along the axis from the plasmas. Multiple collisions can decelerate ions and a degree of ion reflection may occur [**64**] so that particles acquire momentum directed along the spot axis then arrive at the surface through backward flux, **Figure 14b**.

Two colliding air breakdown plasmas created with high energy, nanosecond laser pulses demonstrated reflection and transmission of expanding shock waves depending on pulse energy and plasma seed separations [70]. One might ask if oxidation during ultrafast laser ablation is significant here, but we have also observed the Jets in pure Nitrogen at 1 atm (10 ps, not reported here) with the aid of a gas cell. We estimate that oxidation of Fe atoms to Fe_3O_4 could at most add %10 additional energy, not nearly enough to explain the phenomena here.

Time-resolved plasma emission provided critical plasma properties, including lifespan and temperatures, whereas plasma collisions across regions indicated plume stagnation, leading to bi-directional jets. The impact of quasi-stationary shock waves is evident, compressing air and elevating debris from the surface between locations. Time-resolved shadowgraphy utilizing 600 fs pulses revealed micron-sized particles, which form the jets, exclusively when the ablation areas were in close proximity. The re-deposited jet debris comprises agglomerated nanoparticle chains. Our observations of directed deposition at low fluence and jets during high-fluence, multi-spot ultrafast laser ablation of stainless steel in air represent a potentially intriguing new research domain, establishing a foundation for further inquiries to develop a comprehensive understanding of these intricate phenomena. Our research and its results are deemed significant for a more profound comprehension of the interaction of multi-spot arrays in laser patterning [71,72] and improved laser deposition of thin films [53,55]. The phenomena of plasma collisions and shocks examined herein are pertinent to astrophysics and physics research, including magnetic reconnection and phenomena such as bi-directional jet formation and particle acceleration observed at ultrahigh laser intensities (10^{14}–10^{15} W cm^{-2}) during two-spot ablation with high transient magnetic fields.

AT A GLANCE

We present new findings on the directed re-deposition of ablation debris during the ultrafast laser micro-structuring of stainless steel in air, utilizing several beams in close proximity to the surface. This intriguing behavior is observed with both 10 ps and 600 fs NIR laser pulses at a 5 kHz repetition rate. The geometry of ablation spots can be modified using beam-splitting optics or a phase-only spatial light modulator. At low fluence (F ~ 1.0 J cm^{-2}) and pulse exposure of a few hundred pulses, the debris appears as concentrated narrow "filaments" connecting the ablation spots, while at higher fluence, (F ~ 5.0 J cm^{-2}) energetic jets of material emanated symmetrically along the axes of symmetry, depositing debris well beyond the typical re-deposition radius with a single spot. The patterns of backward debris re-deposition on the surface are presumably associated with clashing shock waves and plasma plumes interacting with the surrounding air, resulting in stagnation when the locations are in close proximity. The 2D surface debris patterns reflect the intricate 3D interactions occurring across extensive periods during ablation, ranging from picoseconds to microseconds.

REFERENCES

Krausz, F.; Stockman, M.I. Attosecond metrology: From electron capture to future signal processing. Nat. Photonics 2014, 8, 205–213. [Google Scholar] [CrossRef]

Austin, D.R.; Johnson, A.S.; McGrath, F.; Wood, D.; Miseikis, L.; Siegel, T.; Hawkins, P.; Harvey, A.; Mašín, Z.; Patchkovskii, S. Extracting sub-cycle electronic and nuclear dynamics from high harmonic spectra. Sci. Rep. 2021, 11, 1–8. [Google Scholar] [CrossRef] [PubMed]

Wille, H.; Rodríguez, M.; Kasparian, J.; Mondelain, D.; Yu, J.; Mysyrowicz, A.; Sauerbrey, R.; Wolf, J.-P.; Woeste, L. Teramobile: A mobile femtosecond-terawatt laser and detection system. EPJ Appl. Phys. 2002, 20, 183–190. [Google Scholar] [CrossRef]

Tochitsky, S.; Welch, E.; Panagiotopoulos, P.; Polyanskiy, M.; Pogorelsky, I.; Kolesik, M.; Koch, S.W.; Moloney, J.V.; Joshi, C. Filamentation of long-wave infrared pulses in the atmosphere. JOSA B 2019, 36, G40–G51. [Google Scholar] [CrossRef]

So, P.T.; Dong, C.Y.; Masters, B.R.; Berland, K.M. Two-photon excitation fluorescence microscopy. Annu. Rev. Biomed. Eng. 2000, 2, 399–429. [Google Scholar] [CrossRef] [Green Version]

Varnavski, O.; Goodson, T., III. Two-photon fluorescence microscopy at extremely low excitation intensity: The power of quantum correlations. J. Am. Chem. Soc. 2020, 142, 12966–12975. [Google Scholar] [CrossRef]

Abouzeid, H.; Ferrini, W. Femtosecond-laser assisted cataract surgery: A review. Acta Ophthalmol. 2014, 92, 597–603. [Google Scholar] [CrossRef]

Russo, R.E.; Mao, X.; Gonzalez, J.J.; Mao, S.S. Femtosecond laser ablation ICP-MS. J. Anal. At. Spectrom. 2002, 17, 1072–1075. [Google Scholar] [CrossRef]

Diwakar, P.K.; Harilal, S.S.; LaHaye, N.L.; Hassanein, A.; Kulkarni, P. The influence of laser pulse duration and energy on ICP-MS signal intensity, elemental fractionation, and particle size distribution in NIR fs-LA-ICP-MS. J. Anal. At. Spectrom. 2013, 28, 1420–1429. [Google Scholar] [CrossRef] [PubMed] [Green Version]

Pozebon, D.; Scheffler, G.; Dressler, V. Recent applications of laser ablation inductively coupled plasma mass spectrometry (LA-ICP-MS) for biological sample analysis: A follow-up review. J. Anal. At. Spectrom. 2017, 32, 890–919. [Google Scholar] [CrossRef]

Le Drogoff, B.; Margot, J.; Chaker, M.; Sabsabi, M.; Barthelemy, O.; Johnston, T.; Laville, S.; Vidal, F.; Von Kaenel, Y. Temporal characterization of femtosecond laser pulses induced plasma for spectrochemical analysis of aluminum alloys. Spectrochim. Acta B 2001, 56, 987–1002. [Google Scholar] [CrossRef]

Zhu, Y.; Zhou, P.-W.; Zhong, S.-C.; Peng, Q.-X.; Zhu, L.-G. A multi-spot laser induced breakdown spectroscopy system based on diffraction beam splitter. Rev. Sci. Instrum. 2019, 90, 123105. [Google Scholar] [CrossRef] [PubMed]

Ihlemann, J.; Wolff, B.; Simon, P. Nanosecond and femtosecond excimer laser ablation of fused silica. Appl. Phys. A 1992, 54, 363–368. [Google Scholar] [CrossRef]

Allahyari, E.; Nivas, J.J.; Avallone, G.; Valadan, M.; Singh, M.; Granata, V.; Cirillo, C.; Vecchione, A.; Bruzzese, R.; Altucci, C. Femtosecond laser surface irradiation of silicon in air: Pulse repetition rate influence on crater features and surface texture. Opt. Laser Technol. 2020, 126, 106073. [Google Scholar] [CrossRef]

Rajab, F.H.; Liauw, C.M.; Benson, P.S.; Li, L.; Whitehead, K.A. Picosecond laser treatment production of hierarchical structured stainless steel to reduce bacterial fouling. Food Bioprod. Process. 2018, 109, 29–40. [Google Scholar] [CrossRef]

Xu, S.; Ding, R.; Yao, C.; Liu, H.; Wan, Y.; Wang, J.; Ye, Y.; Yuan, X. Effects of pulse durations and environments on femtosecond laser ablation of stainless steel. Appl. Phys. A 2018, 124, 1–8. [Google Scholar] [CrossRef]

Hauschwitz, P.; Jochcová, D.; Jagdheesh, R.; Cimrman, M.; Brajer, J.; Rostohar, D.; Mocek, T.; Kopeček, J.; Lucianetti, A.; Smrž, M. Large-beam picosecond interference patterning of metallic substrates. Materials 2020, 13, 4676. [Google Scholar] [CrossRef] [PubMed]

Kautek, W.; Krüger, J. Femtosecond Pulse Laser Ablation of Metallic, Semiconducting, Ceramic, and Biological Materials. In Laser Materials Processing: Industrial and Microelectronics Applications; Springer: Berlin/Heidelberg, Germany, 1994; pp. 600–611. [Google Scholar]

Schille, J.; Schneider, L.; Kraft, S.; Hartwig, L.; Loeschner, U. Experimental study on double-pulse laser ablation of steel upon multiple parallel-polarized ultrashort-pulse irradiations. Appl. Phys. A 2016, 122, 1–11. [Google Scholar] [CrossRef]

Dyer, P.; Sidhu, J. Spectroscopic and fast photographic studies of excimer laser polymer ablation. Int. J. Appl. Phys. 1988, 64, 4657–4663. [Google Scholar] [CrossRef]

Yalçın, Ş.; Tsui, Y.Y.; Fedosejevs, R. Pressure dependence of emission intensity in femtosecond laser-induced breakdown spectroscopy. J. Anal. At. Spectrom. 2004, 19, 1295–1301. [Google Scholar] [CrossRef]

Yalçin, S.; Tsui, Y.Y.; Fedosejevs, R. Images of femtosecond laser plasma plume expansion into background air. IEEE Trans. Plasma Sci. 2005, 33, 482–483. [Google Scholar] [CrossRef]

Farid, N.; Harilal, S.; Ding, H.; Hassanein, A. Dynamics of ultrafast laser plasma expansion in the presence of an ambient. Appl. Phys. Lett. 2013, 103, 191112. [Google Scholar] [CrossRef]

Lichtman, D.; Ready, J. Laser beam induced electron emission. Phys. Rev. Lett. 1963, 10, 342. [Google Scholar] [CrossRef]

Bechtel, J.; Smith, W.L.; Bloembergen, N. Two-photon photoemission from metals induced by picosecond laser pulses. Phys. Rev. B 1977, 15, 4557. [Google Scholar] [CrossRef]

Mao, S.S.; Mao, X.-L.; Greif, R.; Russo, R.E. Simulation of infrared picosecond laser-induced electron emission from semiconductors. Appl. Surf. Sci. 1998, 127, 206–211. [Google Scholar] [CrossRef]

Russo, R.; Mao, X.; Liu, H.; Yoo, J.; Mao, S. Time-resolved plasma diagnostics and mass removal during single-pulse laser ablation. Appl. Phys. A 1999, 69, S887–S894. [Google Scholar] [CrossRef]

Mao, S.S.; Mao, X.; Greif, R.; Russo, R.E. Initiation of an early-stage plasma during picosecond laser ablation of solids. Appl. Phys. Lett. 2000, 77, 2464–2466. [Google Scholar] [CrossRef]

Mao, S.S.; Mao, X.; Greif, R.; Russo, R.E. Simulation of a picosecond laser ablation plasma. Appl. Phys. Lett. 2000, 76, 3370–3372. [Google Scholar] [CrossRef]

Zeng, X.; Mao, X.; Greif, R.; Russo, R. Experimental investigation of ablation efficiency and plasma expansion during femtosecond and nanosecond laser ablation of silicon. Appl. Phys. A 2005, 80, 237–241. [Google Scholar] [CrossRef]

Nica, P.; Gurlui, S.; Osiac, M.; Agop, M.; Ziskind, M.; Focsa, C. Investigation of femtosecond laser-produced plasma from various metallic targets using the Langmuir probe characteristic. Phys. Plasmas 2017, 24, 103119. [Google Scholar] [CrossRef]

Wen, S.-B.; Mao, X.; Greif, R.; Russo, R.E. Expansion of the laser ablation vapor plume into a background gas. I. analysis. Int. J. Appl. Phys. 2007, 101, 023114. [Google Scholar] [CrossRef]

Wu, J.; Wei, W.; Li, X.; Jia, S.; Qiu, A. Infrared nanosecond laser-metal ablation in atmosphere: Initial plasma during laser pulse and further expansion. Appl. Phys. Lett. 2013, 102, 164104. [Google Scholar] [CrossRef]

Harilal, S.; Farid, N.; Freeman, J.; Diwakar, P.; LaHaye, N.; Hassanein, A. Background gas collisional effects on expanding fs and ns laser ablation plumes. Appl. Phys. A 2014, 117, 319–326. [Google Scholar] [CrossRef]

Pereira, A.; Delaporte, P.; Sentis, M.; Marine, W.; Thomann, A.L.; Boulmer-Leborgne, C. Optical and morphological investigation of backward-deposited layer induced by laser ablation of steel in ambient air. Int. J. Appl. Phys. 2005, 98, 064902. [Google Scholar] [CrossRef]

Harilal, S.S.; Brumfield, B.E.; Cannon, B.D.; Phillips, M.C. Shock wave mediated plume chemistry for molecular formation in laser ablation plasmas. Anal. Chem. 2016, 88, 2296–2302. [Google Scholar] [CrossRef]

Bulgakova, N.M.; Panchenko, A.N.; Zhukov, V.P.; Kudryashov, S.I.; Pereira, A.; Marine, W.; Mocek, T.; Bulgakov, A.V. Impacts of ambient and ablation plasmas on short-and ultrashort-pulse laser processing of surfaces. Micromachines 2014, 5, 1344–1372. [Google Scholar] [CrossRef] [Green Version]

Taylor, R.; Leopold, K.; Singleton, D.; Paraskevopoulos, G.; Irwin, R. The effect of debris formation on the morphology of excimer laser ablated polymers. Int. J. Appl. Phys. 1988, 64, 2815–2818. [Google Scholar] [CrossRef]

Miotello, A.; Kelly, R.; Braren, B.; Otis, C.E. Novel geometrical effects observed in debris when polymers are laser sputtered. Appl. Phys. Lett. 1992, 61, 2784–2786. [Google Scholar] [CrossRef]

Küper, S.; Brannon, J. Ambient gas effects on debris formed during KrF laser ablation of polyimide. Appl. Phys. Lett. 1992, 60, 1633–1635. [Google Scholar] [CrossRef]

Burghardt, B.; Scheede, S.; Senczuk, R.; Kahlert, H.-J. Ablation plume effects on high precision excimer laser-based micromachining. Appl. Phys. A 1999, 69, S137–S140. [Google Scholar] [CrossRef]

Garrelie, F.; Champeaux, C.; Catherinot, A. Study by a Monte Carlo simulation of the influence of a background gas on the expansion dynamics of a laser-induced plasma plume. Appl. Phys. A 1999, 69, 45–50. [Google Scholar] [CrossRef]

Ravi-Kumar, S.; Lies, B.; Zhang, X.; Lyu, H.; Qin, H. Laser ablation of polymers: A review. Polym. Int. 2019, 68, 1391–1401. [Google Scholar] [CrossRef]

Jacob, J.; Shanmugavelu, P.; Balasubramaniam, R. Investigation of the performance of 248 nm excimer laser assisted photoresist removal process in gaseous media by response surface methodology and artificial neural network. J. Manuf. Process. 2019, 38, 516–529. [Google Scholar] [CrossRef]

Elaboudi, I.; Lazare, S.; Belin, C.; Talaga, D.; Labrugère, C. Underwater excimer laser ablation of polymers. Appl. Phys. A 2008, 92, 743–748. [Google Scholar] [CrossRef]

Péreira, A.; Cros, A.; Delaporte, P.; Itina, T.; Sentis, M.; Marine, W.; Thomann, A.; Boulmer-Leborgne, C. Formation of Iron Oxide Nanoparticles by Pulsed Laser Ablation. In Proceedings of Atomic and Molecular Pulsed Lasers VI; SPIE: Bellingham, WA, USA, 2006; p. 62630Q. [Google Scholar]

Pereira, A.; Delaporte, P.; Sentis, M.; Cros, A.; Marine, W.; Basillais, A.; Thomann, A.; Leborgne, C.; Semmar, N.; Andreazza, P. Laser treatment of a steel surface in ambient air. Thin Solid Films 2004, 453, 16–21. [Google Scholar] [CrossRef]

Förster, D.J.; Faas, S.; Gröninger, S.; Bauer, F.; Michalowski, A.; Weber, R.; Graf, T. Shielding effects and re-deposition of material during processing of metals with bursts of ultra-short laser pulses. Appl. Surf. Sci. 2018, 440, 926–931. [Google Scholar] [CrossRef]

Kraft, S.; Schille, J.; Mauersberger, S.; Schneider, L.; Loeschner, U. Time-resolved visualization of high-repetition ultrashort pulse laser ablation. Appl. Opt. 2020, 59, 452–458. [Google Scholar] [CrossRef]

Kraft, S.; Schille, J.; Mauersberger, S.; Schneider, L.; Loeschner, U. Pump-Probe Imaging for Process Control and Optimization in High-Speed Laser Micro Machining. In Laser-Based Micro-and Nanoprocessing XIV; SPIE: Bellingham, WA, USA, 2020; p. 112681H. [Google Scholar]

Vreugdenhil, M.; Van Oosten, D.; Hernandez-Rueda, J. Dynamics of femtosecond laser-induced shockwaves at a water/air interface using multiple excitation beams. Opt. Lett. 2018, 43, 4899–4902. [Google Scholar] [CrossRef]

Baasanjav, D.; Hernandez-Rueda, J.; Mosk, A.P.; Van Oosten, D. Optical method for micrometer-scale tracerless visualization of ultrafast laser induced gas flow at a water/air interface. Appl. Opt. 2020, 59, 5205–5209. [Google Scholar] [CrossRef]

Gupta, S.L.; Thareja, R.K. ZnO thin film deposition using colliding plasma plumes and single plasma plume: Structural and optical properties. Int. J. Appl. Phys. 2013, 114, 224903. [Google Scholar] [CrossRef]

Pandey, P.K.; Thareja, R.K.; Singh, R.P.; Costello, J.T. Deposition of nanocomposite Cu–TiO_2 using heterogeneous colliding plasmas. Appl. Phys. B 2018, 124, 1–8. [Google Scholar] [CrossRef]

Donnelly, T.; O'Connell, G.; Lunney, J.G. Metal nanoparticle film deposition by femtosecond laser ablation at atmospheric pressure. Nanomaterials 2020, 10, 2118. [Google Scholar] [CrossRef]

Serra, P.; Piqué, A. Laser-induced forward transfer: Fundamentals and applications. Adv. Mater. Technol. 2019, 4, 1800099. [Google Scholar] [CrossRef] [Green Version]

Liu, J. Simple technique for measurements of pulsed Gaussian-beam spot sizes. Opt. Lett. 1982, 7, 196–198. [Google Scholar] [CrossRef]

Zhao, W.; Wang, W.; Jiang, G.; Li, B.Q.; Mei, X. Ablation and morphological evolution of micro-holes in stainless steel with picosecond laser pulses. Int. J. Adv. Manuf. Technol. 2015, 80, 1713–1720. [Google Scholar] [CrossRef]

Taylor, G.I. The formation of a blast wave by a very intense explosion I. Theoretical discussion. Proc. R. Soc. Lond. 1950, 201, 159–174. [Google Scholar]

Singh, S.; Argument, M.; Tsui, Y.; Fedosejevs, R. Effect of ambient air pressure on debris redeposition during laser ablation of glass. Int. J. Appl. Phys. 2005, 98, 113520. [Google Scholar] [CrossRef]

Liu, H.; Mao, X.; Yoo, J.; Russo, R. Early phase laser induced plasma diagnostics and mass removal during single-pulse laser ablation of silicon. Spectrochim. Acta B 1999, 54, 1607–1624. [Google Scholar] [CrossRef]

Kramida, A.; Ralchenko, Y.; Reader, J. NIST Atomic Spectra Database (ver. 5.3); NIST: Gaithersburg, MA, USA, 2015. [Google Scholar]

Harilal, S.; Bindhu, C.; Tillack, M.; Najmabadi, F.; Gaeris, A. Internal structure and expansion dynamics of laser ablation plumes into ambient gases. Int. J. Appl. Phys. 2003, 93, 2380–2388. [Google Scholar] [CrossRef] [Green Version]

Cargill, P.; Goodrich, C.; Papadopoulos, K. Interaction of two collisionless shocks. Phys. Rev. Lett. 1986, 56, 1988. [Google Scholar] [CrossRef] [PubMed]

Fallon, C.; Hayden, P.; Walsh, N.; Kennedy, E.; Costello, J. The effect of wedge angle on the evolution of a stagnation layer in a colliding plasma experiment. J. Phys. Conf. Ser. 2014, 548, 12036. [Google Scholar] [CrossRef] [Green Version]

Fallon, C.; Hayden, P.; Walsh, N.; Kennedy, E.; Costello, J. Target geometrical effects on the stagnation layer formed by colliding a pair of laser produced copper plasmas. Phys. Plasmas 2015, 22, 93506. [Google Scholar] [CrossRef]

Gibbon, P. Introduction to plasma physics. arXiv 2020, arXiv:2007.04783. [Google Scholar]

Yang, Z.; Wei, W.; Han, J.; Wu, J.; Li, X.; Jia, S. Experimental study of the behavior of two laser produced plasmas in air. Phys. Plasmas 2015, 22, 073511. [Google Scholar] [CrossRef]

Rambo, P.; Denavit, J. Interpenetration and ion separation in colliding plasmas. Phys. Plasmas 1994, 1, 4050–4060. [Google Scholar] [CrossRef]

Guthikonda, N.; Manikanta, E.; Chelikani, L.; Sai Shiva, S.; Sree Harsha, S.; Ikkurthi, V.; Prem Kiran, P. Interaction of two counterpropagating laser induced plasmas and shock waves in air. Phys. Plasmas 2020, 27, 23107. [Google Scholar] [CrossRef]

Jin, Y.; Perrie, W.; Harris, P.; Allegre, O.; Abrams, K.; Dearden, G. Patterning of Aluminium thin film on polyethylene terephthalate by multi-beam picosecond laser. Opt. Lasers Eng. Opt. Laser Eng. 2015, 74, 67–74. [Google Scholar] [CrossRef]

Tang, Y.; Perrie, W.; Rico Sierra, D.; Li, Q.; Liu, D.; Edwardson, S.P.; Dearden, G. Laser–material interactions of high-quality ultrashort pulsed vector vortex beams. Micromachines 2021, 12, 376. [Google Scholar] [CrossRef]

Taylor, J. Relaxation and magnetic reconnection in plasmas. Rev. Mod. Phys. 1986, 58, 741. [Google Scholar] [CrossRef]

Totorica, S.; Abel, T.; Fiuza, F. Particle acceleration in laser-driven magnetic reconnection. Phys. Plasmas 2017, 24, 41408. [Google Scholar] [CrossRef] [Green Version]

Nilson, P.; Willingale, L.; Kaluza, M.; Kamperidis, C.; Minardi, S.; Wei, M.; Fernandes, P.; Notley, M.; Bandyopadhyay, S.; Sherlock, M. Bidirectional jet formation during driven magnetic reconnection in two-beam laser–plasma interactions. Phys. Plasmas 2008, 15, 92701. [Google Scholar] [CrossRef]

CHAPTER

SQUARED FOCAL INTENSITY DISTRIBUTIONS FOR LASER MATERIAL PROCESSING APPLICATIONS

12

1. INTRODUCTION: AN OVERVIEW

In contemporary laser material processing, beam shaping is garnering more focus, as it enables the modulation of focal intensity distribution to optimize the interaction between laser radiation and the material [1,2,3]. This enhances the flexibility, precision, and efficacy of laser processing. Furthermore, the variety of attainable structural types is broadened [4,5,6], and surface characteristics can be precisely modified [7,8,9]. Top-hat profiles exhibit a consistent intensity distribution over the beam's cross-section within a designated operating plane. A diverse array of beam-shaping techniques can generate them, including apertures [10], refractive optical elements [11,12,13], diffractive optical elements [10,14], and beam integrators or homogenizers [10,15]. A spatial light modulator is a versatile optical device that enables the generation of diverse intensity profiles by the dynamic alteration of the amplitude, phase, and polarization of the input laser beam [16,17,18]. The benefit of refractive beam shapers lies in their uncomplicated design and exceptional conversion efficiency. They exhibit reduced sensitivity to wavelength variations, are simpler to produce, and can endure greater laser intensities in comparison to diffractive components. This work introduces a highly flexible and compact refractive freeform beam-shaping device ideal for laser material processing applications. Extensive structuring of surfaces at the micro- and nano-meter scale typically necessitates the scanning of a focussed laser beam across the material's surface. The prevalent Gaussian intensity distribution results in a specific inhomogeneity of the deposited energy, which subsequently affects the processing outcome. An initial enhancement was seen in this context by the utilization of a beam shaper that facilitates the generation of a rotationally symmetric top-hat profile [19]. Nevertheless, owing to the rectangular symmetry of the unidirectional scanning method with a defined overlap between subsequent scan lines, the application of a squared top-hat appears to be particularly beneficial for uniform energy coupling. This squared top-hat is generated from the collimated Gaussian intensity profile of the input beam by integrating the examined refractive beam shaper with focusing optics. The latter may serve as either a basic focusing lens or a sophisticated f-Theta objective, commonly employed in laser-based surface structuring. Consequently, a primary aim of this study was to assess whether the x-y symmetry of the focal distribution can enhance material processing when utilizing a scanner and an f-Theta lens. This factor is crucial for contemporary laser systems and their elevated repetition rates, since optimal use guarantees enhanced processing efficiency. The beam-shaping element was empirically evaluated in several applications utilizing stainless steel as the substrate material. This encompasses surface nano-structuring by laser-induced periodic surface structures (LIPSS), enabling the engineering of surfaces with diverse functional properties [7,20]. Moreover, the ablation-based micro-structuring of the sample surface utilizing femtosecond lasers (fs-lasers) has been established as a standard processing method [21].

2. MATERIALS AND METHODS

2.1. Focal Beam-Shaping and Characterization

The most common approach for shaping focal intensity distributions is based on Fourier optics, where the focusing optics transforms the incoming beam profile into the desired shape. This is well-explained for rotational symmetry in **[19,22]**. However, for non-rotationally symmetric focal intensity distributions, the symmetry of the beam-shaping element needs to be broken, while the focusing optics remains the same. Therefore, the beam-shaping element becomes a freeform surface. As a phase plate, it imprints a phase modulation on the incoming plan wave of the collimated beam, which stays collimated as this modulation is in the order of λ (PV) and thus very small. The focusing optics performs the Fourier transformation of this modified wave front and therefore determines the width of the generated profiles inverse proportional to the numerical aperture (NA). Consequently, this approach is modular with regards to the focusing optics. The phase plate (a|SqAiryShape) itself is a freeform distribution based on a super Gaussian profile:

$$z(h) = He^{-2\left(\frac{4h^2}{D^2}\right)^{\frac{M}{2}}} \quad (1)$$

of order M = 8, diameter D and plateau height H in x-y-symmetry. The element is made of fused silica via CNC-based grinding and polishing. It can therefore withstand high-power laser intensities as well as ultrashort laser pulses, as employed in the experimental studies of the present work.

As a measure of how much the beam shaping will be affected by diffraction effects, the β-factor:

$$\beta = \frac{2\pi RD}{f_L \lambda} \quad (2)$$

can bc used [23], with R being the half diameter of the entrance beam, D being the half diameter of the focal distribution, f_L being the focal length and λ being the wavelength. Using the specific parameters of the present study (R = 10 mm, D = 25 µm, f_L = 100 mm and λ = 1030 nm) leads to β = 15.25. Having such a large value ensures stable beam-shaping profiles, which are essential for well-controlled material processing.

The squared focal intensity distribution was characterized in order to compare the experimental results with simulations. For this purpose, a supercontinuum laser (SC-400-4, Fianium, Southampton, UK) was used as a light source, from which λ = 1064 nm was selected via a spectrometer (LLTF Contrast, Fianium, Southampton, UK), coupled to a fiber, collimated (ACM25-20-L-C-1064, asphericon, Jena, Germany) and expanded (BAM25-175-D-C-1064, asphericon, Jena, Germany) to a collimated beam with 10 mm diameter. This beam fulfilled the input conditions for the freeform beam-shaping element. All following optical components used for characterization of the beam shaping are depicted in **Figure 1**. The beam-shaping element (a|SqAiryShape) was placed in the collimated beam, which was subsequently focused by a plano-convex lens with 300 mm focal length (#38-631, Edmund Optics, Mainz, Germany). The focal region was then imaged onto a camera (SP928, Ophir, Germany) by using a microscope objective (Aspheric Lens Kit 5720-B, Newport, Irvine, CA, USA). The setup was calibrated to obtain the real physical size of the focal distribution.

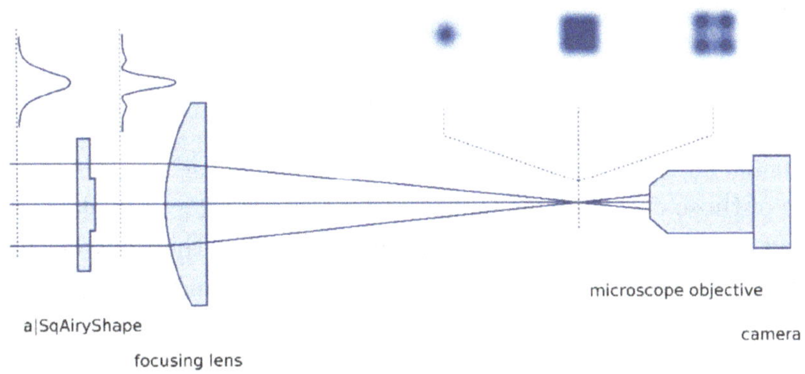

Figure 1. Experimental setup for characterization of the freeform beam-shaping element.

In the simulation, only the freeform beam-shaping element and the plano-convex focusing lens were considered. The focal intensity distribution was calculated via the point spread function with CodeV (Synopsis). This calculation is based on the squared Fourier transform of the pupil function [24]. The latter includes all aberrations and thus the phase manipulation introduced by the beam-shaping element.

Close to the focal point, various intensity distributions, such as a squared top-hat or squared donut, emerge. The following laser processing experiments were carried out with the squared top-hat, which occurs closest to the focal point. This is the top-hat shown in **Figure 2**. Comparison with the simulation, especially through the cross-sections in the bottom of **Figure 2**, reveals close agreement between measurement and simulation in terms of shape and size of the focus spot.

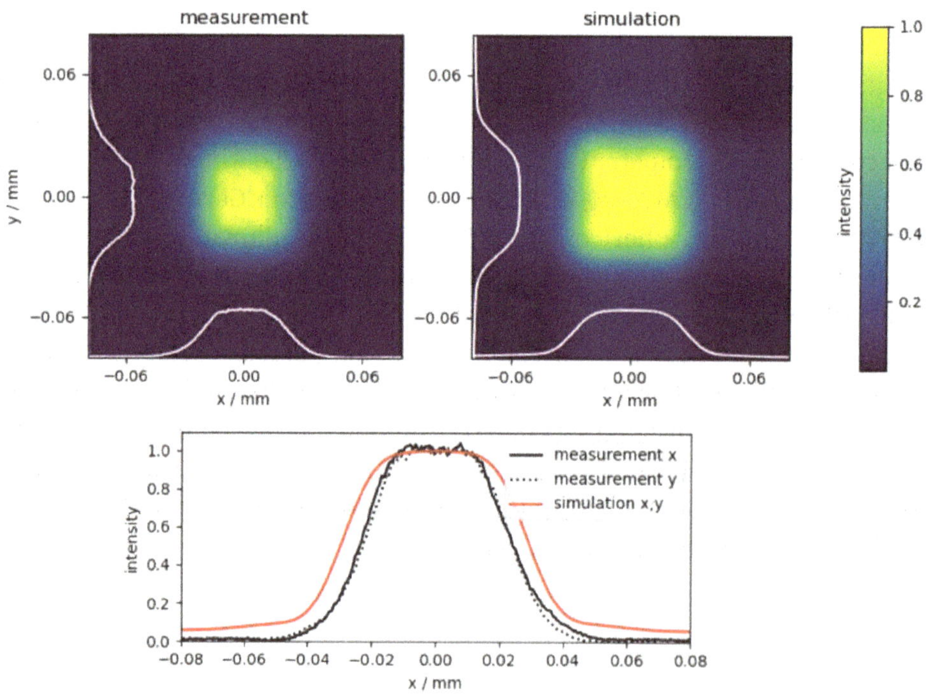

Figure 2. Color maps and cross-sections of the squared focal intensity distribution of the top-hat closest to the focal plane (λ = 1064 nm, f = 300 mm) normalized to the maximum intensity for comparison between measurement and simulation.

A sequence of those cross-sections taken along the optical axis is plotted in **Figure 3**. The upper panel (**Figure 3a**) shows the beam waist of the Gaussian input beam without the freeform beam-shaping element. With the beam-shaping element (panel b), the Gaussian distribution is notably flattened to a top-hat. Note that the beam shaping only occurs close to the focal point (z = 0 mm), whereas the profile remains almost Gaussian for larger distances. The blue-shaded distribution corresponds to the one shown in **Figure 2**.

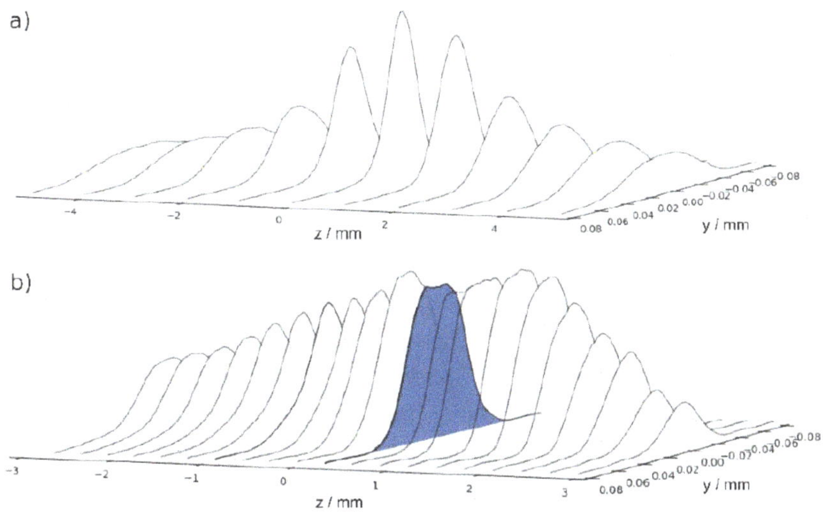

Figure 3. Profile cross-sections measured along the optical axis (z) in the focal region (a) without and (b) with freeform beam-shaping element.

2.2. Material Processing Setup and Sample Preparation

The operation of the beam-shaping element was investigated by means of surface structuring experiments using the experimental setup illustrated in Figure 4. The linearly polarized Gaussian output beam (M^2 ~ 1.08) of the diode-pumped Yb:KYW thin-disc fs-laser system (JenLas D2.fs, Jenoptik, Germany) with a central wavelength $\lambda = 1025$ nm is characterized by a pulse duration $\tau = 300$ fs (FWHM) and pulse energies $E_{imp} \leq 40$ µJ at a repetition frequency $f_{rep} = 100$ kHz. Before passing the beam shaper, the output beam was widened by a beam expander (5×). The shaped beam was then focused onto the sample surface by a galvanometer scanner (IntelliScan14, Scanlab, Puchheim, Germany). The focal length of the associated f-Theta objective (JENar, Jenoptik, Jena, Germany) was $f_L = 100$ mm.

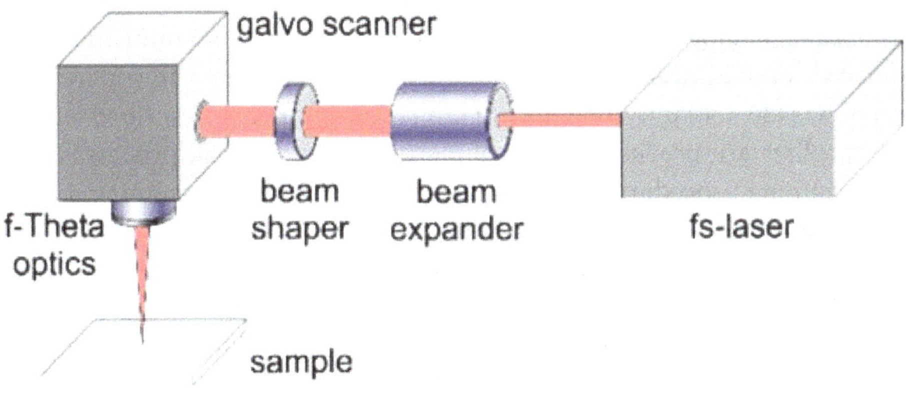

Figure 4. Experimental setup used for surface structuring experiments.

The structuring experiments were carried out on commercially available austenitic stainless-steel (X2CrNiMo17-12-2, Outokumpu, Krefeld, Germany) at normal incidence and under air atmosphere. Prior to laser irradiation, the sample surface was manually grounded and polished to a mirror finish, with an average surface roughness R_a ~4 nm. This procedure involves the use of SiC abrasive paper of 800, 1200 and 2400 grit and 6, 3 and 1 µm polycrystalline diamond suspension, respectively. Besides, the sample surfaces were ultrasonically cleaned in acetone and isopropanol both before and after laser structuring. The morphology of the samples was characterized by scanning electron microscopy (SEM) (SigmaVP, Carl Zeiss, Oberkochen, Germany) at an accelerating voltage of 5 kV using a secondary electron detector. The spatial periods and the orientation of the LIPSS were quantified by 2D-Fourier transform analysis (2D-FT) of the SEM micrographs. White light interference microscopy (WLIM) (CCI HD, Taylor Hobson, Weiterstadt, Germany) was utilized to evaluate the surface topography. The microscope was equipped with a 50× objective.

3. RESULTS AND DISCUSSION

3.1. Single-Spot Experiments

The first part of the experiments was to investigate whether the theoretical calculations for the focal intensity distribution correlate with the geometry of the ablated region within the focal spot on the material surface. For this purpose, the surface was irradiated with 10 single fs-laser pulses with a temporal pulse spacing of 10 µs (f_{rep} = 100 kHz) and a single-pulse energy of E_{imp} = 6.5 µJ. Without the beam-shaping element in the beam path, the utilized f-Theta objective results in a focal spot diameter of $2w_f$ = (0.5 ± 24) µm, determined using the method of Liu [25]. This corresponds to a fs-laser peak fluence of $F = 2E_{imp}/(\pi w_f^2)$ = 2.9 J/cm² at the pulse energy used.

Figure 5 shows SEM micrographs of the ablation spots obtained at different distances (z-positions) of the stainless-steel surface from the focusing objective using the freeform beam-shaping element. At the reference position (z = 0 mm) and at z = 0.6-mm, squared top-hat intensity profiles are available that result in square material ablation of size (30 × 30) µm² and (40 × 40) µm², respectively. As described later, these intensity profiles are predestined for numerous laser material processes that are based on scanning procedures with x-y-symmetry (Figure 6). Assuming a perfectly rectangular fluence distribution of both top-hat profiles, the larger focal dimensions together with the pulse energy E_{imp} = 6.5 µJ result in a lower fs-laser peak fluence of F = 0.72 and F = 0.41 J/cm² respectively, when compared to the initial Gaussian focal spot. It should be noted that the peak fluence, F, is identical to the average fluence, F_{av}, of the fluence distribution. The larger fluence at z = 0 compared to z = -0.6 mm is indicated by the stronger melt formation in the ablation spot fabricated with the smaller top-hat profile.

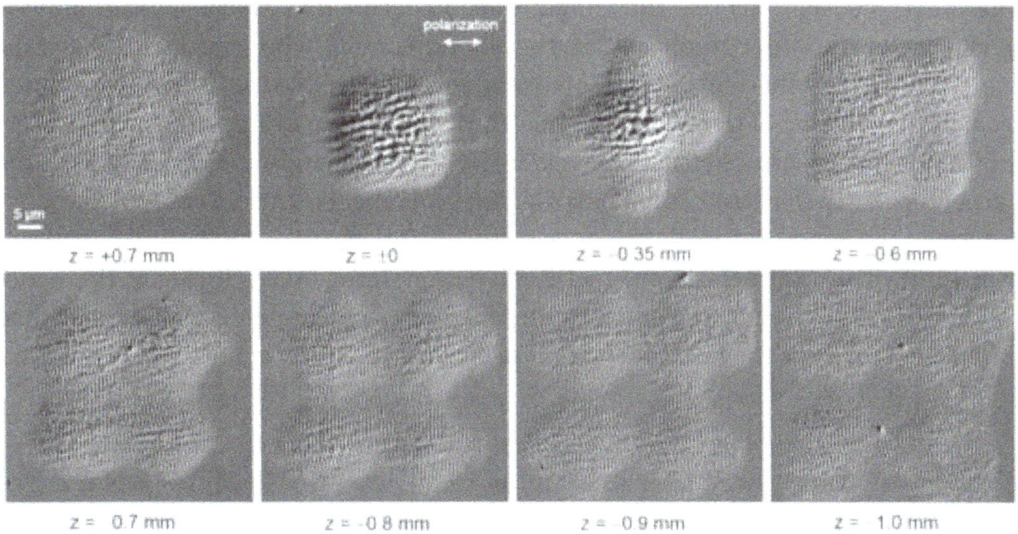

Figure 5. SEM micrographs of the stainless-steel surface after irradiation of 10 linearly polarized single fs-laser pulses at different z-positions of the sample surface within the focal intensity distribution. For comparability, the single-pulse energy was kept constant at E_{imp} = 6.5 µJ.

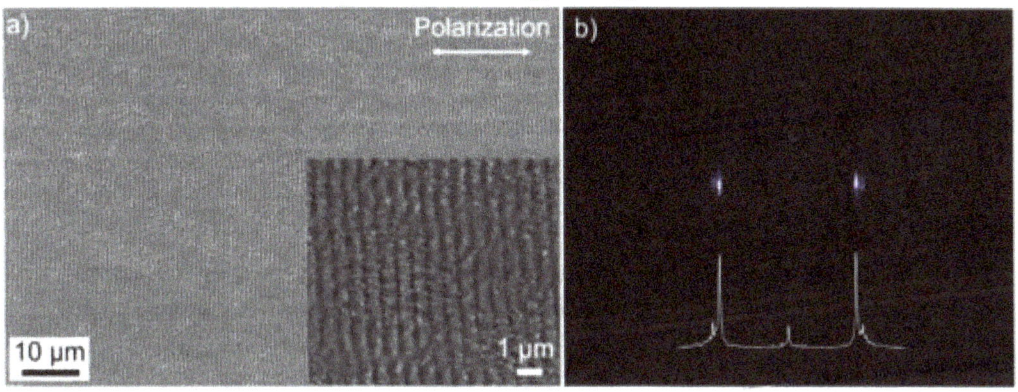

Figure 6. LIPSS structuring of stainless-steel surfaces using the squared top-hat profile (z = 0 mm) of the linearly polarized fs-laser beam ($2w_f$ ~30 µm, E_{imp} = 3.5 µJ, v = 1 m/s, Δx = 10 µm): (a) SEM micrograph and (b) 2D-FT spectrum and its corresponding cross-section calculated from the SEM micrograph quantifying the homogeneity and orientation of the LIPSS.

In addition to the top-hat profiles, the beam-shaping element provides other intermediate stages, which are characterized by more exotic intensity distributions. At z = -1.0 mm, a donut-shaped focal intensity distribution with square beam cross-section can be observed, characterized by the zero-intensity center on the beam axis. In contrast, at z = -0.35 mm, the intensity distribution is almost completely reversed, which is why a stronger intensity can be observed at the center. Both examples are intensity distributions suitable for the generation of novel surface structures (e.g., bio-inspired surfaces) [5].

The SEM micrographs of the single-spot experiments reveal a well-defined modulated surface topography within the ablation crater, whose periodicity is in the sub-μm range and therefore much smaller when compared to the diffraction-limited focal spot diameter of a few tens of μm. These periodic patterns are termed laser-induced periodic surface structures (LIPSS), which for metals are typically oriented perpendicular to the linear beam polarization [**26**]. The most widely accepted theory describes the formation of LIPSS as a result of interference of the incident laser radiation with surface electromagnetic waves generated by scattering at the nano-rough surface [**27,28**]. These scattering processes are possible despite the polished initial surface, since the LIPSS formation process is generally a multi-pulse process.

Figure 5 illustrates that the precise shape of the LIPSS pattern within the focal point is contingent upon the fluence distribution in the beam cross-section, specifically the intensity profile. Melt formation in the concentrated areas of the beam cross-section offers a specific contribution. It is obvious that a notably uniform LIPSS distribution may be achieved with the careful selection of the z-position, indicating a consistent fluence in the beam profile. This is a crucial element for uniform large-area structuring using LIPSS [29]. The capacity to control the LIPSS distribution within the focus spot by choosing a certain beam profile improves the current state-of-the-art, hence augmenting the versatility of the direct-writing structuring method. The fluence values calculated for the top-hat profile are comparable to the fluence required for ablation-induced LIPSS formation on stainless steel under analogous irradiation conditions. [30,31].

3.2. Laser Surface Structuring on the Nanoscale

Based on the demonstrated LIPSS formation within the focal spot, the nanostructures can be scaled to large surface areas by a relative movement between the focused fs-laser beam and the sample surface. This is particularly achievable with metals due to the linear absorption of the fs-laser radiation. In this context, the specific optical properties make stainless-steel a well-suited candidate for the production of highly regular LIPSS [**31**]. This concerns in particular the exact alignment and a small dispersion of the spatial frequencies of the LIPSS, which are coherently linked to each other despite the unidirectional scanning procedure [**30,31**].

The SEM micrograph in Figure 6a shows the stainless-steel surface obtained from scanning the squared top-hat profile (z = 0 mm) with a scan speed of v = 1 m/s and a line spacing of Δx = 10 μm across the sample surface. The single-pulse energy was reduced to E_{imp} = 3.5 μJ, i.e., to F = 0.39 J/cm^2, in order to avoid the extensive melt formation observed in Figure 5. The calculated 2D-FT spectrum (Figure 6b) confirms the high regularity of the LIPSS aligned perpendicular to the linear beam polarization by the very narrow characteristic peaks. The spatial periods of the LIPSS were determined to be around (950 ± 20) nm.

The obtained regularity is comparable with our previous work on rotationally symmetric beam shaping [19], although the following main differences should be discussed:

1. As already demonstrated in [30,31], highly regular LIPSS can also be generated with conventional Gaussian beam profiles. However, the radial intensity dependence and the circular spot area require a larger overlap of the laser pulses in both x- and y-directions to realize an almost homogeneous energy deposition. The optimal process parameters for Gaussian beam processing were derived in [19] to be v = 0.67 m/s, Δx = 6 μm, E_{imp} = 2.6 μJ, F = 1.15 J/cm² at f_{rep} = 100 kHz and $2w_f$ = 24 μm. These values result in a processing rate of about 32 s/cm² and in a laser pulse number $N_{eff\_2D} = (\pi w_f^2 \cdot f_{rep})/(v \cdot \Delta x)$ ~11.3, that effectively hit the focal spot area [28].
2. The main difference between the ideal squared top-hat and the Gaussian profile results from the relative movement during scanning. Figure 7a illustrates that for the Gaussian profile at a certain position y at the surface, every single pulse of the total required number (N_{eff_2D}) contributes to structuring with different fluence values. Consequently, the specific fluence of some of the individual pulses can be significantly smaller than the ablation and LIPSS formation threshold of the material.

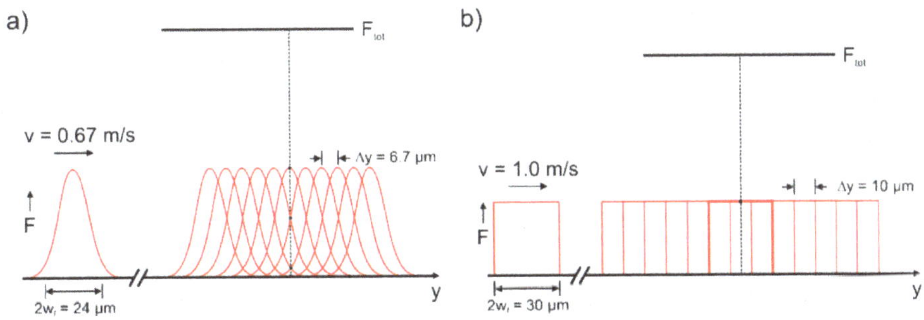

Figure 7. Overlap and contribution of individual pulses during scanning of a single line using (a) the Gaussian focal intensity distribution and (b) an ideal top-hat beam.

1. In the case of the top-hat profile (z = 0) with its square shape, the slightly larger beam diameter (here given by the beam width in x- and y-directions) and steeper flanks, LIPSS structures with almost equal properties can be achieved. Nevertheless, the structuring can be performed with larger scanning speed (v = 1 m/s) and line separation (Δx = 10 μm). As described above, the pulse energy must be increased slightly for this to exceed the required LIPSS formation threshold. The optimum parameters, however, result in N_{eff_2D} = 9 and a processing rate of about 14 s/cm², which is more than a factor of 2 faster than structuring with the conventional Gaussian beam.

The accumulated fluence of the ideal top-hat profile with $F_{tot} = N_{eff\_2D} \cdot F = 3.51\ J/cm^2$ is smaller than for the Gaussian intensity distribution ($F_{tot} = 6.47\ J/cm^2$), since during structuring with the top-hat, each single pulse contributes equally to F_{tot} (Figure 7b).

3.3. Surface Micro-Structuring

In order to investigate the influence of the intensity distribution on the micro-structuring of surfaces, channels were created using the top-hat profile (z = 0 mm) and 10 over-scans with v = 0.1 m/s. Earlier studies with rotationally symmetrical top-hat and donut-shaped beam profiles already showed that steeper channel walls and a better surface quality (especially at the bottom) can be achieved in comparison to the Gaussian profile [19].

These findings can also be observed in Figure 8, which shows 3D-WLIM micrographs of channels produced with the top-hat profile using different pulse energies, E_{imp}. It becomes evident for moderate pulse energies that the homogeneous fluence distribution across the beam profile leads to a very high quality of both the walls and the bottom of the channel (Figure 8a,b). The corresponding arithmetical roughness values of the channel surface were measured via WLIM to be $S_a = 0.1072$ µm for $E_{imp} = 2.5$ µJ and $S_a = 0.1844$ µm for $E_{imp} = 4.6$ µJ. As indicated by the cross-sections in Figure 8d, the corresponding depth of the channels increases with increasing pulse energy, from about 6.2 µm at $E_{imp} = 2.5$ µJ (Figure 8a) to about 12.5 µm at $E_{imp} = 7$ µJ (Figure 8c). In addition, a slight increase in the channel width can be observed. However, the WLIM micrographs indicate that at very high pulse energies and thus large laser peak fluences, an inhomogeneous and rough channel bottom results from extensive ablation and the melt formation already detected in the single-spot experiments. The irregularity is expressed in particular in hole-like depressions that are not adequately displayed by the perspective of the 3D micrograph in Figure 8c. However, a qualitative measure is given by the grey-shaded area in the cross-section of Figure 8d that illustrates the remarkably larger variation, especially at the channel bottom, when compared to smaller E_{imp}. This is also confirmed by the larger surface roughness that was measured to be $S_a = 0.5174$ µm for $E_{imp} = 7$ µJ.

Structuring with the donut-shaped intensity profile at z = -0.9 mm allows to create channels with steep walls and a w-shaped cross-sectional geometry that is determined by the specific intensity distribution (see Figure 5). According to the cross-sectional height profile of the channel (Figure 9c), 10 over-scans with v = 0.1 m/s and $E_{imp} = 4.6$ µJ result in a width of about 40 µm and a maximum ablation depth of about 10 µm. The channel walls are characterized by their very homogeneous and regular surfaces in the form of superimposed nanoscale LIPSS (Figure 9a).

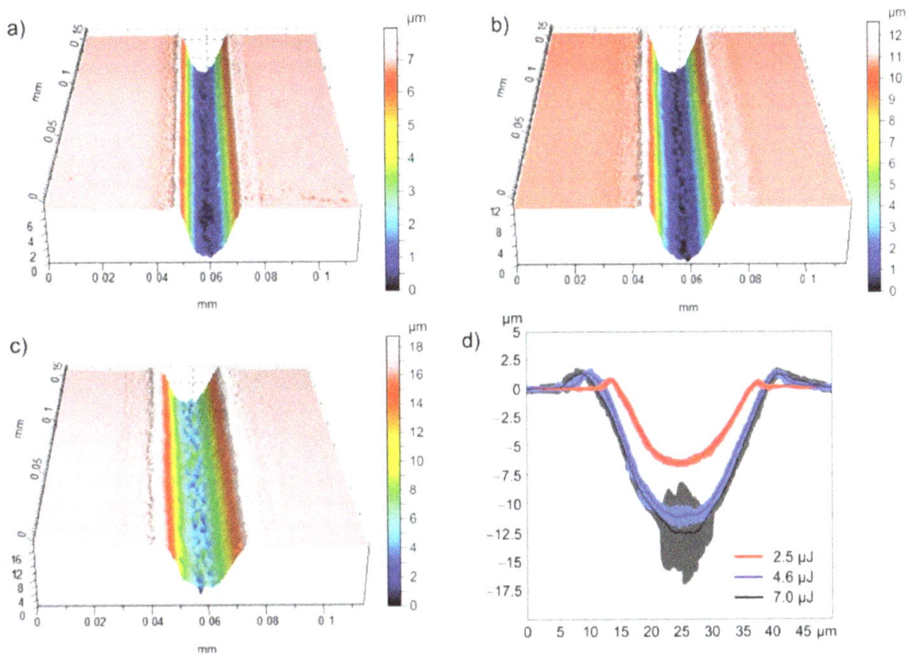

Figure 8. WLIM micrographs of the stainless-steel surface illustrating the 3D geometry of channel-like structures obtained from 10 over-scans with v = 0.1 m/s using the squared top-hat profile at z = 0 mm and different single-pulse energies: (a) E_{imp} = 2.5 µJ, (b) E_{imp} = 4.6 µJ and (c) E_{imp} = 7 µJ. A comparison between the cross-sections of the channels is illustrated in (d). Here, the solid line corresponds to the mean profile and the shadow illustrates the variation of the profile series measured along the channel.

The high quality is confirmed by the fact that the modulation depth of LIPSS is typically in the order of 200 nm [30]. Contrary to a rotationally symmetrical donut, the used squared intensity profile provides the zero-intensity center and two axes in x- and y-directions with almost zero intensity. Consequently, the areas with lower (zero) intensity are not passed over by highly intensive beam parts during scanning in these directions, resulting in a channel with a less ablated central area (Figure 9b). This may be utilized, for instance, to periodically disseminate these w-shaped structures throughout extensive surface regions according to an appropriate scan line distance. Depending on the focal length of the focusing optics, this would yield a microscale periodic grating with periods ranging from 10 to 20 µm. Comparable surface structures with similar dimensions can be achieved, for instance, by Direct Interference Laser Patterning, which is utilized to create surfaces with functional attributes (optics, tribology, biology, wetting) [32].

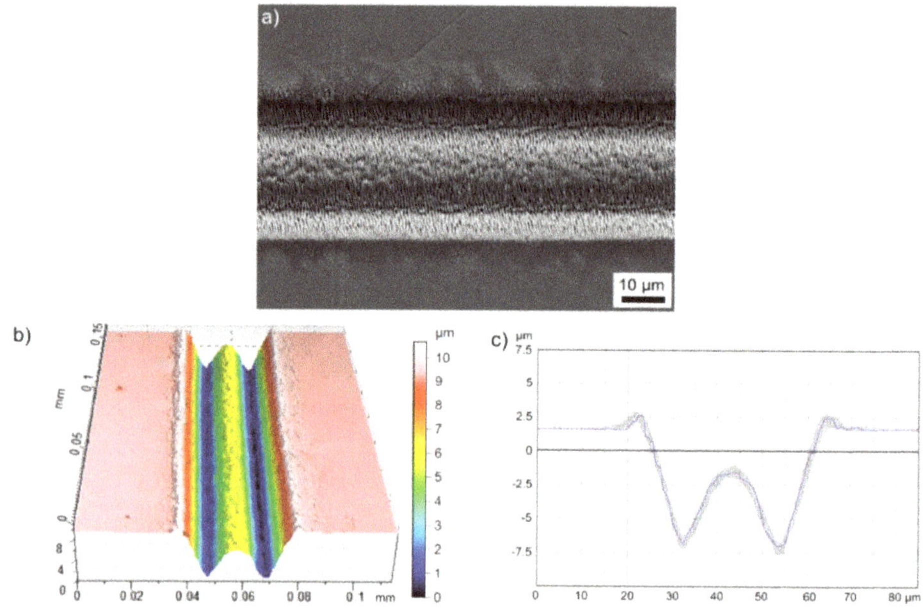

Figure 9. Micro-channel fabricated on stainless-steel using the squared donut-like intensity distribution at z = 0.9- mm, 10 over-scans with v = 0.1 m/s and a single-pulse energy of 4.6 µJ: (a) SEM micrograph, (b) WLIM micrograph and (c) cross-sectional height profile. The blue line in (c) corresponds to the mean profile and the grey shadow illustrates the profile series measured along the channel in (b).

The results indicate that the donut and top-hat profiles produced by the beam-shaping element are beneficial for creating micro-channels with certain wall inclination degrees and customized cross-sectional geometries. This option is also accessible (in situ) during the structuring process, as the individual intensity profiles can be manipulated quite easily by altering the z-position of the sample surface.

4. CONCISE SUMMARY

This work examines a multifaceted refractive beam shaper that transforms the initial Gaussian intensity distribution of the femtosecond laser beam into various squared beam profiles at distinct z-positions. Numerical calculations for the beam-shaping unit were validated experimentally by spatial beam profile measurements and surface structuring at the micro- and nano-scale. The squared top-hat profiles in the focal region have been shown to substantially decrease processing time during large-area LIPSS structuring, while maintaining high regularity. The application of the beam shaper for creating channel-like structures through successive line scans demonstrated that varying focal intensity distributions produced precise channels with adjustable cross-sectional geometries.

AT A GLANCE

Customized intensity profiles within the laser beam's focused spot provide significant possibilities for precise control over the interaction between laser radiation and materials, hence enhancing processing outcomes. This study presents a unique refractive beam-shaping element that generates various squared intensity distributions derived from the Gaussian output beam of the employed femtosecond (fs) laser. The usefulness of the beam-shaping element for femtosecond laser material processing with a conventional f-Theta lens is illustrated through examples of micro- and nano-scale surface structuring of stainless steel. The experimental structuring results align well with beam profile measurements and numerical simulations of the beam-shaping device. The experimental results demonstrate enhanced laser processing characterized by a markedly decreased processing time for surface nano-structuring and the capability to regulate ablation geometry in the production of micro-channels.

REFERENCES

1. Malinauskas, M.; Zukauskas, A.; Hasegawa, S.; Hayasaki, Y.; Mizeikis, V.; Buividas, R.; Juodkazis, S. Ultrafast laser processing of materials: From science to industry. *Light-Sci. Appl.* **2016**, *5*, e16133. **[Google Scholar] [CrossRef] [Green Version]**
2. Raciukaitis, G.; Stankevicius, E.; Gecys, P.; Gedvilas, M.; Bischoff, C.; Jager, E.; Umhofer, U.; Volklein, F. Laser processing by using diffractive optical laser beam shaping technique. *J. Laser Micro Nanoen.* **2011**, *6*, 37–43. **[Google Scholar] [CrossRef]**
3. Sugioka, K.; Cheng, Y. A tutorial on optics for ultrafast laser materials processing: Basic microprocessing system to beam shaping and advanced focusing methods. *Adv. Opt. Technol.* **2012**, *1*, 353. **[Google Scholar] [CrossRef]**
4. Allegre, O.J.; Jin, Y.; Perrie, W.; Ouyang, J.; Fearon, E.; Edwardson, S.P.; Dearden, G. Complete wavefront and polarization control for ultrashort-pulse laser microprocessing. *Opt. Express* **2013**, *21*, 21198–21207. **[Google Scholar] [CrossRef]**
5. Skoulas, E.; Manousaki, A.; Fotakis, C.; Stratakis, E. Biomimetic surface structuring using cylindrical vector femtosecond laser beams. *Sci. Rep.* **2017**, *7*, 45114. **[Google Scholar] [CrossRef] [Green Version]**
6. Ouyang, J.; Perrie, W.; Allegre, O.J.; Heil, T.; Jin, Y.; Fearon, E.; Eckford, D.; Edwardson, S.P.; Dearden, G. Tailored optical vector fields for ultrashort-pulse laser induced complex surface plasmon structuring. *Opt. Express* **2015**, *23*, 12562–12572. **[Google Scholar] [CrossRef] [PubMed]**
7. Müller, F.A.; Kunz, C.; Gräf, S. Bio-inspired functional surfaces based on laser-induced periodic surface structures. *Materials* **2016**, *9*, 476. **[Google Scholar] [CrossRef]**
8. Gräf, S. Formation of laser-induced periodic surface structures on different materials: Fundamentals, properties and applications. *Adv. Opt. Technol.* **2020**, *9*, 11–39. **[Google Scholar] [CrossRef]**
9. Stratakis, E.; Bonse, J.; Heitz, J.; Siegel, J.; Tsibidis, G.; Skoulas, E.; Papadopoulos, A.; Mimidis, A.; Joel, A.-C.; Comanns, P. Laser engineering of biomimetic surfaces. *Mater. Sci. Eng. R Rep.* **2020**, *141*, 100562. **[Google Scholar] [CrossRef]**
10. Dickey, F.M. Theory and techniques for laser beam shaping. *Int. Congr. Appl. Lasers Electro.-Opt.* **2001**, *2001*, 1393–1402. **[Google Scholar]**
11. Frieden, B.R. Lossless conversion of a plane laser wave to a plane wave of uniform irradiance. *Appl. Opt.* **1965**, *4*, 1400–1403. **[Google Scholar] [CrossRef]**

12. Rhodes, P.W.; Shealy, D.L. Refractive optical systems for irradiance redistribution of collimated radiation—Their design and analysis. *Appl. Opt.* **1980**, *19*, 3545–3553. **[Google Scholar] [CrossRef] [PubMed]**

13. Hoffnagle, J.A.; Jefferson, C.M. Design and performance of a refractive optical system that converts a gaussian to a flattop beam. *Appl. Opt.* **2000**, *39*, 5488–5499. **[Google Scholar] [CrossRef] [PubMed]**

14. Veldkamp, W.B.; Kastner, C.J. Beam profile shaping for laser radars that use detector arrays. *Appl. Opt.* **1982**, *21*, 345–356. **[Google Scholar] [CrossRef]**

15. Bollanti, S.; Di Lazzaro, P.; Murra, D. More about the light beam shaping by the integration method. *Eur. Phys. J.-Appl. Phys.* **2004**, *28*, 179–186. **[Google Scholar] [CrossRef]**

16. Häfner, T.; Strauss, J.; Roider, C.; Heberle, J.; Schmidt, M. Tailored laser beam shaping for efficient and accurate microstructuring. *Appl. Phys. A Mater. Sci. Process.* **2018**, *124*, 111. **[Google Scholar] [CrossRef] [Green Version]**

17. Li, J.; Kuanga, Z.; Vilara, J.G.; Deardena, G.; Perriea, W.; Edwardson, S. Dynamic ultrafast laser beam shaping for material processing at the imaging plane using a spatial light modulator. In Proceedings of the 9th International Conference on Photonic Technologies LANE, Gurth, Germany, 19–22 September 2016. **[Google Scholar]**

18. Jin, Y.; Allegre, O.J.; Perrie, W.; Abrams, K.; Ouyang, J.; Fearon, E.; Edwardson, S.P.; Dearden, G. Dynamic modulation of spatially structured polarization fields for real-time control of ultrafast laser-material interactions. *Opt. Express* **2013**, *21*, 25333–25343. **[Google Scholar] [CrossRef]**

19. Möhl, A.; Kaldun, S.; Kunz, C.; Müller, F.A.; Fuchs, U.; Gräf, S. Tailored focal beam shaping and its application in laser material processing. *J. Laser Appl.* **2019**, *31*, 042019. **[Google Scholar] [CrossRef] [Green Version]**

20. Bonse, J.; Kirner, S.V.; Höhm, S.; Epperlein, N.; Spaltmann, D.; Rosenfeld, A.; Krüger, J. Applications of laser-induced periodic surface structures (lipss). *Proc. SPIE* **2017**, *10092*, 100920N. **[Google Scholar]**

21. Sugioka, K.; Meunier, M.; Piqué, A. *Laser Precision Microfabrication*; Springer: Berlin, Germany, 2010. **[Google Scholar]**

22. Cordingley, J.J. Method for Severing Integrated-Circuit Connection Paths by a Phase-Plate-Adjusted Laser Beam. U.S. Patent 5,300,756, 22 October 1991. **[Google Scholar]**

23. Dickey, F.M. *Laser Beam Shaping: Theory and Techniques*, 2nd ed.; CRC Press, Taylor & Francis Group: Boca Raton, FL, USA, 2014. **[Google Scholar]**

24. Hopkins, H.H.; Yzuel, M.J. The computation of diffraction patterns in the presence of aberrations. *Opt. Acta Int. J. Opt.* **1970**, *17*, 157–182. **[Google Scholar] [CrossRef]**

25. Liu, J.M. Simple technique for measurements of pulsed gaussian beam spot sizes. *Opt. Lett.* **1982**, *7*, 196–198. **[Google Scholar] [CrossRef] [PubMed]**

26. Bonse, J.; Krüger, J.; Höhm, S.; Rosenfeld, A. Femtosecond laser-induced periodic surface structures. *J. Laser Appl.* **2012**, *24*, 042006. **[Google Scholar] [CrossRef]**

27. Sipe, J.E.; Young, J.F.; Preston, J.S.; van Driel, H.M. Laser-induced periodic surface structure. I. Theory. *Phys. Rev. B* **1983**, *27*, 1141–1154. **[Google Scholar] [CrossRef]**

28. Bonse, J.; Gräf, S. Maxwell meets marangoni-a review of theories on laser-induced periodic surface structures. *Laser Photon. Rev.* **2020**, *14*, 2000215. **[Google Scholar] [CrossRef]**

29. Kunz, C.; Engel, S.; Müller, F.A.; Gräf, S. Large-area fabrication of laser-induced periodic surface structures on fused silica using thin gold layers. *Nanomaterials* **2020**, *10*, 1187. **[Google Scholar] [CrossRef]**

30. Gräf, S.; Kunz, C.; Undisz, A.; Wonneberger, R.; Rettenmayr, M.; Müller, F.A. Mechano-responsive colour change of laser-induced periodic surface structures. *Appl. Surf. Sci.* **2019**, *471*, 645–651. **[Google Scholar] [CrossRef]**

31. Gnilitskyi, I.; Derrien, T.J.Y.; Levy, Y.; Bulgakova, N.M.; Mocek, T.; Orazi, L. High-speed manufacturing of highly regular femtosecond laser-induced periodic surface structures: Physical origin of regularity. *Sci. Rep.* **2017**, *7*, 8485. **[Google Scholar] [CrossRef] [Green Version]**

32. Müller, D.W.; Fox, T.; Grützmacher, P.G.; Suarez, S.; Mücklich, F. Applying ultrashort pulsed direct laser interference patterning for functional surfaces. *Sci. Rep.* **2020**, *10*, 3647. **[Google Scholar] [CrossRef] [PubMed] [Green Version]**

CHAPTER

STRUCTURING GLASS WITH ULTRAFAST BESSEL BEAMS AT A HIGH ASPECT RATIO

13

1. INTRODUCTION: AN OVERVIEW

Cutting and drilling glass presents a significant technological challenge, especially when the width of the alteration is considerably less than its length. This constitutes high aspect ratio structuring. In this context, ultrafast laser processing is highly appealing because to the infrared wavelength and the exceptionally high strength of the pulses, which facilitate energy deposition in three dimensions, including within the bulk of glass [1]. Laser-induced micro/nano-structures may include index changes, nanogratings, or voids [2]. These alterations can be integrated with chemical etching, resulting in numerous advancements in the domains of photonics, microfluidic devices, and mechanics [3,4]. Due to the somewhat sluggish nature of point-by-point scanning, various strategies have been devised to enhance the overall pace of micro-structuring. Moreover, elongated voids cannot be generated through scanning, as a nano-void created by an initial pulse would be obliterated by the microexplosion generated by the subsequent pulse adjacent to it [5]. Bessel beams have been identified as superior options for manipulating the creation of high aspect ratio elongated structures within transparent materials [6,7]. Bessel beams are generated through a conical superposition of plane waves, wherein the central hotspot retains its diameter throughout the whole length of the interference zone [8,9]. Crucially, when the focusing angle of the Bessel beam is sufficiently elevated, the conical energy flow intersects the high-intensity zone alone at the center hotspot. This significantly diminishes the impact of the Kerr effect, allowing for the processing of longitudinally-invariant structures within dielectrics [10,11,12]. High aspect ratio laser structuring is especially advantageous for stealth dicing, a technique that employs a sequence of pulses during rapid sample translation to generate a complete alteration plane [13]. A mechanical, chemical, or thermal stress is subsequently employed to fracture the glass along the altered plane. The primary advantage of this method is the capability to cut at exceptionally high velocities, generally ranging from 10 centimeters to 1 meter per second. Filamentation, or ultrafast pulse self-channeling in glass, was initially employed for stealth dicing applications [14,15]. Bessel beams additionally facilitate the straightforward regulation of material damage initiation using simple geometric considerations, as nonlinear effects are diminished. They were employed to sever glass and sapphire [16,17,18,19,20]. Given that beam length and focusing angle can be separately modified, it is feasible to generate extended Bessel beams with an identical angle suitable for cutting thick glass. This, however, necessitates elevated pulse energies as the energy is distributed across the whole focal line of the beam [21]. Bessel beams, utilizing millijoule energy, were employed to sever glass up to 1 cm in thickness [22,23]. Recently, the disruption of the symmetry of the Bessel beam was employed to enhance the precision and efficacy of directing fractures [24,25,26,27]. Regulating void formation with ultrafast laser pulses is essential for enhancing the stealth dicing of glass.Here, we show that, depending on the material, voids do not form for the same parameters. We compare three glass types: fused silica, Corning®® Eagle XG®®, and Corning®® Gorilla®® glass. We examine the conditions under which different structures are generated using single-shot Bessel beams, contingent upon pulse energy, pulse duration, and sample position. We noted that these three criteria govern the creation of diverse structures within glass. The future advancement of applications like laser cutting should consequently function within the appropriate parameter range.

2. MATERIALS AND METHODS

The experimental configuration (refer to Figure 1) consists of a chirped pulse amplified (CPA) Ti-Sapphire laser source with a central wavelength of 800 nm, and a reflective axicon (Cailabs) paired with a telescope system to enhance the cone angle [28]. The grating compressor of the amplifier can extend the laser pulses from 100 femtoseconds to 3 picoseconds. The experimental characterization of the beam is shown in Figure 2. The Bessel beam has a cone angle of 19.5○° in air, which corresponds to a central spot diameter of 0.9 μμm Full Width at Half Maximum (FWHM). The Bessel beam length is ~220 μm FWHM. One can notice the high degree of circular symmetry of the Bessel beam. All samples were 700 μm thick.

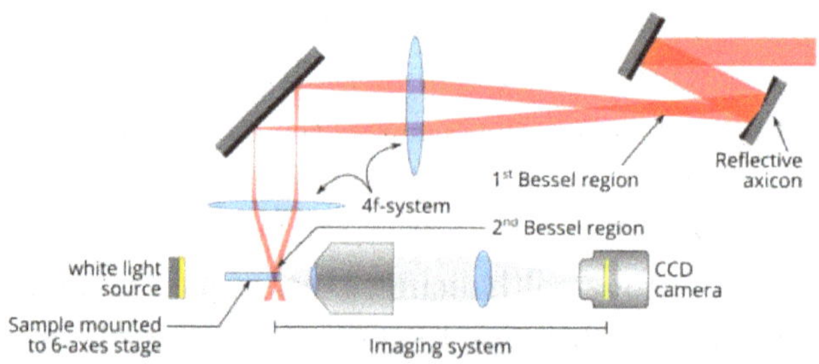

Figure 1. Experimental setup. Ultrafast laser beam is shaped by a reflective axicon (Cailabs). It is then demagnified by a telescopic arrangement in a 2f-2f configuration. Laser-written structures are observed using an incoherent white light source.

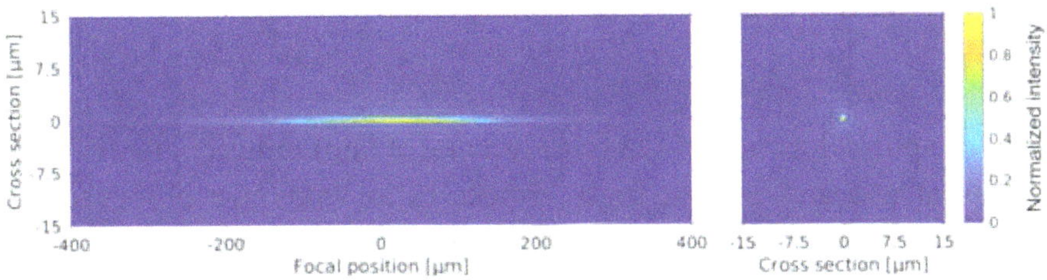

Figure 2. Characterization of experimental Bessel beam. (left) Longitudinal cross-cut of fluence distribution as a function of relative propagation distance in air. Please note aspect ratio of axes. (right) Cross-section of fluence distribution at center of beam.

After processing, the laser-modified regions were characterized by scanning the sample with white light, brightfield illumination, in an optical microscope arrangement with a numerical aperture of 0.8 and aberration-correction. Stitching the microscope images provided high resolution images of the whole high aspect ratio structure. Figure 3 shows the result of single shot illumination in fused silica at a pulse energy of 52 µJ, for increasing pulse durations. A periodic background pattern can be observed and simply arises from the illumination inhomogeneity associated to the stitching operation.

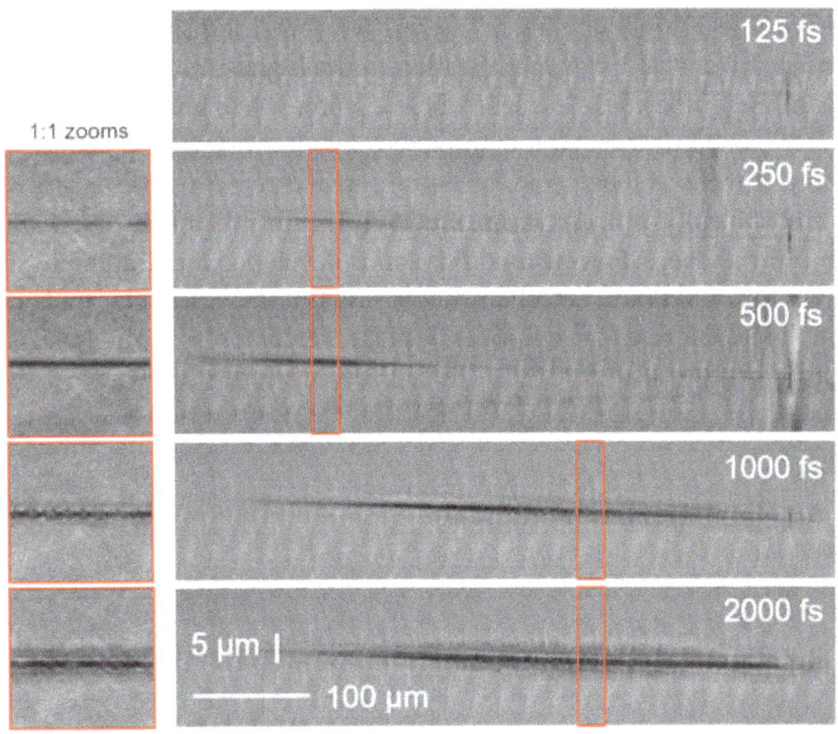

Figure 3. Microscope images of damages produced in fused silica glass for several pulse durations, for same beam position inside sample and for a constant single shot energy of 52 µJ. Please note image aspect ratios (left insets show 1:1 zooms of panels to their right). Beam propagates from left to right.

3. RESULTS

3.1. Microstructure Types

In Figure 3, at a pulse duration of 125 fs, only a line of faint index change is observable. In the cases of 250 and 500 fs, high- aspect ratio voids are observed, they are characterized by black marks. We readily see that the void structure is shorter than the Bessel beam length. The void is present only in the moderate intensity region while in the highest intensity region, the void structure has almost vanished. From 1 ps, the black mark is split around its center into small dots: these are void bubbles (see scanning electron microscopy image in Figure 4 of reference [29]). This process is called bubbling hereafter. Faint bubbling appears here in the central region of the Bessel beam and is slightly stronger for the case of 2 ps.

Importantly, the morphology of the material transformation can vary along the beam. An example is shown in Figure 4 where we report the results for the interaction of a 3 ps Bessel beam in Gorilla glass for different focusing positions inside the sample. In this example, the material modification undergoes minor variations with the position. We can observe three sections with different morphologies from left to right: (i) a void with faint bubbling in the relatively low intensity region of the Bessel beam (onset), (ii) a wide index modification in the highest intensity segment of the Bessel beam, and (iii) a void. The central panel allows us to understand the ordering of the material modification type with increasing intensity: index modification on a small diameter (central lobe), void formation, structure with more or less randomly distributed bubbles, wide index index modification on a diameter much larger than the beam central lobe. The wide modification is developed longitudinally step-wise, suggesting a threshold-based mechanism. It erased all structures like void or bubbles. Note that the bubbles appear elliptical because we compressed the horizontal axis scale by a factor 5.

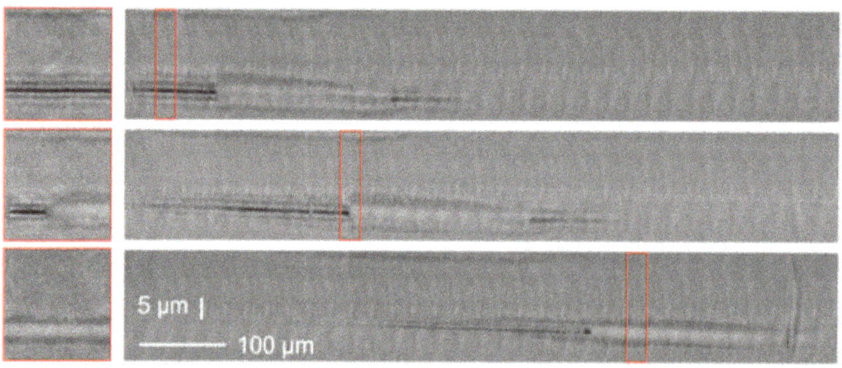

Figure 4. Brightfield microscopy images of damages produced in Gorilla glass with 3 ps pulse duration at a constant energy of 52 μJ as in Figure 4. Position of beam in sample was varied from top to bottom. Please note image aspect ratios (left insets show 1:1 zooms of panels to their right).

We further explored, in a parametric study, the evolution of the laser-induced structures as a function of pulse energy, pulse duration, and beam position. We summarized our results in Figure 5 for three positions of the Bessel beam inside the fused silica sample. In Figure 5a, the beam is positioned at the center of the sample such that none of the entrance and exit surfaces are damaged. We classified the modifications in 5 groups, ranging from no damage to strong bubbling. We note that we did not make a separate class for the wide index modification, such as the one shown in Figure 4, because those modifications are conventionally useless for applications. Also, they often occur alongside voids, which we are more interested in. For the drilling and cutting of glass applications, the narrow index modifications and nice voids, shown respectively as gray and green markers, are the most interesting features. When bubbling occurred on a substantial section of the beam, the structure was classified as bubbling.

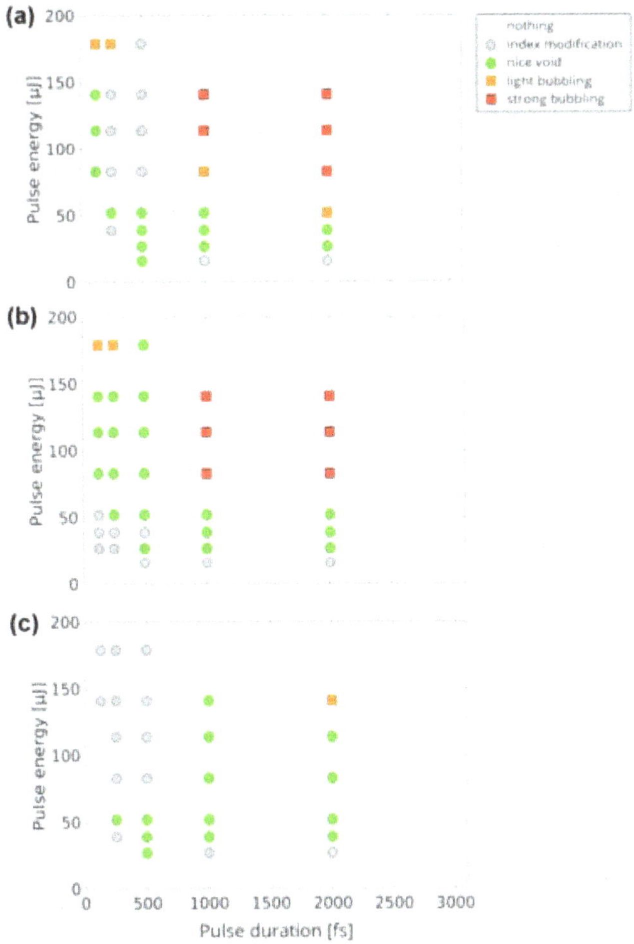

Figure 5. Morphology of single shot-induced damage in fused silica as a function of pulse energy and pulse duration. This is performed for a Bessel beam position: (a) at center of sample, (b) crossing the entrance surface, (c) crossing exit surface.

In **Figure** 5a, we see that the relatively large range of parameters allow for inducing nice voids in fused silica, similarly as in reference [30]. Higher energies and longer pulse durations tend to create bubbling. When the pulse duration is increased, the energy required to open a void is reduced, but bubbling also arises earlier. The energy range over which nice voids can be opened is reduced at longer pulse durations. In **Figure** 5b, the beam crosses the entrance side. In this case, the shortest pulse durations provide conditions where voids can be opened on a larger range of parameters. The presence of the interface increases the ability of void opening after energy deposition, as was already noticed in a former reference [6]. In **Figure** 5c, the beam crosses the exit interface. In this case, the short pulse durations induce voids with very small diameters. A much larger range of energies and pulse durations enable void formation. In almost all the cases tested here, nice voids could be produced. We hypothesize that our results can be understood by the following: the presence of the interface allows the phenomenon of ablation cooling [31] where a fraction of the energy deposited inside the material can be evacuated via the open channel at the surface. This allows for a fast quenching of the material relaxation which increases the capability of "freezing" a void before the compressed melt part surrounding the void can close it. In the case where the beam crosses the entrance surface (**Figure** 5b), the dynamics of plasma formation is slightly different from the two other cases because a dense plasma can form on the surface which can then shield light preventing void formation in the very first micrometres from the surface. This could prevent the effect of the ablation cooling described above.

3.2. Other Glasses

In **Figure** 6, we show the results for Eagle XG glass (**Figure** 6a) and Gorilla glass (**Figure** 6b) when the beam is enclosed within the sample, as in **Figure** 5a. For these two materials, the situation highly differs from the case of fused silica. Overall, it is much more difficult to create voids, particularly for short pulses, where no modification is apparent below a threshold pulse duration of 1 ps.

The operating window for void formation is restricted to a small energy range before bubbling sets in. This window is even smaller in the case of Gorilla glass. For most of the pulse durations in Gorilla glass, above the energy for the bubbling regime, only faint structures are found, whereas this phenomenon occurs in Eagle XG only for 1.5 ps and never in fused silica. It could be attributed to a slower relaxation in Gorilla glass than in that of the other media.

We attribute the differences between the different glasses in terms of void opening to the difference in softening point. Indeed, it exceeds 1500 ∘°C for fused silica, while it is only of 971 ∘°C for Eagle XG, and 853 ∘°C for Gorilla glass. Therefore, assuming a similar profile of laser-deposited energy, the molten region is wider in the case of Eagle XG and Gorilla glass than in the case of fused silica. This would lead to a faster quenching of the relaxation in fused silica, with stronger gradients, in contrast with the other two glasses.

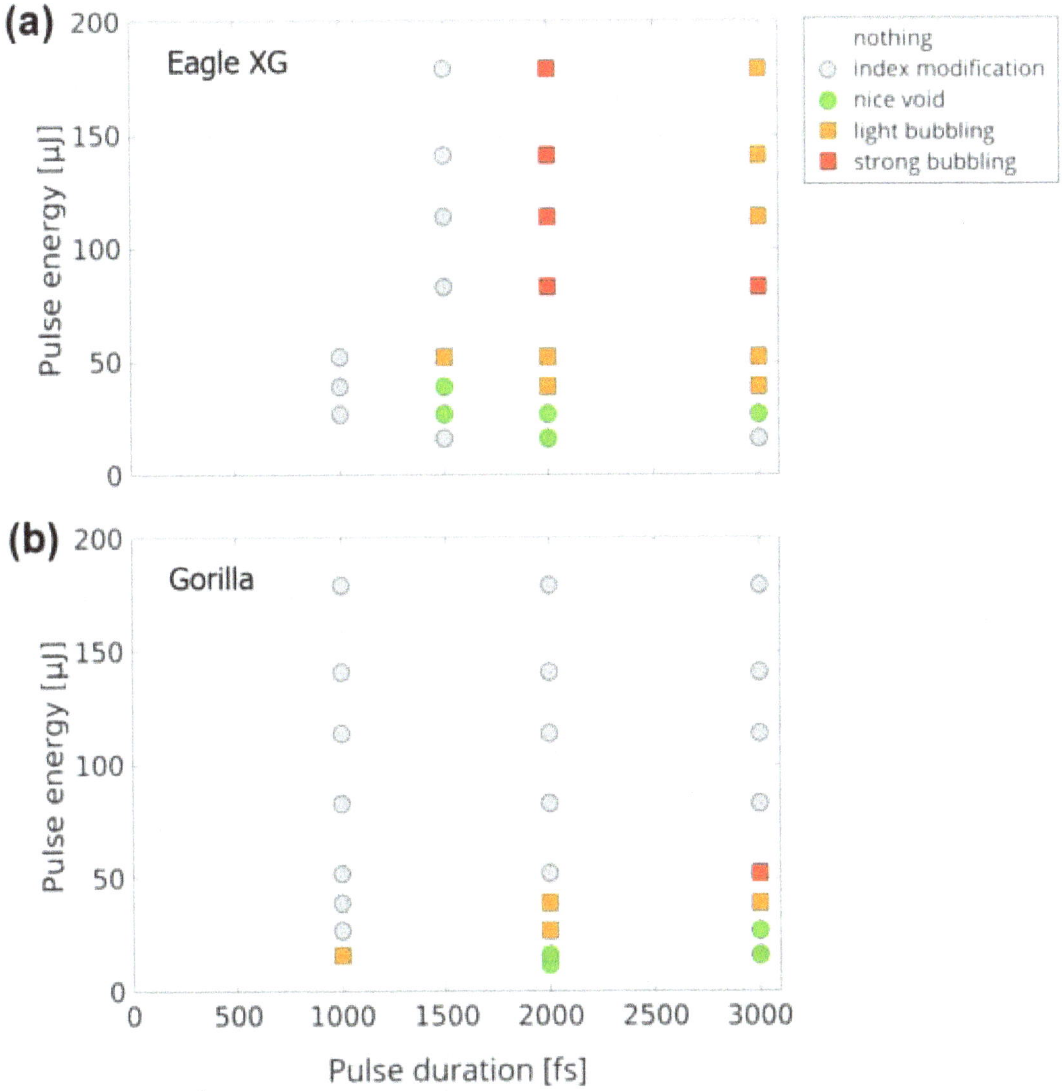

Figure 6. Morphology of single shot-induced damage in (a) Eagle XG and (b) Gorilla glass as a function of pulse energy and pulse duration. Beam is centered inside sample.

In Eagle XG and Gorilla glass, beam positioning close to the entrance and exit surfaces yield to similar tendencies as in the case of fused silica. While no major differences could be observed with **Figure 6** when the beam crosses the entrance surface, we observed a much larger operating window for void opening when the beam crosses the exit surface for both glass types.

3.3. Large Heat-Affected Zone for Picosecond Pulses

Figure 7 shows the result of a 3 ps laser shot in Eagle XG. One can notice that the heat affected zone is very wide, typically above 8 μm in diameter. Importantly, in this regime, the bubbles are not aligned: it seems that they have randomly moved from the optical axis during the relaxation stage. This well supports the interpretation of a large molten layer around the void at large pulse energies and duration.

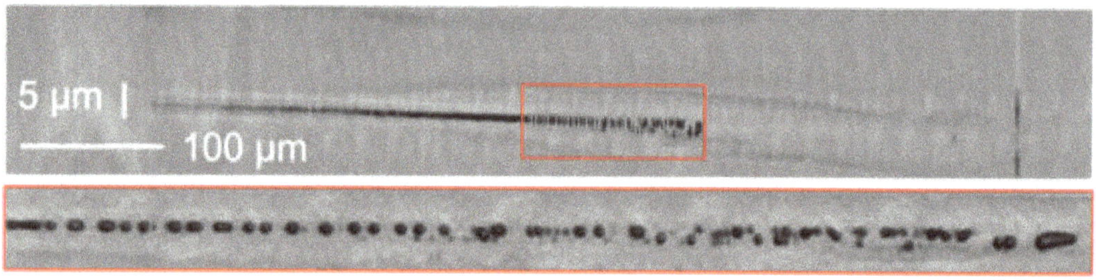

Figure 7. Strong bubbling in Eagle XG glass. Pulse duration is 3 ps, pulse energy 141 μJ. Please note image aspect ratio. Bottom inset shows a 1:1 zoom of marked area in panel above.

4. CONCLUSIONS

We observed that void formation in transparent glasses strongly depends on energy and pulse duration. In fused silica, voids can be formed for a relatively large set of parameters. By increasing pulse energy and pulse duration, the high aspect ratio void channel splits into bubbles with a diameter of approximately 1 μm. This structure can even vanish for higher energies. In Eagle XG and Gorilla glass, only very long pulse durations could yield void formation, but for a very limited energy range. For a pulse duration of 2 ps and 16 μJ pulse energy, a void can be formed in all three materials, with a length of 150 ± 10 μm in Eagle XG, of 170 ± 10 μm in fused silica and the highest length is for Gorilla glass with 230 ± 10 μm. We observe that void formation can occur over a broader range of energies when the beam traverses the exit surface for all materials. We analyze the progression of the morphological changes in the damage caused by ultrashort Bessel beams in these glasses as follows. At sufficiently high energy, a void may develop within glass, either following a micro-explosion [32,33] or through cavitation in a molten region [34]. We deduce that this gap may be temporarily encircled by a molten layer, the thickness of which is contingent upon the type of glass and the duration of the pulse. Extended pulse durations and elevated energy typically augment the thickness of the molten layer. A low softening point, as shown in Eagle XG and Gorilla, further enhances the molten layer. Thicker molten layers enhance the frequency of bubbling or may even seal the hole. The creation of bubbles is understood as a redistribution of material during the cooling phase, involving the division or temporary closure of the voids that are generated. We perceive the function of the exit surface as a cooling mechanism. The evacuation of a portion of the heated material down the void channel created at the center of the Bessel beam facilitates the quenching of relaxation and preserves the void structure, even under conditions where the void would otherwise be partially or completely reclosed by the surrounding molten material. Recent findings on double pulse illumination suggest that this technique may partially diminish the molten layer and enhance the temperature differential between the optical axis and the adjacent lobes [35]. These findings elucidate the development of high aspect ratio void channels inside various glasses. We anticipate they will influence future technologies for ultrafast laser material processing of glass.

AT A GLANCE

Regulating the development of high aspect ratio void channels within glass is crucial for applications such as high-speed glass dicing. This study examines void creation utilizing ultrafast Bessel beams in a single-shot illumination context. We delineate the morphology of the damage in relation to pulse energy, pulse length, and beam positioning within fused silica, Corning Eagle XG, and Corning Gorilla glass. The extensive parameter set permits void creation within fused silica, whereas the operational range is significantly more limited for Eagle XG and Gorilla glass. The temporary creation of a molten layer surrounding voids allows for the interpretation of morphological change in relation to pulse energy and duration.

REFERENCES

1. Sugioka, K.; Cheng, Y. Ultrafast lasers—Reliable tools for advanced materials processing. *Light Sci. Appl.* **2014**, 3, e149. **[Google Scholar] [CrossRef]**
2. Gattass, R.R.; Mazur, E. Femtosecond laser micromachining in transparent materials. *Nat. Photonics* **2008**, *2*, 219–225. **[Google Scholar] [CrossRef]**
3. Malinauskas, M.; Žukauskas, A.; Hasegawa, S.; Hayasaki, Y.; Mizeikis, V.; Buividas, R.; Juodkazis, S. Ultrafast laser processing of materials: From science to industry. *Light Sci. Appl.* **2016**, 5, e16133. **[Google Scholar] [CrossRef] [PubMed] [Green Version]**
4. Sugioka, K. Progress in ultrafast laser processing and future prospects. *Nanophotonics* **2017**, 6, 393–413. **[Google Scholar] [CrossRef] [Green Version]**
5. Watanabe, W.; Itoh, K. Motion of bubble in solid by femtosecond laser pulses. *Opt. Express* **2002**, *10*, 603. **[Google Scholar] [CrossRef]**
6. Bhuyan, M.K.; Courvoisier, F.; Lacourt, P.A.; Jacquot, M.; Salut, R.; Furfaro, L.; Dudley, J.M. High aspect ratio nanochannel machining using single shot femtosecond Bessel beams. *Appl. Phys. Lett.* **2010**, *97*, 081102. **[Google Scholar] [CrossRef]**
7. Courvoisier, F. Ultrafast Laser Micro-Nano Structuring of Transparent Materials with High Aspect Ratio. In *Handbook of Laser Micro- and Nano-Engineering*; Sugioka, K., Ed.; Springer International Publishing: Cham, Switzerland, 2020; pp. 1–37. **[Google Scholar] [CrossRef]**
8. Durnin, J.; Miceli, J.J.; Eberly, J.H. Diffraction-free beams. *Phys. Rev. Lett.* **1987**, *58*, 1499–1501. **[Google Scholar] [CrossRef]**
9. Duocastella, M.; Arnold, C. Bessel and annular beams for materials processing. *Laser Photon. Rev.* **2012**, 6, 607–621. **[Google Scholar] [CrossRef]**
10. Polesana, P.; Franco, M.; Couairon, A.; Faccio, D.; Trapani, P.D. Filamentation in Kerr media from pulsed Bessel beams. *Phys. Rev. A* **2008**, 77, 043814. **[Google Scholar] [CrossRef] [Green Version]**

11. Garzillo, V.; Jukna, V.; Couairon, A.; Grigutis, R.; Trapani, P.D.; Jedrkiewicz, O. Optimization of laser energy deposition for single-shot high aspect-ratio microstructuring of thick BK7 glass. *J. Appl. Phys.* **2016**, *120*, 013102. **[Google Scholar] [CrossRef]**

12. Lamperti, M.; Jukna, V.; Jedrkiewicz, O.; Trapani, P.D.; Stoian, R.; Itina, T.E.; Xie, C.; Courvoisier, F.; Couairon, A. Invited Article: Filamentary deposition of laser energy in glasses with Bessel beams. *APL Photonics* **2018**, *3*, 120805. **[Google Scholar] [CrossRef] [Green Version]**

13. Kumagai, M.; Uchiyama, N.; Ohmura, E.; Sugiura, R.; Atsumi, K.; Fukumitsu, K. Advanced dicing technology for semiconductor wafer—Stealth dicing. *IEEE Trans. Semicond* **2007**, *20*, 259–265. **[Google Scholar] [CrossRef]**

14. Hosseini, S.A.; Herman, P. Method of Material Processing by Laser Filamentation. U.S. Patent 20130126573 A1, 23 May 2013. **[Google Scholar]**

15. Butkus, S. Rapid Cutting and Drilling of Transparent Materials via Femtosecond Laser Filamentation. *J. Laser Micro Nanoeng.* **2014**, *9*, 213–220. **[Google Scholar] [CrossRef] [Green Version]**

16. Bhuyan, M.K.; Jedrkiewicz, O.; Sabonis, V.; Mikutis, M.; Recchia, S.; Aprea, A.; Bollani, M.; Trapani, P.D. High-speed laser-assisted cutting of strong transparent materials using picosecond Bessel beams. *Appl. Phys. A* **2015**, *120*, 443–446. **[Google Scholar] [CrossRef]**

17. Mishchik, K.; Beuton, R.; Caulier, O.D.; Skupin, S.; Chimier, B.; Duchateau, G.; Chassagne, B.; Kling, R.; Hönninger, C.; Mottay, E.; et al. Improved laser glass cutting by spatio-temporal control of energy deposition using bursts of femtosecond pulses. *Opt. Express* **2017**, *25*, 33271. **[Google Scholar] [CrossRef]**

18. Jenne, M.; Flamm, D.; Ouaj, T.; Hellstern, J.; Kleiner, J.; Grossmann, D.; Koschig, M.; Kaiser, M.; Kumkar, M.; Nolte, S. High-quality tailored-edge cleaving using aberration-corrected Bessel-like beams. *Opt. Lett.* **2018**, *43*, 3164. **[Google Scholar] [CrossRef]**

19. Dudutis, J.; Pipiras, J.; Stonys, R.; Daknys, E.; Kilikevičius, A.; Kasparaitis, A.; Račiukaitis, G.; Gečys, P. In-depth comparison of conventional glass cutting technologies with laser-based methods by volumetric scribing using Bessel beam and rear-side machining. *Opt. Express* **2020**, *28*, 32133. **[Google Scholar] [CrossRef] [PubMed]**

20. Li, Z.; Wang, X.; Wang, J.; Allegre, O.; Guo, W.; Gao, W.; Jia, N.; Li, L. Stealth dicing of sapphire sheets with low surface roughness, zero kerf width, debris/crack-free and zero taper using a femtosecond Bessel beam. *Opt. Laser Technol.* **2021**, *135*, 106713. **[Google Scholar] [CrossRef]**

21. Mitra, S.; Chanal, M.; Clady, R.; Mouskeftaras, A.; Grojo, D. Millijoule femtosecond micro-Bessel beams for ultra-high aspect ratio machining. *Appl. Opt.* **2015**, *54*, 7358. **[Google Scholar] [CrossRef] [Green Version]**

22. Meyer, R.; Froehly, L.; Giust, R.; Hoyo, J.D.; Furfaro, L.; Billet, C.; Courvoisier, F. Extremely high-aspect-ratio ultrafast Bessel beam generation and stealth dicing of multi-millimeter thick glass. *Appl. Phys. Lett.* **2019**, *114*, 201105. **[Google Scholar] [CrossRef] [Green Version]**

23. Feuer, A.; Thomas, J.U.; Freitag, C.; Weber, R.; Graf, T. Single-pass laser separation of 8 mm thick glass with a millijoule picosecond pulsed Gaussian–Bessel beam. *Appl. Phys. A* **2019**, *125*, 332. **[Google Scholar] [CrossRef]**

24. Meyer, R.; Jacquot, M.; Giust, R.; Safioui, J.; Rapp, L.; Furfaro, L.; Lacourt, P.A.; Dudley, J.M.; Courvoisier, F. Single-shot ultrafast laser processing of high-aspect-ratio nanochannels using elliptical Bessel beams. *Opt. Lett.* **2017**, *42*, 4307. **[Google Scholar] [CrossRef] [Green Version]**

25. Meyer, R.; Giust, R.; Jacquot, M.; Dudley, J.M.; Courvoisier, F. Submicron-quality cleaving of glass with elliptical ultrafast Bessel beams. *Appl. Phys. Lett.* **2017**, *111*, 231108. **[Google Scholar] [CrossRef]**

26. Jenne, M.; Flamm, D.; Chen, K.; Schäfer, M.; Kumkar, M.; Nolte, S. Facilitated glass separation by asymmetric Bessel-like beams. *Opt. Express* **2020**, *28*, 6552. **[Google Scholar] [CrossRef] [PubMed]**

27. Flamm, D.; Grossmann, D.G.; Sailer, M.; Kaiser, M.; Zimmermann, F.; Chen, K.; Jenne, M.; Kleiner, J.; Hellstern, J.; Tillkorn, C.; et al. Structured light for ultrafast laser micro- and nanoprocessing. *Opt. Eng.* **2021**, *60*, 025105. **[Google Scholar] [CrossRef]**

28. Boucher, P.; Hoyo, J.D.; Billet, C.; Pinel, O.; Labroille, G.; Courvoisier, F. Generation of high conical angle Bessel–Gauss beams with reflective axicons. *Appl. Opt.* **2018**, *57*, 6725. **[Google Scholar] [CrossRef] [PubMed]**

29. Velpula, P.K.; Bhuyan, M.K.; Courvoisier, F.; Zhang, H.; Colombier, J.P.; Stoian, R. Spatio-temporal dynamics in nondiffractive Bessel ultrafast laser nanoscale volume structuring. *Laser Photon. Rev.* **2016**, *10*, 230–244. **[Google Scholar] [CrossRef]**

30. Bhuyan, M.K.; Velpula, P.K.; Colombier, J.P.; Olivier, T.; Faure, N.; Stoian, R. Single-shot high aspect ratio bulk nanostructuring of fused silica using chirp-controlled ultrafast laser Bessel beams. *Appl. Phys. Lett.* **2014**, *104*, 021107. **[Google Scholar] [CrossRef]**

31. Kerse, C.; Kalaycıoğlu, H.; Elahi, P.; Çetin, B.; Kesim, D.K.; Önder, A.; Yavaş, S.; Aşık, M.D.; Öktem, B.; Hoogland, H.; et al. Ablation-cooled material removal with ultrafast bursts of pulses. *Nature* **2016**, *537*, 84–88. **[Google Scholar] [CrossRef]**

32. Juodkazis, S.; Nishimura, K.; Tanaka, S.; Misawa, H.; Gamaly, E.G.; Luther-Davies, B.; Hallo, L.; Nicolai, P.; Tikhonchuk, V.T. Laser-Induced Microexplosion Confined in the Bulk of a Sapphire Crystal: Evidence of Multimegabar Pressures. *Phys. Rev. Lett.* **2006**, *96*, 166101. **[Google Scholar] [CrossRef] [Green Version]**

33. Beuton, R.; Chimier, B.; Quinoman, P.; de Martínez, P.G.A.; Nuter, R.; Duchateau, G. Numerical studies of dielectric material modifications by a femtosecond Bessel–Gauss laser beam. *Appl. Phys. A* **2021**, *127*, 334. **[Google Scholar] [CrossRef]**

34. Bhuyan, M.K.; Somayaji, M.; Mermillod-Blondin, A.; Bourquard, F.; Colombier, J.P.; Stoian, R. Ultrafast laser nanostructuring in bulk silica, a "slow" microexplosion. *Optica* **2017**, *4*, 951. **[Google Scholar] [CrossRef]**

35. del Hoyo, J.; Meyer, R.; Furfaro, L.; Courvoisier, F. Nanoscale confinement of energy deposition in glass by double ultrafast Bessel pulses. *Nanophotonics* **2020**, *10*, 1089–1097. **[Google Scholar] [CrossRef]**

CHAPTER 14

FORECASTING OPTIMAL PROCESS PARAMETERS FOR DIRECT LASER INTERFERENCE PATTERNING UTILIZING CENTRAL COMPOSITE DESIGN

1. INTRODUCTION: AN OVERVIEW

The functionalization of technical surfaces through the creation of deterministic topographies currently serves as an innovation driver in contemporary materials engineering. Nature exemplifies superior surface engineering, with the design of textured surfaces frequently adhering to a biomimetic approach inspired by natural patterns [1,2]. Consequently, emulating natural designs aids in comprehending the function of surface microstructures and correlating topographies with macroscopic surface attributes. For example, these precisely defined and highly directed structures created at the micron and sub-micron size on the surfaces of contemporary industrial products present a distinct opportunity for substantial enhancement of product performance [3]. This update pertains to several applications, including antifouling [4], wetting management [5,6], tribology [7,8,9,10], enhancement of electrical conductivity [11], cell adhesion [12], and modification of surface visual appearance [13,14]. The replication of these multifaceted surface structures is a major technical problem today owing to their intricacy. In this context, non-contact manufacturing processes, such as laser-based microprocessing, emerged as a highly effective method for replicating natural surfaces, as they offer both the necessary technological and economic advantages while ensuring the production of high-resolution features [15]. Currently, the foremost laser patterning techniques are direct laser writing, laser-induced periodic surface structures, and direct laser interference patterning [15]. In direct laser writing, a focused laser beam is scanned across the material surface using pulse-to-pulse techniques, with resolution constrained by optical diffraction at the focal point. Laser-induced periodic surface structures yield recurring patterns by self-organization processes, achieving feature sizes that are below the diffraction limit and comparable to or significantly smaller than the wavelength. Conversely, direct laser interference patterning (DLIP) utilizes the principle of coherent light wave interference to create periodic structures on a surface by directly transferring the pattern shape to the material when adequate laser energy per unit area (fluence) is applied [17,18,19,20]. The interference patterns are generated by dividing a coherent laser beam into many beams and subsequently overlapping them on the surface of the samples (see Figure 1a,b). Consequently, as a result of constructive and destructive interference, a distinct intensity pattern is produced, and the dimensions of the periodic structures can be regulated by altering the angle between the interfering beams and the laser wavelength, in accordance with Equation (1):

$$\Lambda_2 = \frac{\lambda}{2\sin(\theta)} \quad (1)$$

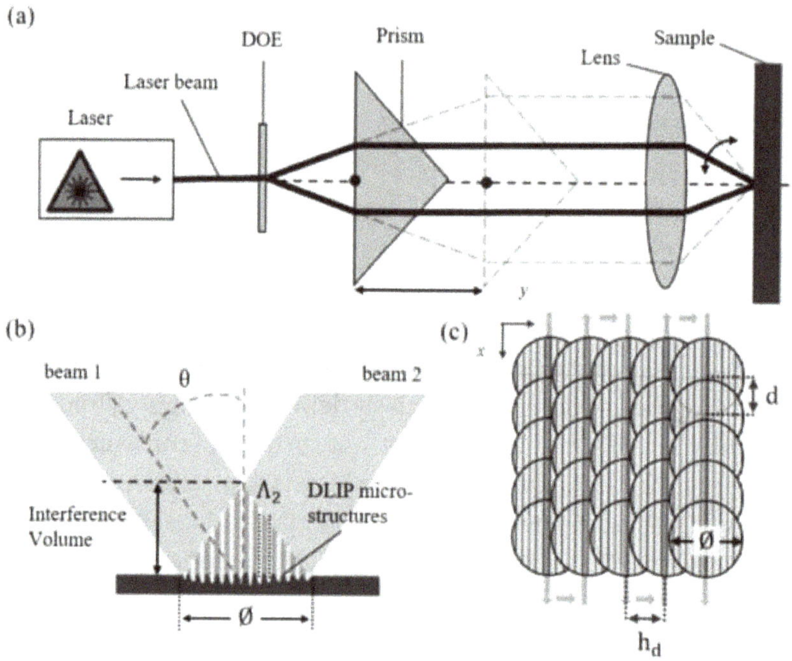

Figure 1. (a) Schematic representation of DLIP optical setup; (b) schematic view of the interaction zone (with a diameter Ø) of two beams intercepting at angle 2θ and forming an interference pattern with a particular period of Λ2Λ2; (c) distribution of pulses during the DLIP process: Ø, d, and h_d denote the laser beam diameter (at level $1/e^2$), pulse overlap, and hatch distance, respectively.

Despite the simple optical arrangement required for DLIP for fabricating periodic structures, it is also important to assure in several cases a uniform distribution of these structures to guarantee, overall, a specific surface functionality.

As reported in several studies, the quality of the final surface morphology strongly depends on laser-material interaction and, thus, on the used laser beam profile and the structuring strategy [21,22]. During the DLIP process, laser-material interaction occurs predominantly at the positions corresponding to the interference maxima, inducing various metallurgical processes, such as melting, ablation, and recrystallization [20]. During nanosecond-pulsed laser processing of metals, structuring mechanisms are mainly based on recoil vapor pressure as well as on Marangoni convection, which have an effect on the overall picture of melt flow [22,23,24]. In most of the commercial laser systems, the laser beam profile has a Gaussian distribution (TEM00TEM00 profile), which necessitates the use of advanced processing strategies, such as consecutively overlapped irradiation of interfering laser pulses in order to form homogeneous structures on large surface areas. For example, Aguilera et al. [22] used this technology and varied different processing parameters in order to produce homogeneous structures on a large surface area. Experiments were carried out in such a way that one process parameter (factor) was applied and varied through different values (levels), then the response was analyzed, whereas the other factors remain unaffected. This procedure is called one variable at a time [25]. For instance, in the mentioned study, discrete values of overlap distance, hatch distance, and fluence were chosen for specific special periods, and it was found that for fluence $F = 1.42\ J/cm^2$, 98.5% overlap and hatch distance of 20 μm, for 120 μm beam diameter spot, the structured surface was uniform and, thus, obtaining a homogeneous periodic pattern [22]. However, this approach is time-consuming since several variables have to be to screened independently, resulting in a large number of experiments. Furthermore, the number of experiments increases exponentially with the increase of the number of the factors and their levels.

Additionally, in one variable at a time method, it is impossible to single out the effects of factors interaction that can only be observed when varying multiple factors at the same time. Furthermore, by applying one variable at a time approach, it is possible to miss a process window with optimal settings of factors that will give the desired response. Therefore, these factors should not be examined independently. They should be taken into account simultaneously and must be investigated together since one factor might depend on the level of the other factor. In this case, a more effective approach is a statistical design of the experiment, which aims to decrease the number of experiments and to study the effect of interactions between different factors. For this purpose, many experimental design methods, namely, Plackett and Burman, factorial, Box–Behnken, and central composite design (CCD), have been developed [26,27,28]. Among these experimental methods, CCD developed by Box and Wilson is a very efficient experimental design method to reduce the number of experiments in the studies with a large number of factors and levels [28]. CCD has more advantages compared to other experimental design methods. For instance, it provides high-quality predictions in studying linear and quadratic interaction effects of factors influencing a system. Whereas interactions, unobserved in Plackett and Burman's experimental methods and Box–Behnken has less coverage than in the case of CCD [28,29]. Therefore, CCD has been widely used in the fields of engineering and science [29,30,31,32,33,34]. The CCD consists of three main parts and of $2^k + 2k + m$ runs.

The factorial part of CCD is a two-level design with 2^k factorial points at the corners of a cube denoting its design in space as shown in Figure 2. For a cube design, the number 2 in the last expression results from the amount of levels, and k is set to 3, representing the number of factors. The other part of CCD is fixed at the center of the design space and consists of m center points (see Figure 2), which represent the middle levels of all the factors investigated. The replication of these points allows estimation of experimental error, detection of curvature in the fitted data, and checking the adequacy of the model. Consequently, the replication of the entire experimental design is not required [28].

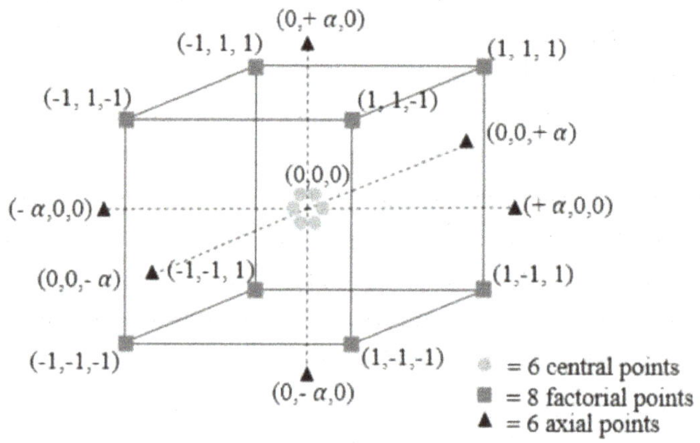

Figure 2. Central composite design with three factors and five levels each.

The last part comprising CCD is to define the axial points. There are 2k axial points in a CCD, and they establish new extreme levels (the lowest and the highest level) for each factor. The distance between the axial and center points is denoted by ± α± value, where α=$2k/4$≅1,68179 for k = 3. This value gives rotatability to the design, which ensures that the variance of the model prediction is constant at all points equidistant from the design center [35]. It makes the CCD method able to explore the wide process space and to capture a strong curvature for studying the effects of the interactions between the design factors on the model [36,37,38].

In this contribution, we present the optimization and fabrication of homogeneous periodic surface microstructures on bearing steel (100Cr6) using the design of experiments approach and employing a two-beam-DLIP setup with an infrared (IR) nanosecond laser. Since it has been shown that the improvement of the structure homogeneity is more dependent on the strategy used during the experiments than on the pulse duration of the laser source, the aim is to optimize the DLIP process parameters such as laser peak-fluence, pulse overlap, and hatch distance (see **Figure 1c**) with respect to the structures' height and surface texture homogeneity by performing the CCD method. The surface topography is characterized using confocal microscopy and scanning electron microcopy analysis. Furthermore, a 3D-characterization method for measuring the pattern homogeneity was applied based on morphological filtering [**39,40,41,42**], which allows a holistic view of the surface properties, and a new qualification approach of DLIP surface structures was introduced. The method presented here is of significant relevance to assure, in the future, a certain performance over the whole treated area as well as to permit in relevant industrial processes to quantitatively describe the produced topography in terms of homogeneity. It is required, for instance, for quality management.

2. MATERIALS AND METHODS

2.1. Materials

Laser texturing studies were conducted on a hardened bearing steel surface (100Cr6), which is frequently utilized in the production of automobile components.The test samples were sectioned into circular substrates having a diameter of 40 mm and a thickness of 10 m. Each test surface was ground, resulting in surface roughness S_a of 0.74 μm (DIN-ISO 25178). Before the laser treatment, all samples were cleaned using isopropanol.

2.2. Nanosecond Two-Beam-DLIP Setup and Process Strategy

The laser experiments were carried out using a two-beam DLIP-μFab system (developed by Fraunhofer IWS, Dresden, Germany), which produces the interference spot that contains a line-like intensity pattern. The system is equipped with a pulsed Q-switched Nd:YLF laser (Laser-export Tech- 1053 Basic, Moscow, Russia), operating at 1053 nm wavelength and providing 12 ns pulses with pulse energy up to 290 μJ at 1 kHz. The laser emits the fundamental transverse mode (TEM00TEM00) with a laser beam quality factor of $M^2 < 1.2$. This system also includes a compact DLIP optical head, where the main beam is split into two beams using a diffractive optical element (DOE), then the beams are parallelized by a prism and finally overlapped using a lens with a focal distance of 40 mm (see Figure 1a). Such an optical configuration provides the fully-automatic control of the spatial period of the interference profile between 1.45 μm and 8.50 μm by varying the incidence angle of the beams (θ) on the sample, which, in turn, is realized by moving the prism with respect to the DOE [43].

In this study, different pulse peak-fluences FpF (ranging from 4.97 J/cm² up to 7.07 J/cm²) were used to determine the energy range needed for uniform structuring of bearing steel surface. The peak-fluence on the interference spot was calculated according to Equation (2):

$$F_p = \frac{2n \cdot P}{f \cdot \pi \cdot w^2}, \quad (2)$$

where n is a number of interfering beams, P and f is power and repetition rate of the laser, respectively, and w denotes a beam radius (at level $1/e^2$). In this work, the beam radius w was determined using the D-squared method and was equal to 52.5 μm [44,45].

To structure larger areas, the sample was moved by XY-stages (PRO Series, Aerotech Ltd., Tadley, UK) with linear speeds ranging from 0.22 cm/s to 1.60 cm/s. The arrangement of the pulse position in the structuring process is illustrated schematically in Figure 1c. This process occurs consecutively such that first overlap in the x-direction (feed direction) occurs then substrate is moved in the y-direction by hatch distance, making an overlap also in y-direction. Notable that x-direction was chosen as feed direction since it is parallel to the orientation of the line-like pattern, and thus guarantees a well-defined periodic structure. The pulse-to-pulse overlap in x-direction is denoted as P_O (%) and represented as a function of pulse-to-pulse separation distance d ($d = v_{scan} \cdot f$, where v_{scan} is the scanning speed of x-axis and f is the used laser pulse repetition rate) and the laser beam diameter Ø using Equation (3):

$$P_O\,[\%] = \left(1 - \frac{d}{Ø}\right) \cdot 100\% \quad (3)$$

In the experiments, the pulse-to-pulse overlap was varied between 82.56% and 98.52%. On the other hand, the hatch distance h_d (distance between the vertical lines) was controlled by shifting the sample with the y-axis in such a way that h_d was always kept multiple of the spatial period. In addition, hatch distance h_d was varied from 25.5 μm up to 68 μm. The laser pulses overlap in y-direction is also introduced, and it can be represented as a function of hatch distance h_d and the laser beam diameter Ø (at level $1/e^2$) using Equation (4):

$$H_D\,[\%] = \left(1 - \frac{h_d}{Ø}\right) \cdot 100\% \quad (4)$$

2.3. Central Composite Design Method

In the present study, the experimental plan with the variation of parameters and statistical analyses of the experimental data was carried out using MINITAB 18 statistical package. The experiments were designed based on the CCD method. In Figure 2, each axis corresponds to a factor while each point on the cube represents certain levels. The three selected factors are fluence (X_1), pulse overlap (X_2), and hatch distance overlap (X_3). It is recommended that six center points are taken in a CCD with three factors [35].

As was stated before, the center point is replicated to find the experimental error, and so the replication of the entire experimental design is usually not required. However, during pretests model variability (i.e., how well the regression model fits the experimental data) determined by the R^2 factor was <%80, which is less than the acceptable value according to [38]. Therefore, it was decided to make a replication of the entire experimental design in order to increase the statistical significance of the measurements and improve the models' regression fit. As a result, the factorial, center and axial points in a CCD method build up an experimental design with five levels for each factor and three replicates, making 60 runs in total ($3(2^k + 2k + m) = 3(8 + 6 + 6) = 60$). The designed 60 experiments of the DLIP process were conducted in random order to exclude any bias in the response variables and to avoid a systematic error associated with the specific factor combinations as it was suggested in [28,37].

The experimental plan with the coded and uncoded levels of design factors is presented in **Table 1**. The low, middle, and high levels of each factor are coded as -1, 0, and +1, respectively, while the lowest and the highest levels are coded as -1.5 and +1.5.

Table 1. Factors and their adopted (uncoded) values at different coded levels.

Factors (*Y*)	Symbol	Coded: -1.5	-1	0	+1	+1.5
		Uncoded Values of Coded Levels:				
Peak-Fluence: F (J/cm²)	X_1	4.97	5.32	6.02	6.72	7.07
Pulse Overlap: P_o (%)	X_2	82.56	85.22	90.54	95.86	98.52
Hatch Distance: H_D (%)	X_3	40	46	58	70	76

The mathematical relationship of *Y* response on the corresponding factors is expressed by the following second-order polynomial equation [36,37]:

$$Y = \beta_0 + \sum_{i=1}^{k} \beta_i X_i + \sum_{i=1}^{k} \beta_{ii} X_i^2 + \sum_{i=1}^{k}\sum_{j>i}^{k} \beta_{ij} X_i X_j + \varepsilon \, ; \, i = 1, 2, \ldots, k; \, j = 2, \ldots, k; \, i \neq j, \tag{5}$$

where *Y* is the observed response value (structures height, structure-height-error, and waviness percentage); X_i and X_j are the coded values of factors, β_0 is the constant, β_i, β_j, and β_{ij} are the linear, quadratic, and interaction coefficients respectively, k is the number of the factors, and ε is the error term. MINITAB was also used to generate the Pareto charts, response surface, contour plots of factors, as well as the optimization plots. Excluding the control factors, each test was carried out under the same experimental conditions in the ambient environment without post-treatment.

2.4. Surface Characterization

The 3D surface topography of structured samples was characterized using confocal and coherence scanning interferometry microscopy (Sensofar, S Neox non-contact 3D Surface Profiler, Barcelona, Spain) employing a 50× magnification objective, with a lateral and vertical resolution of 340 nm and 4 nm, respectively. Using this objective, a total area of 351 μm × 264 μm could be recorded in each measurement. Afterward, using the software MountainsMap® 7.4 (Digital Surf, Besançon, France), the surface profiles of the recorded topographies are extracted applying morphological filters (ISO16610-14), and the topographical 3D roughness parameters are calculated by the recognized measurement (ISO 25178-2). In addition, topographical measurements have been carried out using a scanning electron microscopy at an operating voltage of 15 kV (JEOL, JSM 6610LV, Tokyo, Japan).

To describe the homogeneity of the fabricated structures, two topographical parameters were used, namely, structure height error and waviness. Waviness shows how the structure height is changing at distances larger than the structure period. This kind of structure inhomogeneity is caused by too large a hatch distance between the laser pulses [22]. However, when the structure hatch distance is small enough, the waviness of the structure becomes close to zero. At this point, the homogeneity of the structure is described better by the structure height error parameter.

2.5. Development of a Topographical Analysis Method

To perform a homogeneity analysis of the structured surface, an analysis algorithm, based on the extraction of surface envelopes from the measured surface (S) using morphological filters (MF), was developed. A morphological filter is based on two morphological operations, dilation, and erosion, that work using a structuring element (SE) of a given size [41,42]. In this case, the SE was set to the size of the structure period Λ. By applying the dilation (dMF) and erosion (eMF) morphological filtering the upper (U) and lower (L) structure envelopes were achieved, respectively (profile inset Figure 3a). The U and L envelopes represent the distribution of heights and depths of structure hills and valleys, respectively. Accordingly, $H=|U-L|$ represent the fabricated structure height distribution. Then, the average structure height (h) and its error (h_{err}) can be calculated by finding an average and root mean square of H, respectively. Equations (6) and (7) represents the mathematical procedures described in this paragraph:

$$U = eMF_{SE=\Lambda}(S);\ L = dMF_{SE=\Lambda}(S);\ H = |U - L|; \quad (6)$$

$$h_{err} = \frac{RMS(H)}{H} \cdot 100\%. \quad (7)$$

In order to separate structure height error from the waviness calculations, special filtering is applied to the measured surface S. Spatial filtering is performed by calculating the fast Fourier transform of the S, then applying spatial filter and finally performing the inverse Fourier transform on the filtered Fourier space data:

$$(\text{i}) : FS = iFFT(S\_Filter(FFT(S))) \quad (8)$$

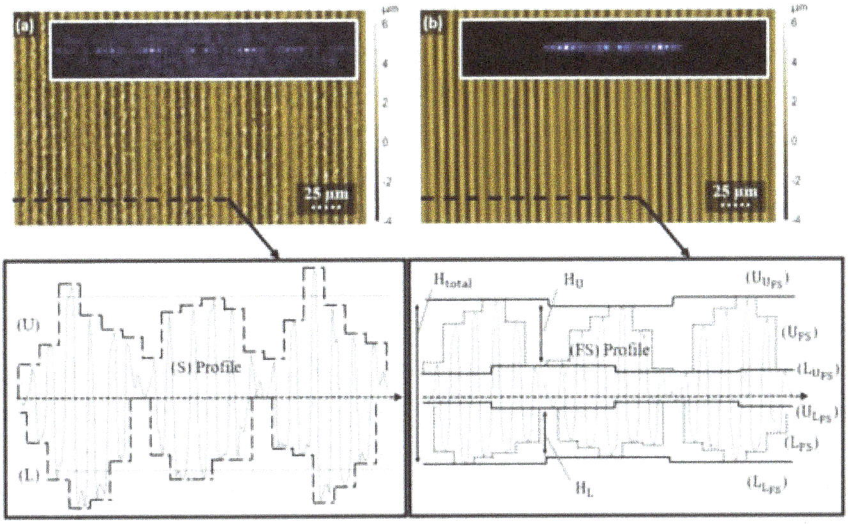

Figure 3. Topography images of (a) measured DLIP surface and (b) filtered surface with corresponding roughness profiles. The insets show Fourier transform of each of the surfaces.

As it was mentioned in Section 2.4, the waviness is proportional to the hatch distance between the laser pulses. Therefore, the waviness is measured by using similar technique described in previous paragraph, but with structuring element of the MF filter equal to the hatch distance that was used to form the structure. Altogether, the calculation of the waviness consist of few steps. First, the upper (*UFS*) and lower (*LFS*) envelops of the filtered surface (FS) are calculated:

$$(\mathrm{ii}) : U_{FS} = eMF_{SE=\Lambda}(FS); \ (\mathrm{iii}) : L_{FS} = dMF_{SE=\Lambda}(FS). \quad (9)$$

Then the upper and lower envelopes of *UFS* and *UFS* are calculated with *SE=h=h*:

$$\begin{aligned} &(\mathrm{iv}) : U_{U_{FS}} = eMF_{SE=h_d}(U_{FS}); \ (v) : U_{U_{FS}} = dMF_{SE=h_d}(U_{FS}); \\ &(\mathrm{vi}) : U_{L_{FS}} = eMF_{SE=h_d}(L_{FS}); \ (\mathrm{vii}) : \ L_{L_{FS}} = dMF_{SE=h_d}(L_{FS}). \end{aligned} \quad (10)$$

Finally, the waviness of the function is expressed by Equation (11):

$$H_U = \left(U_{U_{FS}} - U_{U_{FS}}\right); \ H_L = \left(U_{L_{FS}} - L_{L_{FS}}\right); \ H_{Wav} = H_U + H_L; \ H_{total} = U_{U_{FS}} - L_{L_{FS}}; \\ W[\%] = \frac{H_{wav}}{H_{total}} 100 \quad (11)$$

An example of this procedure is shown in Figure 3, in which explanation is done on 2D profiles instead of 3D for simplicity. It shows the topography image of a DLIP structured surface measured by confocal microscope in Figure 3a,b represents its filtered Fourier space. The inserts in both images show the fast Fourier transforms of the corresponding topographies and visualizes how the smallest scale elements from surface (a) that form the noise (unreal measured peaks, solidified debris, dirt particles, etc.) are filtered by using the special filtering method (ISO25178), leaving only the surface waviness and the smaller line-like pattern. Additionally, the extracted profile pictures of selected areas are shown for both cases, showing the different topography feature indicators described in the previous section.

3. RESULTS AND DISCUSSIONS

3.1. Statistical Analysis of Results

The surface of the steel sample was patterned using the two-beam DLIP configuration. The μsec processing strategy was already described in Section 2.2. Areas with a size of 3 × 3 mm were structured with different processing parameters. The varied process parameters were (i) hatch distance, (ii) pulse overlap, and (iii) peak-laser-fluence (see Table 1).

The scanning electron micrographs of three exemplary patterned surfaces are shown in Figure 4b,d. The surfaces were processes with the following process parameters: $F = 6.02\ J/cm^2$, P_O = %90.54, H_D = %58 (Figure 4b), F = 6.02 J/cm2, P_O = %98.52, H_D = %58 (Figure 4c) and $F = 6.72\ J/cm^2$, P_O = %98.52, H_D = %70 (Figure 4d). The reference topography with initial surface roughness is presented in Figure 4a. Due to the high roughness of the initial samples' surface, the grinding stripes are still be visible after the laser process (see features perpendicular to the line-like structure in Figure 4b). In this case, low laser fluence and pulse overlap were used, which leads to the formation of shallow structures with height in the range of surface roughness. Differently, at higher fluence values and/or increased pulse overlap, the initial surface roughness is flattened as it can be seen from Figure 4c,d. Furthermore, the high magnification images depicted in the insets of Figure 4c,d clearly show that during nanosecond-pulsed laser treatment besides ablation phenomena, redeposition of the molten material driven by Marangoni convection and recoil vapor pressure [22,24] takes part in the structuring mechanism, creating structures with higher aspect ratios (height to spatial period ratio) [46].

In addition, to the periodic microstructures produced by the interference pattern, also waviness of the structured topography is very pronounced for the used hatch distances 42.5 μm, which corresponds to H_D = %58 in Figure 4b,c. This structure topography can be explained since the interference patterns are obtained using two overlapped beams with a Gaussian intensity distribution (TEM00TEM00), which results in an interference pattern also with a Gaussian distribution [47,48]. Therefore, due to the Gaussian beam profile, the fluence in the center of the spot is higher, which lead to high cumulative energy and, thus, strong ablation and a significant quantity of molten material occur. Furthermore, since the sample is moved in the x direction (parallel to the orientation of the interference lines) and the pulses are overlapped, which leads to deeper structures at the central area as well as shallow structures at the periphery, producing the larger repetitive structure. Moreover, the repetitive length of waviness modulation is equivalent to the hatch distance used as reported in [22].

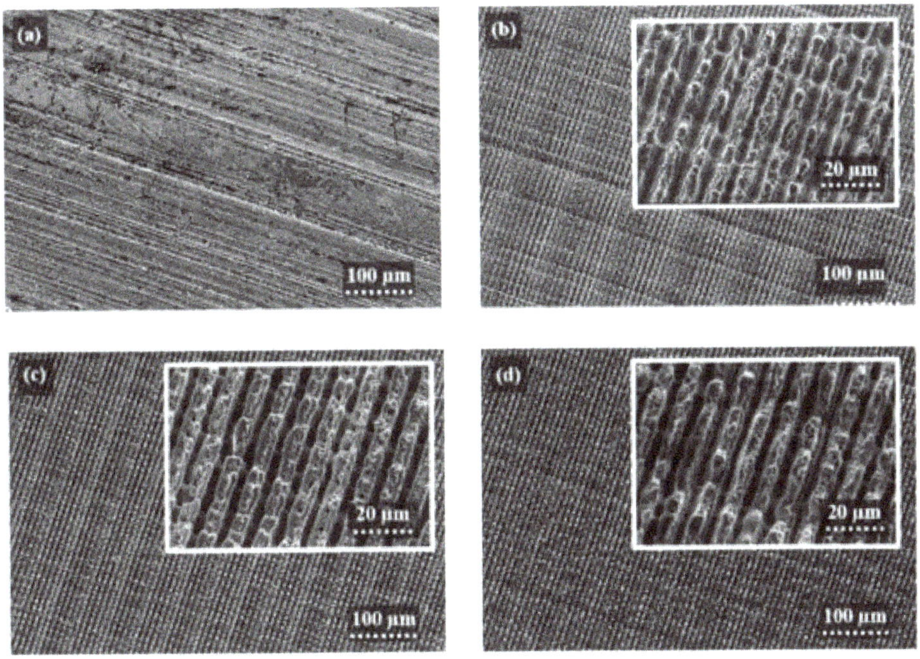

Figure 4. Scanning electron micrographs of initial surface topography (a), and the nanosecond DLIP structures produced on 100Cr6 steel using spatial period Λ = 8.50 μm, and the corresponding processing parameters: F = 6.02 J/cm², P_0 = %90.54, H_D = %58 (b), F = 6.02 J/cm², P_0 = %98.52, H_D = %58 (c), and F = 6.72 J/cm², P_0 = %98.52, H_D = %70 (d). The insets show the corresponding magnification of the topographies.

For determining the general surface texture homogeneity, the experimental results obtained by using the topographical analysis method described in **Section 2.5** were analyzed statistically. Analysis of variance (ANOVA) was performed to identify the significance of the factors and their interactions as well as for estimating the adequacy of the model. Pareto charts of the ANOVA standardized effect estimates are presented in **Figure 5**. A Pareto chart is very useful for reviewing a large number of factors and for presenting the magnitude and the importance of the effects. In the plotted diagrams, a vertical reference line indicates the minimum magnitude of statistically significant effects, which depends on the significance level denoted by α and set in our model to 0.05 (5% of risk) by convention described in [**49**]. In addition, the bars that cross the reference line are statistically significant.

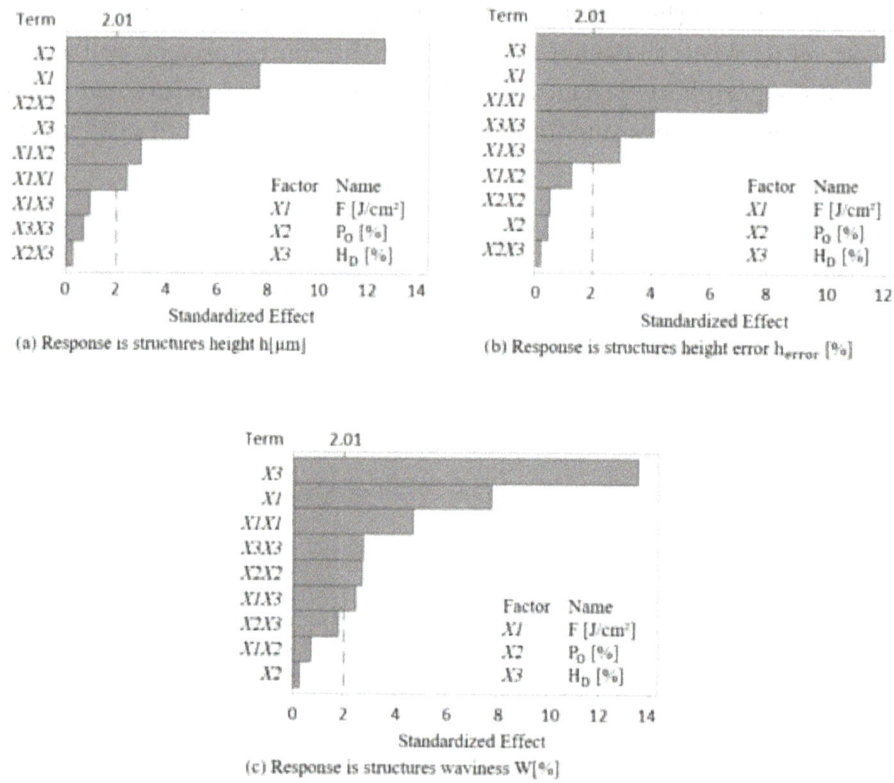

Figure 5. Pareto charts of the standardized effect of (a) structures height, (b) structures height error, and (c) surface waviness.

For instance, in Figure 5a where different standardized effects in the structure height response are shown, it is worth to point out that pulse overlap (X_2) as well as laser-fluence (X_1) have the highest effect on the formation of structure height. Other significant effects according to the developed model are square interaction of each of pulse overlap ($X_{2×2}$) and laser-fluence (X_1X_1) as well as their two-way interaction (X_1X_2). Other relevant single effect which is worth to mention is the hatch distance (X_3), whereas the square interaction of hatch distance (X_3X_3) as well as the two-way interactions between hatch distance with each of fluence (X_1X_3) and pulse overlap (X_2X_3) are not significant at the 0.05 level with the current model terms.

In the same manner, the Pareto chart of Figure 5b, shows the standardized effects for structure height error. In this case, almost all factors have significant effects at the 0.05 level on the model, except those of single, square and two-way interactions of pulse overlap (X_2, X_2X_2, X_2X_3, and X_1X_2). Finally, in the Pareto chart shows in Figure 5c, it is possible to see that for the standardized effects for waviness, almost all factors have significant effects at the 0.05 level, except the effect of single interaction of pulse overlap (X_2) as well as the effects of each of the two-way interactions between pulse overlap with each of fluence (X_1X_2) and hatch distance (X_2X_3).

Moreover, from the Pareto charts, it is possible to determine the most influencing factors among all the relevant candidates. This is visible when calculating the contribution in percentages of each factor for the different developed models, which is shown in Table 2. The results show that pulse overlap (X_2), followed by fluence (X_1), have the highest contribution to the structure height, whereas in the case of height error and waviness, the hatch distance (X_3) followed by fluence (X_1) have the highest contributions.

Table 2. Contribution of significant main factors (in %), their interactions and quadratic effects of factors from the model for each of the responses, with X1 = Fluence, X2 = Pulse overlap and X3 = Hatch Distance.

The regression equations obtained by removing insignificant terms from the model for each of the responses are given as follows:

$$\begin{gathered}\text{Model \#1}: \ S = 0.883 \ and \ R^2 = 84.82\%; \\ h \ (m) = 275.50 + 2.26X_1 + 6.41X_2 + 0.06X_3 - 0.79X_1X_1 \\ -0.03X_2X_2 + 0.15X_1X_2\end{gathered} \tag{12}$$

$$\begin{gathered}\text{Model \#2}: \ S = 7.740 \ and \ R^2 = 87.54\%; \\ h_{error} \ (\%) = 1368 - 328.60X_1 + 0.11X_2 - 9.22X_3 + 22.90X_1X_1 + 0.04X_3X_3 + 0.55X_1X_3\end{gathered} \tag{13}$$

$$\begin{gathered}\text{Model \#3}: \ S = 8.954 \ and \ R^2 = 84.22\% \\ W \ (\%) = 1943 - 167.40X_1 - 27.30X_2 - 1.92X_3 + 15.13X_1X_1 \\ +0.15X_2X_2 + 0.03X_3X_3 - 0.53X_1X_3\end{gathered} \tag{14}$$

The positive and the negative sign in Equations (12)–(14) represent the synergistic and antagonistic effects of the factor on the response for each of the models, respectively. Furthermore, the determination coefficient (R^2) of model #1 is 84.82%, showing a acceptable agreement with the experimental data. Moreover, the standard error of the regression (S) of model #1 is 0.883. This means that this statistical model can explain 84.82% variability in the response and that the average distance of the structure height data points from the fitted line is about 0.883 µm. It is worth mentioning that S is represented in the units of the dependent variable. Likewise, in model #2 the determination coefficient (R^2) and the standard error of the regression (S) are equal to 87.54% (>80%) and 7.74%, respectively. Finally, for model #3 the corresponding determination coefficient and the standard error of the regression are as following R^2 = %80<) %84.22) and S = %8.954. It can be concluded from both values that each of the models is statistically significant.

In order to compare the results from the models with experimental data, the correlation graphs were analyzed showing a good correlation, as shown in Figure 6.

In the plots, the upper and lower limits of the confidence and prediction intervals are displayed as dashed lines. The confidence interval displays the range of 95% of possible values for the mean response, whereas the prediction interval displays the range of 95% of possible values for a single new observation. The R^2 value in each of the correlation graphs, shows that the developed model can explain >%80 variability in the response.

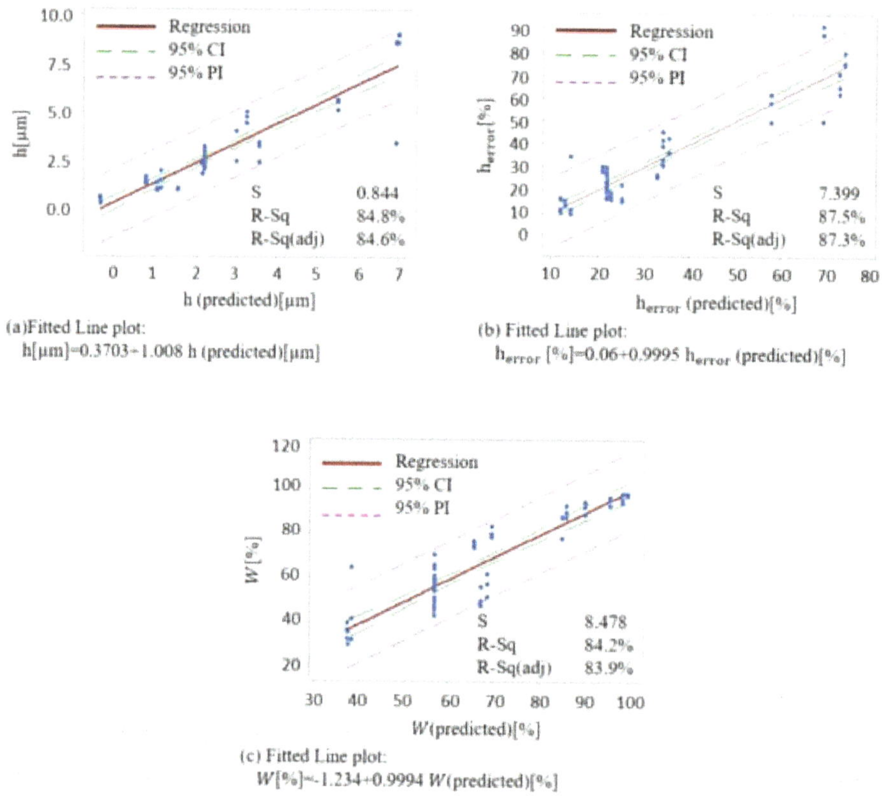

Figure 6. Correlation fitted line plots between experimental and predicted values for each of (a) structures height, (b) structures height error, and (c) surface waviness.

After checking the statistical significance of the model, important interactions of the factors that mostly affect the response were analyzed by using a graphical tool called interaction plots. These interaction plots are shown in Figure 7 and they are plotted from fitted values of predicted responses from the region of interest that consist of central and factorial points, excluding axial ones due to higher magnitude of prediction error. Figure 7a shows that the change in the mean response (height), from a low to a high level of laser fluence factor, depend on the level of the other factor which is pulse overlap. Additionally, the effect of laser fluence on the mean height of the structure is stronger for higher values of pulse overlap (Po), which is visible by the increasing slope of the curves with increasing Po level. The maximum height of the structures is reached when both fluence and pulse overlap are kept at a relatively high-level of 7.07 J/cm^2 and %95.86, respectively. Moreover, Figure 7b,c shows the presence of interaction between the fluence and the hatch distance, which significantly affect the structure height error as well as surface waviness. It is worth mentioning that an increase of both the fluence and the hatch distance tends to decrease the structures height error as well as the surface waviness and thus to improve homogeneity. However, the increase of fluence is effective only until the negative vertex displayed in Figure 7b,c. After this critical point, the curve changes its slope, which means that a further increase of the laser fluence may have an opposite effect in the structure height and will decrease homogeneity. Indeed, a very high value of fluence can negatively affect the quality and the height of the fabricated structures, as the accumulated energy may increase, and the possibility of uncontrolled overmelt occurs, which was already shown in previous investigations [50].

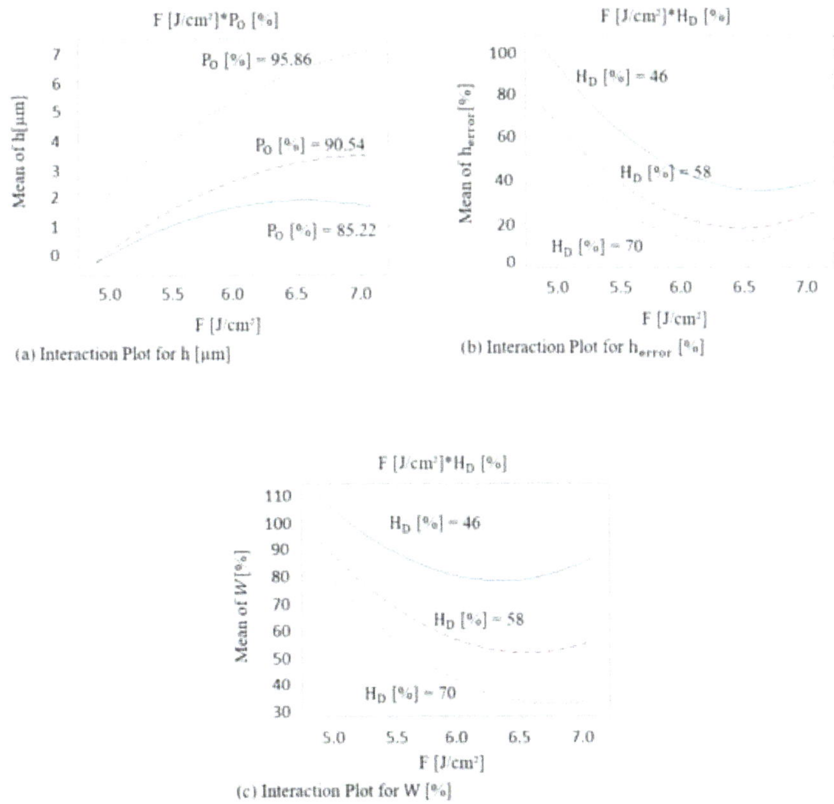

Figure 7. Interaction plots of F (J/cm^2)·POPO (%) for the (a) structures height, and of F (J/cm^2)·HDHD (%) for (b) structure height error, and (c) surface waviness.

Finally, all the effects of the individual main effect of each factor, their quadratic terms, as well as their interactions can be summarized with the help of 2D contour plots, which are generally the graphical representation of the regression equation. Therefore, the accuracy of the contour plot depends on how well the model represents the true relationships between the variables.

Each response surface presents the effect of laser fluence and hatch distance on structure height, height error, and surface waviness, while pulse overlap is held at a fixed specific level of 85.22%, 90.54%, and 95.86%. Each contour plot in Figure 8 has the dominant characteristic of the non-linear surface (fan-shaped, in this case). Such non-linearity implies a strong $X_1 \times X_3$ (fluence with hatch distance) interaction effect, whereas it can be concluded from Figure 8b,c that the pulse overlap does not have a significant effect on the interaction between laser fluence and hatch distance, which indicates that interactions $X_2 \times X_3$ and $X_1 \times X_2$ are negligible. Moreover, the response surface plots exhibit a saddle shape, which means that any increase or decrease of fluence from the saddle peak results in a decrease of each of the measured response factors. For instance, in the case of Po = %85.22, H_D = %60, and for fluences from 6.5 J/cm^2 and higher, the structure height saturates and even slightly reduces, because of material overmelting, as previously reported in [50].

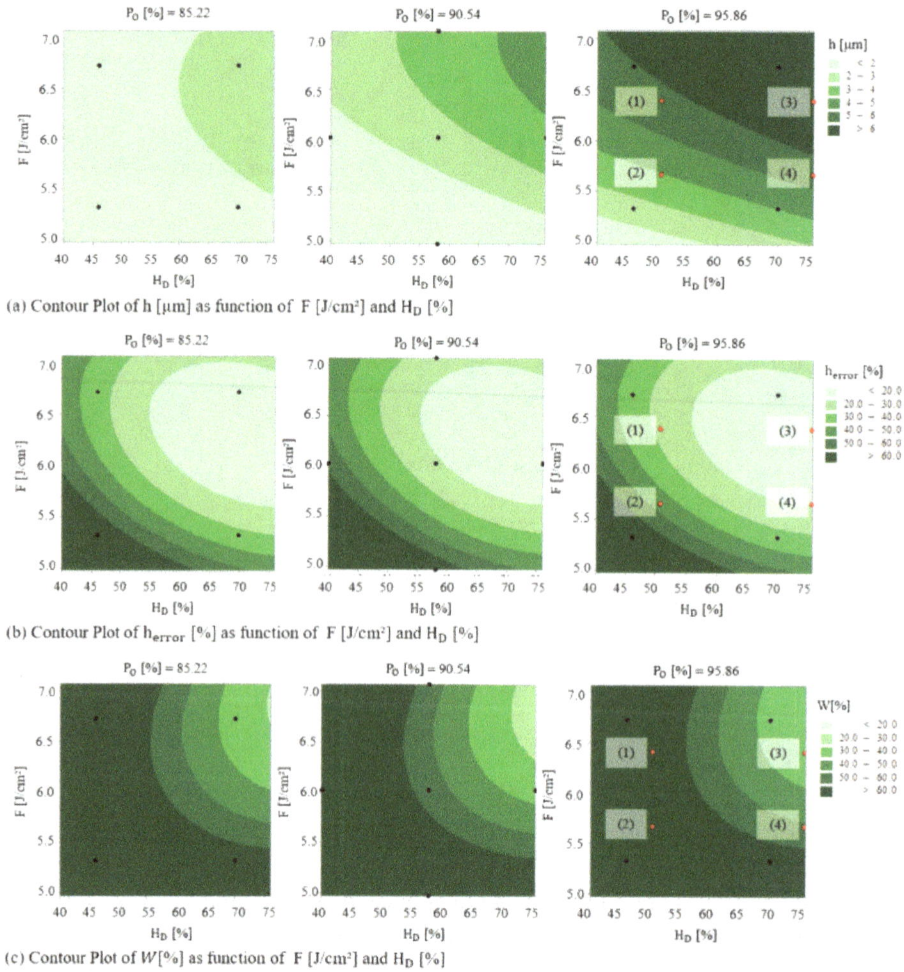

Figure 8. Contour plots of (a) height, (b) height error, and (c) surface waviness.

Similarly, for Po = 90.54%, H_D = %55, and for fluences from 6.5 J/cm^2 and higher, the structure height error reaches its minimum and then again increases. The same behavior is observed for surface waviness by increasing laser fluence more than 6.5 J/cm^2 at Po = %90.54 and H_D = %72. It is attributed to the negative effect of the quadratic term ($X_1 \times X_1$) of the fluence. On the other hand, the quadratic term ($X_3 \times X_3$) of the hatch distance has a positive effect on the response. The best values of response for each of the structure height (>6 μm), height error (<20%), and surface waviness (~24%) are in the upper right corner of the plot, which corresponds with high values of both laser fluence (X_1) and hatch distance (X_3). The lowest values of the structures height (<2 μm), height error (>60%), and surface waviness (>60%) are in the lower-left corner of the plot, which corresponds to low values of both X_1 and X_3.

The tendency that the increase of levels of varied factors of DLIP process improves the response in the form of patterned surface homogeneity can be explained by redistribution of cumulative laser intensity that controls the quantity of molten and ablated material during the movement of the substrate in the x and y directions. Nevertheless, the factors involved in the DLIP process have an optimal level after which a further increase will lead to worse response values, which means that the homogeneity will be damaged and the height of the structure will collapse. This is in agreement with a generally accepted theory for near-surface melt dynamics during laser processes, where the melt flow (based on Marangoni convection) is considered as the main driving force in the microstructure formation besides recoil and plasma pressure [22,24]. In this case, the excess of deposited intensity on the processed surface that leads to uncontrolled melt of the material that is further explained in [50].

Since both error contour plots (Figure 8b,c) showed that the low error zones can be found for high laser-fluence and hatch distance (upper right corner), a correlation graph between the structure height error and surface waviness was realized, permitting to estimate the strength of this relationship. The correlation graph presented in Figure 9 shows that a high correlation exists between the structure height error and surface waviness. The fitted equation for the quadratic model that describes the relationship between h_{error} (%) and W (%) is:

$$W(\%) = 37.16 - 1.185\, h_{error}(\%) + 0.01539\, h_{error}(\%)^2 \quad (15)$$

The R^2 in each of the correlation graphs shows that the developed model can explain >%80 variability in the response. However, this statistically significant relationship does not imply that the height error (h_{error}) causes surface waviness (W (%)). Nevertheless, since the model fits the data well, this equation can be used to predict h_{error} (%) for a value of W (%), or find the settings for W (%) that correspond to a desired value or range of values for h_{error} (%).

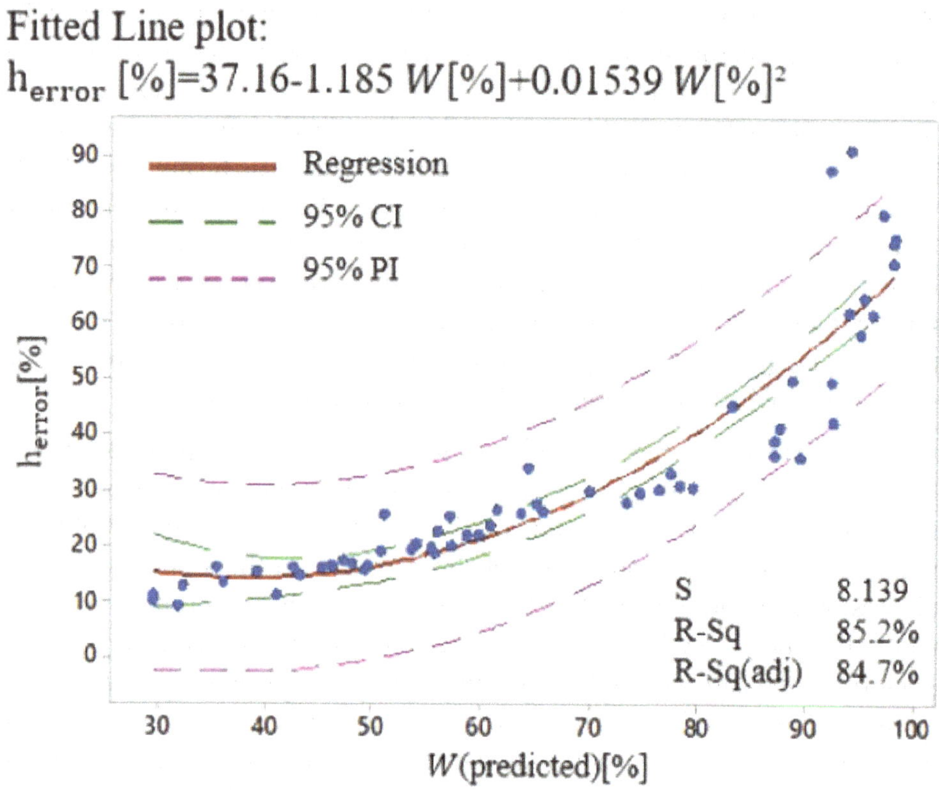

Figure 9. Correlation fitted line plots between height error (h_{error} (%)) and surface waviness (W (%)).

3.2. Model Validation

To confirm the validity and accuracy of the developed model, additional experiments were done in triplicates according to manually chosen parameters (marked by red dots in **Figure 8**). The DLIP structuring runs were conducted in the same conditions as in the previous described experiment. The corresponding topographies with extracted profiles are presented in **Figure 10**.

After the analysis, the relative error between the calculated surface quality parameters and the experimental values was calculated for the structures height, height error, and surface waviness (Figure 11). The results show that the predicted value of the structure height was very close to the experimental results with a relative error varying between 8.5% and 11% for different parameter sets. Additionally, the relative error for surface structures error varied between 10% and 20%. In consequence, the results indicate that the prediction model achieved in the present study is reliable.

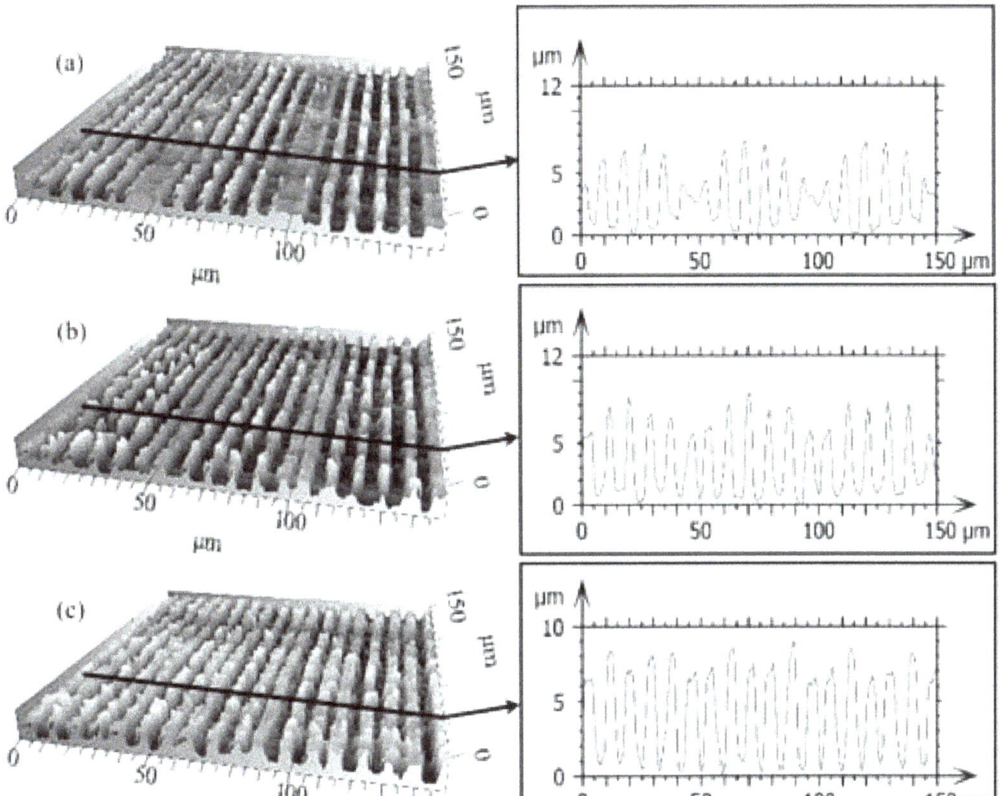

Figure 10. Confocal microscopy pictures of the nanosecond DLIP structures produced on 100Cr6 steel using spatial period Λ = 8.50µm, and the corresponding processing parameters: F = 5.69 J/cm², P_0 = %95.86, H_D = %52 (a), F = 6.4 J/cm², P_0 = %95.86, H_D = %52 (b), F = 5.69 J/cm², P_0 = %95.86, H_D = %76 (c), and F = 6.4 J/cm², P_0 = %95.86, H_D = %76 (d). The insets show the cross-section profiles of the corresponding topographies.

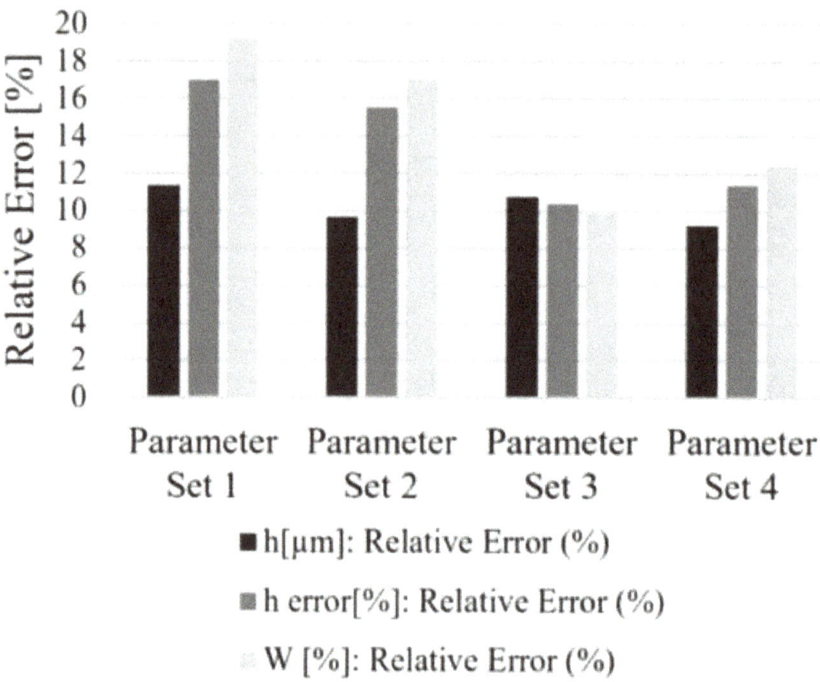

Figure 11. Bar charts summarizing relative error between model-predicted and verification experiments with parameter sets: (1) F = 5.69 J/cm², P_O = %95.86, H_D = ,%52 2)) F = 6.4 J/cm², P_O = %95.86, H_D = 3) ,%52) F = 5.69 J/cm², P_O = %95.86, H_D = %76, and (4) F = 6.4 J/cm², P_O = %95.86, H_D = %76.

3.3. Multi-Objective Optimization

A multiobjective optimization was conducted using Minitab's response optimizer to determine the optimal process parameters that minimize errors in structural height and waviness. The optimization plot illustrates the influence of each process parameter (the model variables; refer to columns in Figure 12a) on the surface texture characteristics (the responses or composite desirability; refer to rows in Figure 12a). The vertical red lines on the graph indicate the present factor settings, while the red numbers at the top of each column denote these current factor level values. Similarly, the horizontal blue lines and their associated blue numbers denote the response values for the present factor level. Furthermore, the composite desirability value represented by D indicates the extent to which the response from the expected factor levels satisfied the initial criteria. Moreover, the optimization plot enables us to interactively modify and tweak the input variable parameters to conduct sensitivity analysis and to seek more favorable or enhanced solutions.

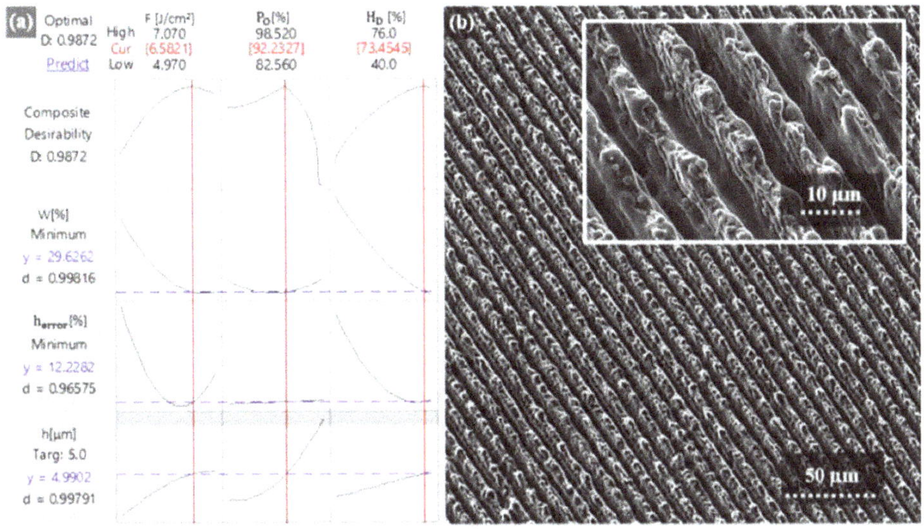

Figure 12. Summary of the multi-objective optimization process where in (a) Minitab Response optimizer is presented and in (b) the Scanning electron micrograph visualizes line-like DLIP microstructures produced on 100Cr6 steel using the processing parameters predicted by the Response optimizer.

The predictive nature of the optimization plot is tested against an experimental example, where the aim was to structure the surface with minimal waviness and line-like microstructures with 5 µm in height which should have minimal height error. In order to reach a desired surface quality, the developed model suggests to use the process parameters of F = 6.58 J/cm^2, P_O = %92.23, and H_D = %73.45 with the composite desirability value equal to %98.7. Afterwards, the predicted and optimized parameters were used in the structuring process. The resulting surface topography is visualized in the scanning electron micrograph presented in Figure 12b. This surface topography is characterized by a 5.43 µm structure height, ~13.78% height error, and ~25.83% waviness, thus, similar well enough to the predicted values of 4.99 µm, 12.23%, and 29.63%, respectively. Consequently, this demonstrates that the developed model is statistically reliable and can be used for prediction and optimization of the processing parameters.

4. CONCISE SUMMARY

This study focuses on the experimental examination and analysis of DLIP operations on bearing steel (100Cr6) utilizing a pulsed nanosecond infrared laser for texturing. The primary objective was to examine the intrinsic link between process factors and the resultant surface texture quality metrics utilizing a central composite design (CCD) methodology. This encompasses the creation of an empirical model articulated by second-order polynomial equations that delineate linear and quadratic interaction effects of laser processing settings on the assessed topographical features. The created model accurately described the association between various parameters and the measured response, with an accuracy of 80%. The analysis of data through Pareto charts indicated that laser fluence and pulse overlap are the most significant process parameters affecting the resultant structure depth. Moreover, it was determined that the laser fluence and hatch distance mostly influence the structural height inaccuracy and surface waviness of the manufactured structure. Furthermore, the model facilitated the identification of cross-correlation between laser fluence and pulse overlap concerning the structures' height response, as well as between laser fluence and hatch distance in connection to both surface waviness and structural height inaccuracy. This indicates that the variation in the mean response from low to high levels of one component is contingent upon the level of the other factor. Moreover, the model facilitates the prediction of appropriate process parameters for the fabrication of target surface textures with defined quality characteristics. Ultimately, the model elucidated how specific undesirable topographical attributes, such as structural height discrepancies and waviness, might be mitigated to enhance the uniformity of periodic surface structures.

AT A GLANCE

This chapter presents the optimization of the direct laser interference patterning process through the application of the design of experiments methodology. Periodic line-like microstructures with a spatial period of 8.50 μm were produced using a two-beam interference apparatus with nanosecond laser pulses, adjusting laser fluence, pulse overlap, and hatch distance. A central composite design with three components and five levels was employed to optimize the necessary number of experiments. The experimental and numerical findings demonstrate the influence of different structuring process parameters on surface uniformity. The evaluated answers include structural height, height error, and pattern waviness. A confocal microscopy and scanning electron microscopy examination of the microstructures on the printed surface was performed. A 3D characterisation method utilizing morphological filtering, providing a comprehensive perspective on surface attributes, was implemented, and a novel qualifying scheme for surface microstructures was established. Empirical models were built and tested to establish correlations between process parameters and performance criteria. Multi-objective optimization was conducted to minimize structural height errors and waviness.

REFERENCES

1. Favret, E.A.; Fuentes Néstor, O. *Functional Properties of Bio-Inspired Surfaces: Characterization and Technological Applications*; World Scientific: Hackensack, NJ, USA, 2009. [**Google Scholar**]
2. Han, Z.; Mu, Z.; Yin, W.; Li, W.; Niu, S.; Zhang, J.; Ren, L. Biomimetic multifunctional surfaces inspired from animals. *Adv. Colloid Interface Sci.* **2016**, *234*, 27–50. [**Google Scholar**] [**CrossRef**] [**PubMed**]
3. Müller, F.; Kunz, C.; Gräf, S. Bio-Inspired Functional Surfaces Based on Laser-Induced Periodic Surface Structures. *Materials* **2016**, *9*, 476. [**Google Scholar**] [**CrossRef**] [**PubMed**]

4. Rosenkranz, A.; Hans, M.; Gachot, C.; Thome, A.; Bonk, S.; Mücklich, F. Direct Laser Interference Patterning: Tailoring of Contact Area for Frictional and Antibacterial Properties. *Lubricants* **2016**, *4*, 2. **[Google Scholar] [CrossRef] [Green Version]**

5. Alamri, S.; Aguilar-Morales, A.I.; Lasagni, A.F. Controlling the wettability of polycarbonate substrates by producing hierarchical structures using Direct Laser Interference Patterning. *Eur. Polym. J.* **2018**, *99*, 27–37. **[Google Scholar] [CrossRef]**

6. Cardoso, J.; Aguilar-Morales, A.; Alamri, S.; Huerta-Murillo, D.; Cordovilla, F.; Lasagni, A.; Ocaña, J. Superhydrophobicity on hierarchical periodic surface structures fabricated via direct laser writing and direct laser interference patterning on an aluminium alloy. *Opt. Lasers Eng.* **2018**, *111*, 193–200. **[Google Scholar] [CrossRef]**

7. Rosenkranz, A.; Stratmann, A.; Gachot, C.; Burghardt, G.; Jacobs, G.; Mücklich, F. Improved Wear Behavior of Cylindrical Roller Thrust Bearings by Three-Beam Laser Interference. *Adv. Eng. Mater.* **2015**, *18*, 854–862. **[Google Scholar] [CrossRef]**

8. Costa, H.; Hutchings, I. Some innovative surface texturing techniques for tribological purposes. *Proc. Inst. Mech. Eng. Part J J. Eng. Tribol.* **2014**, *229*, 429–448. **[Google Scholar] [CrossRef] [Green Version]**

9. Guo, L.; Wong, P.; Gachot, C. Facilitating the Study of the Texturing Effect on Hydrodynamic Lubrication. *Lubricants* **2018**, *6*, 18. **[Google Scholar] [CrossRef] [Green Version]**

10. Kasem, H.; Stav, O.; Grützmacher, P.; Gachot, C. Effect of Low Depth Surface Texturing on Friction Reduction in Lubricated Sliding Contact. *Lubricants* **2018**, *6*, 62. **[Google Scholar] [CrossRef] [Green Version]**

11. Heilmann, S.; Zwahr, C.; Knape, A.; Zschetzsche, J.; Lasagni, A.F.; Füssel, U. Improvement of the Electrical Conductivity between Electrode and Sheet in Spot Welding Process by Direct Laser Interference Patterning. *Adv. Eng. Mater.* **2018**, *20*, 1700755. **[Google Scholar] [CrossRef]**

12. Zwahr, C.; Günther, D.; Brinkmann, T.; Gulow, N.; Oswald, S.; Holthaus, M.G.; Lasagni, A.F. Laser Surface Pattering of Titanium for Improving the Biological Performance of Dental Implants. *Adv. Healthc. Mater.* **2016**, *6*, 1600858. **[Google Scholar] [CrossRef] [PubMed]**

13. Voisiat, B.; Teutoburg-Weiss, S.; Rank, A.; Lasagni, A.F. DLIP holographic structuring: From basic concept to advanced monitoring methods and industrial scale production. In Proceedings of the Laser-Based Micro- and Nanoprocessing XIII, San Francisco, CA, USA, 4 March 2019. **[Google Scholar]**

14. Teutoburg-Weiss, S.; Sonntag, F.; Günther, K.; Lasagni, A.F. Multiple method micromachining laser platform for fabricating anti-counterfeit elements with multiple-scaled features. *Opt. Laser Technol.* **2019**, *115*, 465–476. **[Google Scholar] [CrossRef]**

15. Waugh, D.G.; Lawrence, J. *Laser Surface Engineering: Processes and Applications*; Woodhead: Cambridge, UK, 2016. **[Google Scholar]**

16. Bonse, J.; Krüger, J.; Höhm, S.; Rosenfeld, A. Femtosecond laser-induced periodic surface structures. *J. Laser Appl.* **2012**, *24*, 042006. **[Google Scholar] [CrossRef]**

17. Nakata, Y. Interference laser processing. *Adv. Opt. Technol.* **2016**, *5*, 29–38. **[Google Scholar] [CrossRef]**

18. Indrišiūnas, S.; Voisiat, B.; Gedvilas, M.; Račiukaitis, G. New opportunities for custom-shape patterning using polarization control in confocal laser beam interference setup. *J. Laser Appl.* **2017**, *29*, 011501. **[Google Scholar] [CrossRef]**

19. Lasagni, A.F.; Gachot, C.; Trinh, K.E.; Hans, M.; Rosenkranz, A.; Roch, T.; Eckhardt, S.; Kunze, T.; Bieda, M.; Günther, D.; et al. Direct laser interference patterning, 20 years of development: From the basics to industrial applications. In Proceedings of the Laser-Based Micro- and Nanoprocessing XI, San Francisco, CA, USA, 7 March 2017. **[Google Scholar]**

20. Lasagni, A.F. Advanced Design of Periodical Structures by Laser Interference Metallurgy in the Micro, Nano Scale on Macroscopic Areas. Ph.D. Thesis, Shaker Verlag GmbH, Aachen, Germany, 2007. **[Google Scholar]**

21. He, C.; Steger, M.; Gillner, A. High-Efficiency Nanostructuring using Multi-Beam Interference by Con-secutively Deposited Ultrashort Laser Pulses on Tool Steel. *J. Laser Micro/Nanoeng.* **2018**, *13*, 1–5. [Google Scholar]

22. Aguilar-Morales, A.I.; Alamri, S.; Kunze, T.; Lasagni, A.F. Influence of processing parameters on surface texture homogeneity using Direct Laser Interference Patterning. *Opt. Laser Technol.* **2018**, *107*, 216–227. [Google Scholar] [CrossRef]

23. Chichkov, B.N.; Momma, C.; Nolte, S.; Alvensleben, F.; Tünnermann, A. Femtosecond, picosecond and nanosecond laser ablation of solids. *Appl. Phys. A Mater. Sci. Process.* **1996**, *63*, 109–115. [Google Scholar] [CrossRef]

24. Volkov, A.N.; Zhigilei, L.V. Melt dynamics and melt-through time in continuous wave laser heating of metal films: Contributions of the recoil vapor pressure and Marangoni effects. *Int. J. Heat Mass Transf.* **2017**, *112*, 300–317. [Google Scholar] [CrossRef]

25. Czitrom, V. One-Factor-at-a-Time versus Designed Experiments. *Am. Stat.* **1999**, *53*, 126. [Google Scholar]

26. Siebertz, K.; Bebber, D.V.; Hochkirchen, T. Statistische. In *Versuchsplanung Design of Experiments (DoE)*; Springer: Berlin, Germany, 2017. [Google Scholar]

27. Dean, A.M.; Voss, D.; Danel, D. *Design and Analysis of Experiments*; Springer: Cham, Switzerland, 2017. [Google Scholar]

28. Montgomery, D.C. *Design and Analysis of Experiments*; Wiley: Hoboken, NJ, USA, 2020. [Google Scholar]

29. Tanco, M.; Viles, E.; Pozueta, L. Comparing Different Approaches for Design of Experiments (DoE). In *Lecture Notes in Electrical Engineering Advances in Electrical Engineering and Computational Science*; Springer: Dordrecht, The Netherlands, 2009; pp. 611–621. [Google Scholar]

30. Genna, S.; Tagliaferri, F.; Papa, I.; Leone, C.; Palumbo, B. Multi-response optimization of CFRP laser milling process based on response surface methodology. *Polym. Eng. Sci.* **2017**, *57*, 595–605. [Google Scholar] [CrossRef] [Green Version]

31. Kibria, G.; Sen, A.; Aziz, H.M.T.; Doloi, B.; Bhattacharyya, B. Pulsed Nd:YAG Laser Surface Texturing of Pure Titanium Material. In *Lecture Notes on Multidisciplinary Industrial Engineering Precision Product-Process Design and Optimization*; Springer: Singapore, 2018; pp. 361–390. [Google Scholar]

32. Moradi, M.; Karamimoghadam, M. High power diode laser surface hardening of AISI 4130; statistical modelling and optimization. *Opt. Laser Technol.* **2019**, *111*, 554–570. [Google Scholar] [CrossRef]

33. Witek-Krowiak, A.; Chojnacka, K.; Podstawczyk, D.; Dawiec, A.; Pokomeda, K. Application of response surface methodology and artificial neural network methods in modelling and optimization of biosorption process. *Bioresour. Technol.* **2014**, *160*, 150–160. [Google Scholar] [CrossRef]

34. Diler, E.A.; Ipek, R. An experimental and statistical study of interaction effects of matrix particle size, reinforcement particle size and volume fraction on the flexural strength of Al–SiCp composites by P/M using central composite design. *Mater. Sci. Eng. A* **2012**, *548*, 43–55. [Google Scholar] [CrossRef]

35. Box, G.E.P.; Hunter, J.S. Multi-Factor Experimental Designs for Exploring Response Surfaces. *Ann. Math. Stat.* **1957**, *28*, 195–241. [Google Scholar] [CrossRef]

36. Mathews, P.G. *Design of Experiments with MINITAB*; New Age: New Dehli, India, 2010. [Google Scholar]

37. Khan, R.M. *Problem Solving and Data Analysis Using Minitab: A Clear and Easy Guide to Six Sigma Methodology*; Wiley: Chichester, UK, 2013. [Google Scholar]

38. Wu, C.F.J.; Hamada, M. *Experiments: Planning, Analysis, and Parameter Design Optimization*; Wiley: New York, NY, USA, 2000. [Google Scholar]

39. Dietzsch, M.; Gerlach, M.; Gröger, S. Back to the envelope system with morphological operations for the evaluation of surfaces. *Wear* **2008**, *264*, 411–415. [Google Scholar] [CrossRef]

40. Kumar, J.; Shunmugam, M. A new approach for filtering of surface profiles using morphological operations. *Int. J. Mach. Tools Manuf.* **2006**, *46*, 260–270. [Google Scholar] [CrossRef]

41. Raja, J.; Muralikrishnan, B.; Fu, S. Recent advances in separation of roughness, waviness and form. *Precis. Eng.* **2002**, *26*, 222–235. **[Google Scholar] [CrossRef]**

42. *International standard: Geometrical Product Specifications (GPS)—Filtration—Part 41: Morphological Profile Filters: Disc and Horizontal Line-Segment Filters*; ISO 16610-41:2015; ISO: Geneva, Switzerland, 2015.

43. Roch, T.; Benke, D.; Lasagni, A.F. Method and Arrangement for Forming a Structuring on Surfaces of Components by Means of a Laser Beam 2014. US 9,764,424 B2, 19 September 2017. **[Google Scholar]**

44. Liu, J.M. Simple technique for measurements of pulsed Gaussian-beam spot sizes. *Opt. Lett.* **1982**, *7*, 196. **[Google Scholar] [CrossRef] [PubMed]**

45. Hecht, E. *Optics*; Pearson: Harlow, UK, 2017. **[Google Scholar]**

46. Indrisiunas, S.; Voisiat, B.; Žukauskas, A.; Račiukaitis, G. Direct laser beam interference patterning technique for fast high aspect ratio surface structuring. In Proceedings of the Laser Applications in Microelectronic and Optoelectronic Manufacturing (LAMOM) XX, San Francisco, CA, USA, 4 March 2015. **[Google Scholar]**

47. Alamri, S.; Lasagni, A.F. Development of a general model for direct laser interference patterning of polymers. *Opt. Express* **2017**, *25*, 9603. **[Google Scholar] [CrossRef]**

48. Furlan, V.; Biondi, M.; Demir, A.G.; Pariani, G.; Previtali, B.; Bianco, A. Sub-micrometric surface texturing of AZ31 Mg-alloy through two-beam direct laser interference patterning with a ns-pulsed green fiber laser. *Appl. Surf. Sci.* **2017**, *423*, 619–629. **[Google Scholar]**

49. Moore, D.S.; McCabe, G.P.; Craig, B.A. *Introduction to the Practice of Statistics*; W.H. Freeman, Macmillan Learning: New York, NY, USA, 2017. **[Google Scholar]**

50. El-Khoury, M.; Alamri, S.; Voisiat, B.; Kunze, T.; Lasagni, A.F. Fabrication of hierarchical surface textures using multi-pulse direct laser interference patterning with nanosecond pulses. *Mater. Lett.* **2020**, *258*, 126743. **[Google Scholar]**

CHAPTER 15

COMPETITIVE EFFECTS DURING FEMTOSECOND LASER INDUCED ELEMENT REDISTRIBUTION IN BARIUM AND LANTHANUM MIGRATION-BASED LASER-WRITTEN WAVEGUIDES

1. INTRODUCTION

The alteration of the local composition of glasses treated beneath the surface with focussed femtosecond laser beams has been extensively investigated [1,2]. This phenomenon has recently attracted significant interest [3] owing to its effective application in the fabrication of several high-performance photonic devices [4,5,6]. Femtosecond-induced element redistribution (FLIER) is a general phenomenon observed in multicomponent glasses when subjected to subsurface, high repetition rate, femtosecond laser stimulation. This phenomena can be elucidated by thermo-diffusion processes, namely the Ludwig–Soret effect, which arises from the chemical potential imbalance created in the heated glass volume. The compositional alteration of glass by FLIER simulates the elemental thermal diffusion found in basaltic liquids (silicates) below the Earth›s mantle, but at a much accelerated rate (108 times quicker) and within a precisely delineated micrometric volume [8]. The geometry of the focal volume, local temperature gradients, and heating-cooling cycles significantly influence the final outcome of the FLIER process [3,8]. The effects are especially significant when both relatively heavy elements (e.g., lanthanides) and light elements (e.g., alkali) are included in the glass composition. Nevertheless, FLIER effects have seemingly been neglected in numerous instances when discussing the origin of the refractive index contrast in fs-laser-written waveguides.Thermo-diffusion involves several parameters including thermal and stress gradients, viscosity and its temperature dependence, multiple diffusion coefficients [3], etc. At the end of the day, these parameters condition the local chemical potential of the different glass constituents [12]. Indeed, the prediction of the sign and magnitude of the Soret coefficient of a given component in a multicomponent mixture is still an open problem that requires to consider microscopic parameters and numerical modeling [13]. Still, FLIER can certainly be controlled in practice [14,15] and exploited for producing high-performance photonic devices, as indicated above. Even more, one of the key features of FLIER-based fs-laser writing for photonic applications arises from the potential of ad hoc glass compositional design [4,16]. A given glass composition can be slightly modified by adding small amounts of a modifier made of a relatively heavy ion-oxide (that will act as refractive index carrier) and fast diffusing and lighter (typically alkali) ion-oxide modifier that counter-migrates the index carrier. This gives rise to the formation of regions with increased refractive index and light guiding capabilities. As a matter of fact, by using the electronic polarizabilities [17,18] of the ions involved, it is possible to predict quite accurately the index contrast achievable upon laser writing via FLIER [19]. This principle has been successfully applied to several glass families including phosphates and borates where lanthanides have been used as index carriers [4,5] and, more recently, to silicates [8] using Ba as index carrier.

Apparently, at least in the case of phosphate-based glasses, heavier ions belonging to the same group (i.e., lanthanides (La, Er and Yb)) migrate in the same direction and experience approximately the same amount of enrichment in the higher refractive index region [5]. This apparently applies to the counter-migrating lighter ions (Na, K) too in the low index zone. For the case of Ba, counter-migration of the lighter elements (Na and Si) away from the refractive index increased region has been observed [4,8]. These observations are consistent with what can be called "positive Soret effect" for the heavier constituents of the glass and "negative" for the lighter counter-migrating species, a common observation for thermo-diffusion in glasses or sols. However, not all of the glass constituents necessarily move upon laser excitation. This depends on the material composition [13], the excitation conditions and type of ions involved [20], the shape of the focal volume and the correlated thermal gradients involved [8,21].

In the refractive index structures produced by FLIER, ion cross-diffusion occurs in a hot and viscous high-temperature glass matrix. Therefore, one can expect the presence of either cooperative or competition effects when different types of index carriers (belonging to different groups) are present in the glass composition. However, to the best of our knowledge, there are no studies regarding the simultaneous use of several types of index carriers in the production of waveguides by fs-laser writing via FLIER. This aspect is a very relevant for potential applications of fs-laser processing beyond the production of waveguides. As below indicated, FLIER processes can be detrimental for certain properties of multicomponent glasses and, therefore, a deeper knowledge of the migration behavior of different types of ions upon fs-laser excitation is very desirable. In this work, we aim at analyzing the presence of such cooperative or competition ion migration effects upon FLIER in compositionally pre-designed glasses.

For such a purpose, phosphate glasses modified with the same amount of Na_2O and K_2O, and different amounts of BaO and/or La_2O_3 were used to produce FLIER-based waveguides. The general behavior and performance of the Ba-only and La-only modified glasses is rather similar in terms of migration threshold, laser energy dependence of the size of the modified index structures and index contrast. However, the samples modified with both Ba and La oxides show a strong increase in the migration threshold, and a clear reduction of the size and index contrast of the structures for comparable writing laser energies. These characteristics suggest that significant migratory competition effects arise when index carriers of varying sorts (valence) are incorporated into the glass composition. This competition effect can be exploited to inhibit undesired cross-thermo-diffusion effects in fs-laser writing applications in multicomponent glasses where preserving the glass composition may be a requirement. The maintenance of the local composition of glass during fs-laser irradiation is crucial, particularly in fs-laser-induced crystallization processes for data storage applications [22], the congruent synthesis of non-linear crystals within precursor glasses via fs-laser writing [23,24], or the fabrication of hybrid micro- and nano-structures in semiconductor-doped glasses using ultrafast lasers [25].

2. MATERIALS AND METHODS

The phosphate-based glass samples were obtained through melting and quenching of batches made by mixing $(NH_4)_2HPO_4$, Na_2CO_3, K_2CO_3, BaO, and La_2O_3 (Sigma-Aldrich Spain, Madrid) reagent grade powders. The compositions of the produced samples are shown in **Table 1**.

Table 1. Composition of the designed samples.

Sample	Alkali Oxide Modifiers		Refractive Index Carrier Modifier		Glass Former	Other Modifiers
Oxides Composition in mol %						
	K_2O	Na_2O	BaO	La_2O_3	La_2O_3	Al_2O_3
Ba-glass	15	15	10	0	55	5
La-glass	15	15	0	10	60	0
Ba-La-glass	15	15	5	5	60	0

Three different sample compositions (namely Ba-, La- and Ba-La-glass) were produced. In order to facilitate the alignment procedure during the characterization of the guiding performance of the laser-written structures in the infrared (~1600 nm), all the samples were doped with a very small amount of Er_2O_3, 0.25 wt.%, and Yb_2O_3, 0.5 wt.%, to use the green color up-conversion emission of Er^{3+} when a weak pump laser at 976 nm is coupled in the waveguide [4]. In the Ba-glass sample, a 5 mol% of Al_2O_3 was added to the composition in order to improve its toughness and thus being comparable to the one in the La_2O_3-containing samples. All glasses were formulated for them to have an O/P ratio close to 3, which in phosphate glasses stands for metaphosphate compositions. The batches were slowly calcined up to 400 °C overnight in alumina crucibles; then, melted at 1000 °C (Ba and Ba-La glasses) or 1200 °C (La-glass) for 1 h. The melts were poured onto brass plates and then crashed into small pieces and put into a graphite boat where they were submitted to a homogenization and fining treatment following the procedure applied for phosphate glasses in [26]. For this case, the glasses were remelted at 800 °C (Ba and Ba-La glasses) and 900 °C (La-glass) for 9 h under a constant flow of N_2. The final glass samples were completely dehydroxylated and homogenized without any bubbles or stria and no weight loss was detected that could have originated from volatilization losses. The glass transition temperature (T_g) of the samples was determined from thermal expansion curves recorded in a Netzsch 402 EP dilatometer (Netzsch-Gerätebau GmbH, Selb, Germany) at a heating rate of 5 K·min^{-1}. The T_g of the three glass samples ranged between 510 °C and 550 °C. The observed differences in T_g are consistent with the relatively small compositional differences among the samples.

The samples were then cut (typically in the shape of 10 × 20 × 6 mm^3 plates) and polished to optical quality before optical characterization and fs-laser writing. Ellipsometric spectra were measured at several spatial locations in each sample in order to achieve statistically significant values in the near IR region (800–1700 nm). The acquired Ψ-spectra were fitted using the Cauchy equation (n (λ) = A + (B / λ 2)) for the refractive index *n* as a function of the wavelength (λ). This dispersion equation describes very well the behavior of the refractive index of the samples in the studied spectral region, where they are transparent (absorption coefficient k = 0). Further details regarding the ellipsometry measurement procedure can be found elsewhere [27].

Waveguide-writing via FLIER was performed using a high-repetition-rate, Yb-doped fiber fs-laser (Satsuma HP^2 laser from Amplitude, Pessac, France). The laser generates laser pulses of tunable duration (350 fs–10 ps) at 1030 nm. The laser has a repetition rate that can be varied in the 1 kHz-2 MHz interval with energies up 10 μJ/pulse. In all the experiments described, we used 350 fs-laser pulses at a repetition rate of 500 kHz. The writing beam was slit-shaped (1.2 mm slit width) before being focused with a 0.68 Numerical Aperture (NA) aspheric lens 100 μm underneath the surface. Its polarization was circular in order to minimize the propagation losses of the waveguides. During writing, the sample was translated at a linear speed of 60 µm/s in all cases. A detailed description of the laser writing setup can be found elsewhere [15,28].

Processed samples were grinded and polished to bring the end facets of the laser-written structures to the surface. Optical microscopy observation (in transmission) along the guiding axis of the structures was performed with a Nikon Eclipse system.

After processing, the near field guided modes of the laser-written structures were also imaged by coupling-in light at several different wavelengths in the 980 nm–1640 nm interval with a SMF-28 single mode fiber and using a 50× objective and an infrared camera (Goodrich SUI). The evaluation of the total losses of each structure was made by measuring the output power of the waveguide at 1640 nm with a power meter (Ophir PD300). Propagation losses were calculated as total minus coupling and Fresnel losses, as described elsewhere [6]. The local elemental composition changes were quantified by using Energy-dispersive X-ray Microanalysis (EDX) in a Scanning Electron Microscope (SEM, Leica Cambridge Ltd., Cambridge, UK).

3. RESULTS AND DISCUSSION

3.1. Optical Properties of the Pre-Designed Glass Samples and Index Change Estimates upon FLIER

Table 2 shows the Cauchy coefficients of the refractive index of the samples obtained from the fit of the ellipsometric parameter Ψ measured at three different spatial locations in each sample. The error in the determination of the refractive index values (absolute) has been estimated to be smaller than ± 5 × 10 - 3 [27].

Table 2. Cauchy coefficients of the samples before laser processing and index values at 1590 nm.

Sample	Cauchy Coefficients		Refractive Index (1590 nm)
	A	B	
Ba-glass	1.114	0.0141	1.500
La-glass	1.133	0.0144	1.515
Ba-La-glass	1.125	0.0133	1.510

The spectral dependence of the index (*n*) in the 800–1700 nm range has been plotted in **Figure 1**a for the three glass samples. The measured refractive index values show a very similar dispersion for the three glasses, but the absolute *n* values are distinctly different. The glasses are essentially formed by the same constituents and proportions (~60 mol.% P_2O_5 (glass former) and 30 mol.% of Na_2O and K_2O (modifiers)). The larger electronic polarizability of La_2O_3 compared to BaO provides a larger index to the

La-glass over the whole spectral interval, while the Ba-La-glass shows intermediate values also conditioned by the substitution of 5 mol.% of P_2O_5 by Al_2O_3. The electronic polarizabilities of the constituents of the three glass samples are shown in **Table 3.**

Table 3. Polarizabilities of ions used for local index calculations in the samples upon FLIER.

Polarizability (A^3)	P_2O_5	Na_2O	K_2O	BaO	La_2O_3	Al_2O_3
α_{cation}	0.021	0.175	0.841	1.595	1.048	0.054
α^{2-}	1.350	3.221	1.858	3.652	2.780	1.365
α_{Total}	6.792	3.571	3.540	5.247	10.436	4.203

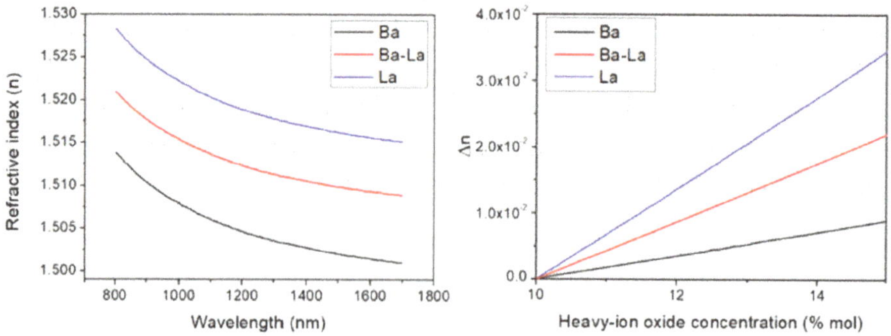

Figure 1. Spectral dependence of the refractive index of the glass samples measured by spectroscopic ellipsometry (a). Index change as a function of the heavy-ion concentration estimated by the electronic polarizability method (b).

By using the electronic polarizability method **[14,15,29]** and the determined refractive index values for wavelengths much longer than the glass matrix resonances, it is possible to accurately estimate the changes in the refractive index of the glass associated to small changes in its local composition caused by FLIER **[16]**. This approach has been shown to work particularly well in the case of phosphate-based glasses **[19,30]**. We have used this approach, using the glass constituent's polarizabilities shown in **Table 3**, to estimate the index changes that take place when the local concentration of the index carrier oxide increases **[19]**. For the calculation, it is assumed that the local concentration of the index carrier oxide (BaO or La_2O_3) increases by the same amount as the concentration of the fast-diffusing ion oxides (Na_2O and K_2O) decreases (e.g., an increase of 2 mol.% of La_2O_3 is accompanied by a corresponding decrease of the Na_2O and K_2O concentrations of 1 mol.% each). For the two heavy ions considered, the Ba- and La-glass index increases nearly linearly with the local concentration increase of the carrier oxide, due to the relatively minor contribution of the counter-migration of K and Na, with a much smaller polarizability. It is found that the Ba-La glass shows again an intermediate behavior. For a local concentration increase of the index carrier of 20% (relative), index changes above or close to 10^{-2} are expected in the La- and La-Ba-glasses which for the Ba-glass are reduced down to 3.6×10^{-3}.

3.2. Characterization and Performance of the Produced Structures

3.2.1. Morphology of the Laser Written Structures

Figure 2 shows several optical micrographs in transmission corresponding to the cross section of the laser-written structures transverse to the sample movement axis (X) for three representative values of the writing pulse energy. As indicated above, after processing, the samples were polished to bring the end facets of the laser-written structures to the surface and, therefore, the cross sections shown correspond to the buried waveguides produced upon laser writing underneath the surface. A top view of the structures (not shown) indicates that the structures (in the form of a continuous line) homogenously spread along the sample movement axis.

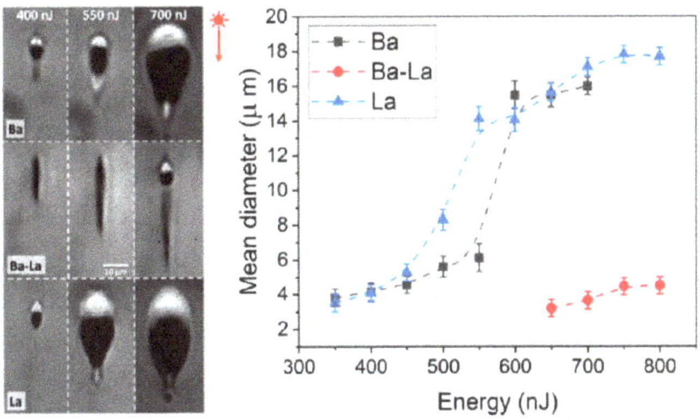

Figure 2. (Left panel) Transmission optical microscopy images of the cross section of structures produced in Ba-, Ba-La-and La-samples at the indicated writing pulse energies. (Right plot) Mean diameter of the structures produced in the three different composition samples as a function of the writing pulse energy.

To facilitate their comparison, the images are shown at the same magnification. In all cases, the structures show an elongated shape along the beam propagation axis due to the combined effect of spherical aberration (SA) and non-linear propagation [21,31]. As a matter of fact, the length of the structures along the beam propagation axis is nearly independent of the sample composition for a given pulse energy. This behavior is related to the relatively small compositional differences among the samples, generating relatively small differences in their linear and non-linear refractive indices, that condition, respectively, the extent of SA and non-linear self-focusing experienced by the laser beam. However, the size and shape of the structures in the transverse dimension (perpendicular to both the laser propagation and sample moving axis) show a strong difference between the samples incorporating a single index carrier element (Ba- and La-glass) or two (Ba-La-glass). In the former case, a bright contrasted region with increased refractive index is clearly visible for writing energies above ~350 nJ/pulse. As confirmed by EDX measurements (see below), the high index region is caused by the migration of Ba or La to the top of the structure while the region underneath, depleted in the carrier element, shows a depressed index and shows a dark contrast [**4,8,15,27,28**]. It is worth noting that Ba or La enrichment can also be observed (to a much lesser extent) at the bottom of the structures produced at the higher energies. This effect has previously been observed both in La- and Ba-modified glasses [**4,8,19**] and is related, as discussed in Ref. [**8**], to the actual shape of the generated thermal gradient. The plot in **Figure 2** shows the mean diameter of the bright contrasted region at the top of the structures as a function of the writing pulse energy. The mean diameter of the structures was computed as the square root of the product of the dimensions of the bright region along the laser propagation axis (Z) and transversally to it (Y) in the images shown in **Figure 2**.

For the Ba- and La-glasses, above ~450 nJ, the size of the structure shows a sudden and strong increase caused by an excessive heat accumulation [**32**] not compensated by heat diffusion [**15**]. The earlier increase of the structure size in the La-glass as a function of the laser energy seems thus related to its expectable lower thermal diffusivity compared with that of the Ba-glass. In contrast, in the Ba-La glass the refractive index increased region is observed only for the higher pulse energies studied (above 650 nJ). For lower values, the laser modified zone adopts the appearance of a dark, non-linear self-focusing filament. Such a difference in the material response is also illustrated quantitatively in the plot of **Figure 2**. While for the Ba- and La- samples, the bright regions are observed for energies as low as 350 nJ, its formation requires ~650 nJ in the Ba-La sample. At this energy and above, its average diameter is ~3–4 μm while the Ba- and La-samples show an average size for the high index zone about four times bigger at a comparable energy. Such morphological differences cannot be attributed to differences in the optical or thermal properties of the Ba-La-glass compared to those of the Ba- and La-glasses, since both optical and thermal properties belong to the group of additive properties of glasses [**33**] and should therefore show (as seen in the optical case, **Figure 1**) intermediate values with respect to those observed in the glasses with a "single index carrier".

The comparison of the morphology of the structures in the three glass compositions analyzed strongly suggests that the laser energy threshold required for generating the cross-migration of the index carrier ions and the alkali species is substantially higher in the Ba-La-sample when compared to the case of the Ba- and La- samples, and that the presence of two heavy index carriers of different valence seems to be hindering their thermo-diffusion against alkali, due to a possible competition effect. In this respect, it is worth noting that a somewhat similar effect of migration inhibition effect has been observed before in the form of two different energy thresholds for inducing the cross-migration of light or heavy ions against alkali in a phosphate glass modified with lanthanides, potassium, silicon and aluminum oxides [20]. The general trend observed there, and later on quantitatively confirmed in [5], was that all the lanthanides showed a similar migration threshold and moved by the same extent and in the same direction. Above such a threshold, the migration of the lighter non-alkali elements was apparently inhibited or minimized below the EDX measurements resolution [20].

3.2.2. Guiding Performance of the Structures

The performance as optical waveguides of the compositional structures generated via FLIER in the different glass samples was assessed for laser wavelengths in the 980 nm–1640 nm interval. **Figure 3** shows the near field image of the propagated modes at 1590 nm for two representative laser pulse energies in the three glass samples. As anticipated from the morphology data, the structures produced in the Ba-La-glass do not support any guided modes for energies below 700 nJ. On the contrary, the structures produced in Ba- and La-glasses show single mode propagation and a relatively small mode field diameter (MFD) for energies between ~400 nJ up to approximately 600 nJ where they start to show multimodal propagation.

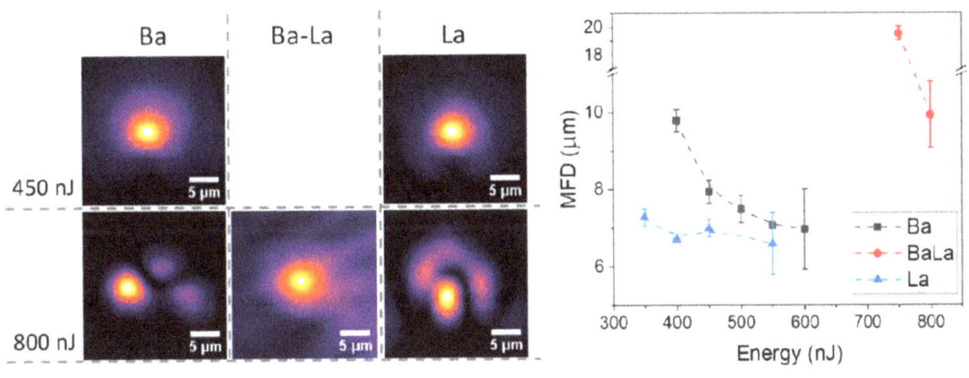

Figure 3. (Left panel) Near field images and (right plot) corresponding mode field diameters (MFD) at 1590 nm for the single-mode structures produced in the different glass compositions.

The plot in **Figure 3** shows the evolution of the MFD (for the single mode waveguides) as a function of the laser pulse energy. The plot indicates that the waveguide formation threshold is slightly smaller for the La-glass (350 nJ) while the MFD remains nearly constant (~7 µm) for pulse energies up to ~550 nJ. At this energy, the transmission of the waveguides turns multimodal, generating a large uncertainty in the MFD estimation. For the Ba-glass, at its waveguide formation threshold (400 nJ), the MFD (~10 µm) is substantially larger than for the La-glass and decreases with the pulse energy down to values close to 7 µm. As for the La-glass, above a certain value (600 nJ), the structures behave as multimodal.

The differences observed as a function of laser pulse energy in the behavior of the waveguides written in the Ba- and La-glasses are consistent with the expected differences in their thermal diffusivity and in the polarizabilities of the index carriers involved. The lower diffusivity of the La-glass and the higher polarizability of La_2O_3 facilitates the formation of a structure capable of supporting a single guide mode for energies slightly lower than those required for the he Ba-glass. However, in the latter case, at the waveguide formation threshold, the index increase in the guiding region is clearly smaller than in the La-glass (larger MFD for the same structure size, **Figure 2**) due to the lower index change expected upon Ba-enrichment. For the same amount of local enrichment of the index carrier (Ba or La), the local index increase should be about three times larger for the La-glass. This can be more clearly seen in **Figure 4**, where we have plotted the index contrast of the single mode structures as a function of the writing pulse energy, as derived from the size of the high index region and its mode field diameter for several wavelengths [19,34]. The figure shows that, already at the waveguide formation threshold, an index contrast above 10^{-2} is induced in the La-glass. This parameter shows a nearly constant behavior in the single mode region (350–550 nJ/pulse). According to the plot in **Figure 2**, for energies in this interval, the index contrast observed is consistent with a local enrichment in La in the guiding region around and above 20% relative to the initial material composition. This might suggest that the transition to a multimodal behavior in the La-glass sample above 550 nJ might be rather conditioned by the size of the structure than by the index contrast achieved, at least in the vicinity of the multimodal transition energy at ~550 nJ.

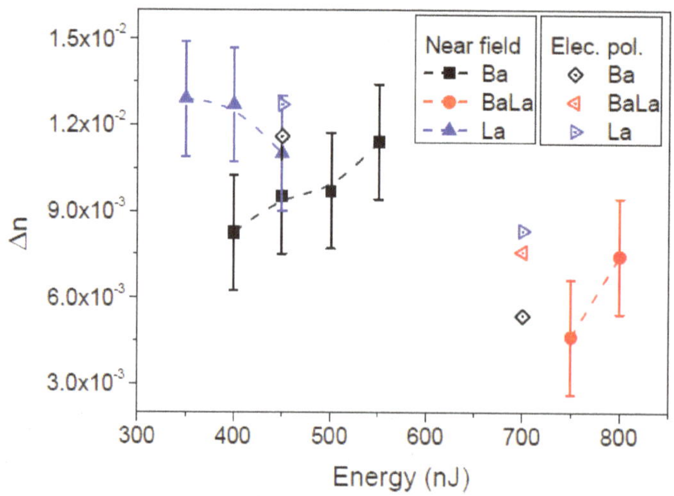

Figure 4. Refractive index contrast at the guiding region estimated from the measured MFD of the single mode waveguides produced in the three glass compositions indicated (Ba-, La- and Ba-La-glass) (solid symbols). The open symbols correspond to the index contrast estimated using experimental EDX local compositional measurements and the electronic polarizability method, as described in the text.

The index contrast of the Ba-glass sample shows nevertheless an increasing behavior with the laser pulse energy. Given the slower increase of the structure size with energy in this case, and the clear diminution of the MFD with the same parameter, the observed behavior suggests a progressive increase in the local content of Ba in the guiding region, that can lead to index contrast values above 10^{-2}. By using the values in **Figure 2**, a local Ba relative enrichment above 40% is foreseen.

Once again, the behavior of the laser-written structures in the Ba-La-glass strongly differs and is not "in between" that observed for the Ba- and La-glass samples. Along with a waveguide formation threshold about twice as big (750 nJ), the index contrast achieved in the guiding region is much smaller, about a factor of two smaller than the one in La-glass. This comparatively poor contrast would be consistent (see **Figure 2**) with a local enrichment around 10% in the index carriers, much smaller than that observed in the Ba- and La-samples at a much higher laser writing energy. This aspect points again to the presence of strong migration competition effects that are even more evident in the measured local compositional changes described below.

Regarding the propagation losses, although the optimization of this parameter is beyond the scope of this work, it must be emphasized that the propagation losses achieved in the structures compare well with previous results of FLIER waveguides based on phosphate- and silicate-based glasses for the Ba- and La-glass samples [**4,5,6,8**]. In the La-glass sample, the losses in the single mode structures are typically in the 0.5–1.0 dB/cm range, while, in the Ba-glass samples, they are somewhat higher, in the 0.7–1.3 dB/cm interval. Much higher losses (2–3 dB/cm) are observed in the structures produced in the Ba-La glass samples. This can be related to both the lower index contrast and much higher writing energies, causing energy coupling fluctuations.

3.2.3. Local Compositional Changes and Migration Competition Effects

As indicated in **Section 2**, the local composition of the laser-induced structures was measured in an SEM by means of EDX. **Figure 5** shows several representative SEM images and compositional maps of structures produced at the same energy in the three samples (700 nJ). Since the images have been recorded on the cross sections of the structures normal to the sample scan axis, they can be easily compared with those shown in **Figure 2**. For all of them, the enrichment in the heavier elements in the guiding region is evidenced by the increased Z-contrast (bright zones) in the backscattered electron SEM images. The correspondence between optical and Z-contrast in FLIER-based optical waveguides has been widely reported (see for instance [**4,6,15**]). The opposite contrast (dark contrasted zones) is consistent with a local depletion of heavy elements, leading to depressed index zone adjacent to the guiding region.

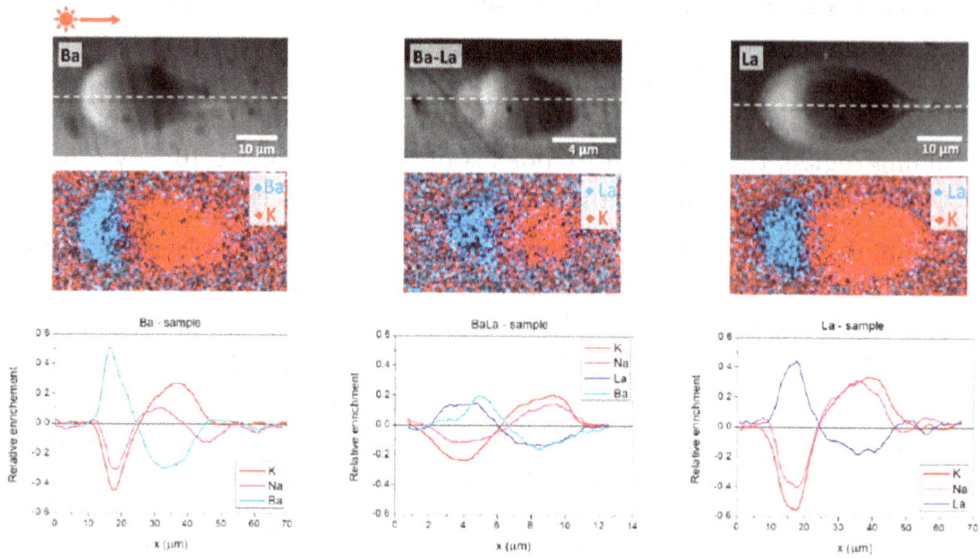

Figure 5. (Top row) SEM backscattered electron images and EDX compositional maps corresponding to Ba, La, and K for structures produced at 700 nJ in the Ba-, La- and Ba-La-glass samples. (Bottom row) Corresponding composition line scans along the dashed line shown in the SEM images.

This can be more clearly appreciated in the compositional maps included, corresponding to the spatial distribution of La, Ba and K in each of the samples (for clarity, the spatial distribution of Na is not included in the maps but is shown in the compositional cross sections in the same figure). For the selected energy (700 nJ), the structures in the Ba- and La-glass show very similar appearance including the similarities in the size and element distribution maps (obviously replacing Ba by La in the corresponding sample). As above indicated, both structures are multimode while for the Ba-La glass the structure shown, just above the FLIER threshold (650 nJ in this sample) does not support a guided mode.

Several additional features are worth noting in the element distribution over the FLIER region. The first one is that the relative enrichment in Ba and La, respectively, in the guiding region of the Ba- and La-glass structures reaches a very similar value, close to 50%, slightly higher in the Ba-glass case. This implies that the guiding region reaches a peak concentration in the index carrier oxide close to 15 mol% in the guiding region. Such an enrichment is accompanied by a more or less similar relative depletion in K and Na, although the K depletion is always bigger than the Na one. The same applies to the longer region depleted in Ba or La, located below the guiding zone, with a depletion around 20–30%, slightly higher in the Ba case. Interestingly, the extent of the region over which compositional changes are observed is also in both glass samples somewhat longer for the K than for the Na case.

The index contrast estimates derived from the structure size (see **Figure 2**) and MFD measurements (see **Figure 3**) have been compared to the values derived from the local composition of the structures obtained by the EDX and the electronic polarizability approach, as described in [19]. The index contrast was computed by using an averaged composition over the index carrier enrichment region by integrating the compositional cross sections of each individual element except oxygen (see **Figure 5**) and assuming oxide stoichiometry. The so-calculated averaged composition was input in the electronic polarizabilities model to compute the averaged local index for two energies (450 and 700 nJ). The obtained values have been plotted in **Figure 4**. It can be seen that there is an excellent correlation between the measured compositional changes and the experimentally determined index contrast in the single mode region where Δn values close to 1.2×10^{-2} are reached for both the Ba- and La-glass samples. At energies above 600 nJ, the index contrast estimated from compositional measurements decreases to values ~7.0×10^{-3}, something expectable from the much bigger size of the structure over which the index carrier is redistributed, also causing the observed multimodal behavior.

In contrast, in the Ba-La-glass sample the relative enrichment in Ba- and La- at the top of the structure at 700 nJ is much smaller, around 15–20% (less than half of what is observed in the Ba- and La-glass samples, compare the linescans shown in **Figure 5**). Since the initial content of the sample in La_2O_3 and BaO is just 5 mol%, the peak concentration of index carrier oxides barely reaches 12 mol%, much lower than in the Ba- or La-glass samples at the same energy. The same applies to the Na and K depletion, that is lower than 20% in average. It is also worth noting that the amount of enrichment and depletion in the Ba- and La- content is similar in the two regions of the structure (bright and dark-contrasted in the SEM image), generating a nearly symmetric compositional structure along the laser beam propagation axis. Indeed, the overall length of the compositionally modified FLIER region in the Ba-La glass (~10 μm) is much smaller than the one in the Ba- and La-glass samples at the same energy (~60 μm). Notice that the length of the laser-affected zone along the laser beam propagation axis is very similar for the three glass compositions, at the same laser pulse energy, as shown in **Figure 2**, and longer than the extension of the FLIER region. The index contrast estimate based on EDX compositional measurements at 700 nJ leads to a value around 5×10^{-3} which is consistent with the fact that the structure, less than ~4 μm in diameter, does not support a guided mode in the near IR.

All these features of the structures produced in the Ba-La-glass, compared to the behavior of the Ba- and La-glass samples, show that the cross-migration of Ba and La against alkalis is hindered by the simultaneous presence of the two index carriers. Whatever the origin of this behavior is, the presence of both (Ba and La) oxide modifiers clearly inhibits their cross-migration against the alkalis over a very broad interval of laser pulse energies (more than 700 nJ) while, above the cross-migration threshold, it reduces the extent of the migration process both in magnitude (smaller compositional changes) and spatial spread (shorter compositionally modified regions).

This kind of blocking effect somewhat resembles the so-called mixed alkali effect (MAE) in glasses [**35,36,37**], although in a very different situation. Please note that in what follows in the comparison with the MAE, we refer not to the presence of Na_2O and K_2O as fast diffusing modifiers in the glass compositions. In this case, we refer to the concurrent effect associated with the presence of both index carriers (Ba and La) that hinders their cross-thermo-diffusion against the alkalis in the FLIER process.

In general, the MAE refers to several phenomena peculiar to mixed-cation glasses, including the huge rise in electrical resistivity occurring on mixing of two mobile ions [**35**]. Indeed, although the MAE, evidenced as a non-additivity of many glass properties, was initially observed as a thermomechanical effect, it has been more thoroughly described in terms of the decrease of several orders of magnitude in the electrical conductivity and ionic diffusivity when one alkali oxide is gradually replaced by another in a series of glasses [**36,37**]. Such a decrease is accompanied by a strong increase of the activation energy of the ionic conductivity [**35**]. It is also worth noting that the MAE has been observed in a wide variety of glass families, including silicate, borate and phosphate-based glasses, something that can be related to the fact that alkalis are usually considered to simply fill the interstitial space between structural rigid polyhedra and convert bridging oxygens to nonbridging oxygens [**37**].

The origin of the MAE has been correlated with the interaction between the alkali cations via structural defects such as non-bridging oxygens [**38**], "memory effects" [**39,40**] or, more recently, to topological effects caused by the network strain associated to different cation radii [**37**]. An important common feature of these "memory" or topological effects is that alkali ions have individualized sites that relax to suit their inhabitants below T_g. According to the thorough literature summary given in [**37**], "when the alkali leaves the site, the site retains "memory" of that ion, making it difficult for ions of the other type to hop into it". The recent use of detailed molecular dynamics (MD) simulations and comparison to MAE experiments enables to confirm that interstices containing mismatched alkali species relax toward different ultimate structures and the competition between these interstices results in stress on the network [**38**].

When comparing our results regarding the cross-migration of Ba and La ions against alkali by thermo-diffusion in the Ba-, La-, and Ba-La-glass samples, there are two features that strongly resemble the MAE. The first is the existence of an energy threshold for the cross-diffusion of species in the FLIER process that resembles the existence of an activation energy for the ionic conductivity in fast ionic conductors [**41**], directly related to the diffusion of cations in the presence of a DC electric field. The second is the strong increase in the cross thermo-diffusion threshold in terms of laser pulse energy when both Ba- and La- oxides are present in the composition, as similarly observed with the ionic diffusion activation energy in the MAE. The atomic number, mass, valence, and ionic radii of the ions involved in the FLIER process are shown in **Table 4**.

Table 4. Atomic number, mass valence and ionic radius (crystalline and effective, assuming a coordination number of six [36]) of the cations involved in the FLIER process.

Element	Na	K	Ba	La
Atomic number	11	19	56	57
Mass (a.m.u.)	22.99	39.10	137.3	138.9
Valence Cation radius (pm)	+1	+1	+2	+3
Crystalline (Effective)	116 (102)	152 (138)	140 (135)	117.2 (103.2)

The table shows that, in spite of their very close atomic number and mass, there is a relatively large difference between the cationic effective radii of Ba^{+2} and La^{+3} (~20%). We might speculate that, even at high temperature, for a low-viscosity glass material that is generated upon high repetition (500 kHz) pulsed laser irradiation, there is still a "memory effect" associated with the type of ion that occupied a given site in the glass. In such a case, we may think of a "memory effect" like the one described in the frame of the dynamic structure model [**39,40**] or like the one associated with the network strain model [**37**]. Whatsoever the origin of the "memory effect" is, this would imply that pathways for ion thermo-diffusion are limited to sites of that ion type, and are dependent on relaxation of those sites in order to form new diffusion pathways. The presence of a cation in an unlike site would be increasingly less likely to form with an increasing size ratio [**37**]. The consequence would be that in the Ba-La glass sample, the sites left by the jump of a given heavy cation would most probably be used by an index carrier of the same type, hindering the overall cross-migration process.

However, when several lanthanides are present in the composition of a phosphate-based glass, they experience similar levels of thermo-diffusion in the same direction, as experimentally shown in references [**5,20**] for La, Er and Yb with the same valence (+3), and ionic radii differing around 10%. Therefore, despite the cation size may play a certain role in limiting the simultaneous diffusion of index carrier elements, it seems that it would not be sufficient by itself to account for the high differences here observed between the Ba- or La-glass samples and the Ba-La-glass. Additionally, it must be considered that the MAE is strongly reduced when the temperature of the glass increases, as it has been observed for the thermal conductivity or the viscosity of phosphate and aluminosilicate glasses **[42,43,44]**, which makes it difficult to justify the existence of a similar effect in a very high temperature melt, such as the one induced upon fs-laser irradiation.

Besides the size (ionic radius) of the element, there is an additional factor that may turn out decisive and that has not been considered before, which is the valence of the index carrier elements. When having single valence index carrier elements, e.g., lanthanides as indicated above, their cross-diffusion with the alkalis takes place all at once without mutual interference, even for those with relatively dissimilar size. At the temperatures involved in the process, the relaxation of the network will be sufficiently fast so that cations with different size but equal valence migrate without constraints. However, the cross-migration of index carrier cations would also be regulated through the electrostatic interaction with the electronic density of the oxygens that belong to the PO_4 network building units. Same valence elements can cross-diffuse with alkalis in order to easily keep the charge neutrality. This makes that Ba^{2+} ions only migrate through positions of nearby Ba^{2+} and that La^{3+} do so via La^{3+} positions to not produce local decompensated charges, resulting in a largely reduced cross-migration. Similarly, to the mixed modifier effect that accounts for non-linear property changes in the presence of different size cations, it is proposed that here applies what could be called a "mixed valence effect" and that would be worth validating in other glass composition systems. Whatsoever the origin of the migration competition effects is, they could potentially be used to inhibit thermo-migration effects in fs-laser writing applications where the preservation of the composition of the glass can be needed as a requirement **[13,20,22,23,24,25]**.

4. CONCISE SUMMARY

We have examined the development of refractive index microstructures by FLIER in specifically developed phosphate glasses. The glasses were altered through the incorporation of minor quantities of Na2O and K2O, along with heavier ion oxides (Ba and La). The aforementioned ions function as index carriers. The laser excitation induces cross-thermo-diffusion of alkali and index carrier cations within the stimulated volume. Areas enriched in BaO and La2O3 and deficient in N2O and K2O are generated within the laser-excited volume. These locations exhibit an elevated refractive index adequate to facilitate guided modes in the near infrared. The glass samples modified with a single index carrier ion oxide (either Ba- or La-oxide) show a relatively low waveguide writing threshold (350–400 nJ) and a light guiding performance comparable to each other. In both cases, index contrast values above 10^{-2}, and low propagation losses are achieved. In contrast, when both types of index carriers (Ba and La cations) are present in the glass composition, a much higher element migration threshold is observed, and the size of the compositionally modified regions is substantially smaller. Indeed, the cross-thermo-diffusion of Ba and La against Na and K is inhibited for laser pulse energies below 750 nJ. The morphological, optical and compositional analysis of the produced structures is consistent with a competition effect in the thermo-diffusion of Ba^{2+} and La^{3+} cations that strongly resembles the mixed alkali effect (MAE). The diffusion competition effect, apparently linked to the differing valence of the index carrier ions, provides an effective compositional strategy to alleviate FLIER effects. This compositional design methodology is crucial for enabling fs-laser processing in multicomponent glasses, where preserving the local composition is vital to maintain specific characteristics of the glass.

AT A GLANCE

Fs-laser induced element redistribution (FLIER) has been the focus of extensive research in recent years. The application of this technology to diverse glass types has already yielded effective optical waveguides, tappers, amplifiers, and lasers. The majority of research on FLIER-based waveguides pertains to structures formed by the cross-migration of alkali metals (Na, K) and lanthanides (mostly La). The latter components function as carriers of the refractive index. Herein, we report the production of Ba-based, FLIER-waveguides in phosphate glass with an index contrast $> 10^{-2}$. Phosphate glasses modified with the same amount of Na_2O and K_2O, and variable amounts of BaO and/or La_2O_3 were used to produce the FLIER-waveguides with Ba and or La acting as index carriers. Ba-only modified glasses show a waveguide writing threshold and light guiding performance comparable to that of La-based structures. Nevertheless, blended Ba-La glasses have a significantly elevated element migration threshold and considerably smaller compositionally altered regions. This pattern aligns with a competitive effect in the cross-migration of both elements (Ba and La) in relation to the alkalis. This phenomenon can be utilized to prevent unwanted element redistribution in fs-laser processing of multicomponent glasses.

REFERENCES

1. Yonesaki, Y.; Miura, K.; Araki, R.; Fujita, K.; Hirao, K. Space-selective precipitation of non-linear optical crystals inside silicate glasses using near-infrared femtosecond laser. *J. Non-Cryst. Solids* **2005**, *351*, 885–892. **[Google Scholar] [CrossRef]**

2. Liu, Y.; Shimizu, M.; Zhu, B.; Dai, Y.; Qian, B.; Qiu, J.; Shimotsuma, Y.; Miura, K.; Hirao, K. Micromodification of element distribution in glass using femtosecond laser irradiation. *Opt. Lett.* **2009**, *34*, 136–138. **[Google Scholar] [CrossRef] [Green Version]**

3. Fernandez, T.T.; Sakakura, M.; Eaton, S.M.; Sotillo, B.; Siegel, J.; Solis, J.; Shimotsuma, Y.; Miura, K. Bespoke photonic devices using ultrafast laser driven ion migration in glasses. *Prog. Mater. Sci.* **2018**, *94*, 68–113. **[Google Scholar] [CrossRef]**

4. Dias, A.; Muñoz, F.; Alvarez, A.; Moreno-Zárate, P.; Atienzar, J.; Urbieta, A.; Fernandez, P.; Pardo, M.; Serna, R.; Solis, J. Femtosecond laser writing of photonic devices in borate glasses compositionally designed to be laser writable. *Opt. Lett.* **2018**, *43*, 2523. **[Google Scholar] [CrossRef] [PubMed]**

5. del Hoyo, J.; Moreno-Zarate, P.; Escalante, G.; Valles, J.A.; Fernandez, P.; Solis, J. High-Efficiency Waveguide Optical Amplifiers and Lasers via FS-Laser Induced Local Modification of the Glass Composition. *J. Light. Technol.* **2017**, *35*, 2955–2959. **[Google Scholar] [CrossRef]**

6. Macias-Montero, M.; Dias, A.; Sotillo, B.; Moreno-Zarate, P.; Ariza, R.; Fernandez, P.; Solis, J. Waveguide tapers fabrication by femtosecond laser induced element redistribution in glass. *J. Light. Technol.* **2020**, *38*, 6578–6583. **[Google Scholar] [CrossRef]**

7. Rahman, M.A.; Saghir, M.Z. Thermodiffusion or Soret effect: Historical review. *Int. J. Heat Mass Transf.* **2014**, *73*, 693–705. **[Google Scholar] [CrossRef]**

8. Montero, M.M.; Muñoz, F.; Sotillo, B.; Hoyo, J.; Ariza, R.; Fernandez, P.; Siegel, J.; Solis, J. Femtosecond laser induced thermophoretic writing of waveguides in silicate glass. *Sci. Rep.* **2021**, *11*, 1–12. **[Google Scholar] [CrossRef]**

9. Osellame, R.; Chiodo, N.; Della Valle, G.; Cerullo, G.; Ramponi, R.; Laporta, P.; Killi, A.; Morgner, U.; Svelto, O. Waveguide lasers in the C-band fabricated by laser inscription with a compact femtosecond oscillator. *IEEE J. Sel. Top. Quantum Electron.* **2006**, *12*, 277–285. **[Google Scholar] [CrossRef]**

10. Fernandez, T.; Gross, S.; Arriola, A.; Withford, M.; Privat, K. Revisiting ultrafast laser inscribed waveguide formation in commercial alkali-free borosilicate glasses. *Opt. Express* **2020**, *28*, 10153–10164. **[Google Scholar] [CrossRef]**

11. Davis, K.M.; Miura, K.; Sugimoto, N.; Hirao, K. Writing waveguides in glass with a femtosecond laser. *Opt. Lett.* **1996**, *21*, 1729–1731. **[Google Scholar] [CrossRef]**

12. Kempers, L.J.T.M. A comprehensive thermodynamic theory of the Soret effect in a multicomponent gas, liquid, or solid. *J. Chem. Phys.* **2001**, *115*, 6330–6341. **[Google Scholar] [CrossRef]**

13. Shimizu, M.; Fukuyo, T.; Matsuoka, J.; Nakashima, K.; Sato, K.; Kiyosawa, T.; Nishi, M.; Shimotsuma, Y.; Miura, K. Determination of thermodynamic and microscopic origins of the Soret effect in sodium silicate melts: Prediction of sign change of the Soret coefficient. *J. Chem. Phys.* **2021**, *154*, 074501. **[Google Scholar] [CrossRef]**

14. Sakakura, M.; Kurita, T.; Shimizu, M.; Yoshimura, K.; Shimotsuma, Y.; Fukuda, N.; Hirao, K.; Miura, K. Shape control of elemental distributions inside a glass by simultaneous femtosecond laser irradiation at multiple spots. *Opt. Lett.* **2013**, *38*, 4939–4942. **[Google Scholar] [CrossRef] [PubMed]**

15. Del Hoyo, J.; Vazquez, R.M.; Sotillo, B.; Fernandez, T.T.; Siegel, J.; Fernández, P.; Osellame, R.; Solis, J. Control of waveguide properties by tuning femtosecond laser induced compositional changes. *Appl. Phys. Lett.* **2014**, *105*, 131101. **[Google Scholar] [CrossRef] [Green Version]**

16. Tan, D.; Wang, Z.; Xu, B.; Qiu, J. Photonic circuits written by femtosecond laser in glass: Improved fabrication and recent progress in photonic devices. *Adv. Photonics* **2021**, 3, 024002. **[Google Scholar] [CrossRef]**

17. Duffy, J.A. The electronic polarisability of oxygen in glass and the effect of composition. *J. Non-Cryst. Solids* **2002**, *297*, 275–284. **[Google Scholar] [CrossRef]**

18. Dimitrov, V.; Komatsu, T. An Interpretation of Optical Properties of Oxides and Oxide Glasses in Terms of the Electronic Ion Polarizability and Average Single Bond Strength. *J. Univ. Chem. Technol. Metall.* **2010**, *45*, 219–250. **[Google Scholar]**

19. Moreno-Zarate, P.; Muñoz, F.; Sotillo, B.; Macias-Montero, M.; Atienzar, J.; Garcia-Pardo, M.; Fernandez, P.; Serna, R.; Solis, J. Role of the La/K compositional ratio in the properties of waveguides written by fs-laser induced element redistribution in phosphate-based glasses. *Materials* **2020**, *13*, 1275. **[Google Scholar] [CrossRef] [Green Version]**

20. Fernandez, T.T.; Sotillo, B.; del Hoyo, J.; Valles, J.-A.; Vazquez, R.M.; Fernandez, P.; Solis, J. Dual Regimes of Ion Migration in High Repetition Rate Femtosecond Laser Inscribed Waveguides. *IEEE Photonics Technol. Lett.* **2015**, *27*, 1068–1071. **[Google Scholar] [CrossRef] [Green Version]**

21. Fernandez, T.T.; Siegel, J.; Hoyo, J.; Sotillo, B.; Fernandez, P.; Solis, J. Controlling plasma distributions as driving forces for ion migration during fs laser writing. *J. Phys. D Appl. Phys.* **2015**, *48*, 155101. **[Google Scholar] [CrossRef] [Green Version]**

22. Jin, M.; Zhou, W.; Ma, W.; Wang, Q.; Liang, X.; Zhang, P.; Xiang, W. The inhibition of CsPbBr3 nanocrystals glass from self-crystallization with the assistance of ZnO modulation for rewritable data storage. *Chem. Eng. J.* **2021**, 129812. **[Google Scholar] [CrossRef]**

23. Zhang, X.; He, X.; Liu, Q.; Poumellec, B.; Lancry, M.; Brisset, F. Abnormal elemental redistribution in silicate glasses irradiated by ultrafast laser. *J. Alloys Compd.* **2017**, *727*, 444–448. **[Google Scholar] [CrossRef]**

24. Stone, A.; Sakakura, M.; Shimotsuma, Y.; Miura, K.; Hirao, K.; Dierolf, V.; Jain, H. Femtosecond laser-writing of 3D crystal architecture in glass: Growth dynamics and morphological control. *Mater. Des.* **2018**, *146*, 228–238. **[Google Scholar] [CrossRef]**

25. Mardilovich, P.; Fletcher, L.B.; Troy, N.W.; Yang, L.; Huang, H.; Risbud, S.H.; Krol, D.M. Ultrafast Laser Fabrication of Hybrid Micro- and Nano-Structures in Semiconductor-doped Borosilicate Glasses. *Int. J. Appl. Glas. Sci.* **2013**, *4*, 87–99. **[Google Scholar] [CrossRef]**

26. Muñoz, F.; Balda, R. A highly efficient method of dihydroxylation and fining of Nd phosphate laser galsses. *Int. J. App. Glass Sci.* **2019**, *10*, 157–161. **[Google Scholar] [CrossRef]**

27. Moreno-Zarate, P.; Gonzalez, A.; Funke, S.; Días, A.; Sotillo, B.; del Hoyo, J.; Garcia-Pardo, M.; Serna, R.; Fernandez, P.; Solis, J. Imaging Ellipsometry Determination of the Refractive Index Contrast and Dispersion of Channel Waveguides Inscribed by fs-Laser Induced Ion-Migration. *Phys. Status Solidi* **2018**, *215*, 1800258. **[Google Scholar] [CrossRef]**

28. Toney Fernandez, T.; Haro-González, P.; Sotillo, B.; Hernandez, M.; Jaque, D.; Fernandez, P.; Domingo, C.; Siegel, J.; Solis, J. Ion migration assisted inscription of high refractive index contrast waveguides by femtosecond laser pulses in phosphate glass. *Opt. Lett.* **2013**, *38*, 5248–5251. **[Google Scholar] [CrossRef] [PubMed] [Green Version]**

29. Duffy, J.A. Electronic polarizability and related properties of the oxide ion. *Phys. Chem. Glasses* **1989**, *30*, 1–4. **[Google Scholar]**

30. Tomihara, Y.; Tricot, G.; Revel, B.; Kishimoto, K.; Okamoto, T.; Saitoh, A. The Relationship among Electronic Polarizability, Photoelasticity, and Refractivity in Ternary Phosphate Glasses. *Phys. Status Solidi Basic Res.* **2020**, *257*, 1–8. **[Google Scholar] [CrossRef]**

31. del Hoyo, J.; de la Cruz, A.R.; Grace, E.; Ferrer, A.; Siegel, J.; Pasquazi, A.; Assanto, G.; Solis, J. Rapid assessment of nonlinear optical propagation effects in dielectrics. *Sci. Rep.* **2015**, *5*, 07650. **[Google Scholar] [CrossRef] [PubMed] [Green Version]**

32. Eaton, S.M.; Zhang, H.; Herman, P.R.; Yoshino, F.; Shah, L.; Bovatsek, J.; Arai, A. Heat accumulation effects in femtosecond laser-written waveguides with variable repetition rate. *Opt. Express* **2005**, *13*, 4708–4716. **[Google Scholar] [CrossRef]**

33. Dragic, P.D.; Ballato, J. 120 Years of Optical Glass Science: From the law of mixtures to mixing the unmixable. *Opt. Photonics News* **2014**, 44–51. **[Google Scholar] [CrossRef]**

34. Hussey, C.D.; Martinez, F. Approximate Analytic Forms For The Propagation Characteristics Of Single-Mode Optical Fibres. *Electron. Lett.* **1985**, *21*, 1103–1104. **[Google Scholar] [CrossRef] [Green Version]**

35. Ingram, M.D. Amorphous Materials: Mixed Alkali Effect. In *Encyclopedia of Materials: Science and Technology*; Elsevier: Amsterdam, The Netherlands, 2001; pp. 220–223. ISBN 978-0-08-043152-9. **[Google Scholar]**

36. Isard, J.O. The mixed alkali effect in glass. *J. Non-Cryst. Solids* **1969**, *1*, 235–261. **[Google Scholar] [CrossRef]**

37. Wilkinson, C.J.; Potter, A.R.; Welch, R.S.; Bragatto, C.; Zheng, Q.; Bauchy, M.; Affatigato, M.; Feller, S.A.; Mauro, J.C. Topological Origins of the Mixed Alkali Effect in Glass. *J. Phys. Chem. B* **2019**, *123*, 7482–7489. **[Google Scholar] [CrossRef] [PubMed]**

38. Greaves, G.N.; Ngai, K.L. Ionic transport properties in oxide glasses derived from atomic structure. *J. Non-Cryst. Solids* **1994**, *172–174*, 1378–1388. **[Google Scholar] [CrossRef]**

39. LaCourse, W.C. A defect model for the mixed alkali effect. *J. Non-Cryst. Solids* **1987**, *95–96*, 905–912. **[Google Scholar] [CrossRef]**

40. Bunde, A.; Ingram, M.D.; Maass, P.; Ngai, K.L. Mixed alkali effects in ionic conductors: A new model and computer simulations. *J. Non-Cryst. Solids* **1991**, *131–133*, 1109–1112. **[Google Scholar] [CrossRef]**

41. Tuller, H.L.; Button, D.P.; Uhlmann, D.R. Fast ion transport in oxide glasses. *J. Non-Cryst. Solids* **1980**, *40*, 93–118. **[Google Scholar] [CrossRef]**

42. Shannon, R.D. Revised effective ionic radii and systematic studies of interatomic distances in halides and chalcogenides. *Acta Crystallogr. A* **1976**, *32*, 751–767. **[Google Scholar] [CrossRef]**

43. Tsuchida, J.E.; Ferri, F.A.; Pizani, P.S.; Martins Rodrigues, A.C.; Kundu, S.; Schneider, J.F.; Zanotto, E.D. Ionic conductivity and mixed-ion effect in mixed alkali metaphosphate glasses. *Phys. Chem. Chem. Phys.* **2017**, *19*, 6594–6600. **[Google Scholar] [CrossRef] [PubMed]**

44. Kim, K.D.; Lee, S.H. Viscosity behavior and mixed alkali effect of alkali aluminosilicate glass melts. *J. Ceram. Soc. Jpn.* **1997**, *105*, 827–832. **[Google Scholar] [CrossRef] [Green Version]**

INDEX

www.ingramcontent.com/pod-product-compliance
Lightning Source LLC
LaVergne TN
LVHW062223200926
848484LV00006B/1462